ETHNOBOTANY AND PLANTS OF TRANS-HIMALAYA

ETHNOBOTANY AND PLANTS OF TRANS-HIMALAYA

Om Prakash Chaurasia
Zakwan Ahmed
Basant Ballabh
Field Research Laboratory
Defence Research & Development Organisation
C/o. 56 APO, Pin : 901205 (India)

2007

SSPH
SATISH SERIAL PUBLISHING HOUSE
403, Express Tower, Commercial Complex, Azadpur, Delhi - 110033 (INDIA)
Phone : 011-27672852 Fax : 91-11-27672046
E-mail : info@satishserial.com, hkjain1975@yahoo.com
Website : www.satishserial.com

Published by :

SATISH SERIAL PUBLISHING HOUSE
403, Express Tower, Commercial Complex, Azadpur, Delhi - 110033 (INDIA)
Phone : 011-27672852 Fax : 91-11-27672046
E-mail : info@satishserial.com, hkjain1975@yahoo.com

ISBN 81-89304-33-X

Printed at
SALASAR IMAGING SYSTEMS, DELHI-35

डॉ. डब्ल्यू. सेल्वामूर्ति
मुख्य नियंत्रक अनुसंधान तथा विकास
Dr. W. Selvamurthy
Chief Controller Research & Development

दूरभाष / Tele : 011-23007201
फैक्स / Fax : 011-23014259, 23007202
भारत सरकार, रक्षा मंत्रालय
Government of India, Ministry of Defence
रक्षा अनुसंधान तथा विकास संगठन
Defence Research & Dev. Org.
201 'ए' ब्लाक, डी आर डी ओ भवन, राजाजी मार्ग,
नई दिल्ली–110011
201 'A' Block, DRDO Bhawan, Rajaji Marg,
New Delhi–110011

FOREWORD

Plants have always been the source of food, medicine and other necessities of life since the beginning of human civilization. As we know India has rich heritage of multifarious uses of plant biodiversity and cold desert trans Himalayan regions are not an exception to it. Majority of Indian cold deserts of trans Himalayas belong to Ladakh (Jammu & Kashmir) and Lahaul-Spiti (Himachal Pradesh). Cold desert parts of the Himalaya are usually recognized as barren and rugged mountains with very sparse vegetation due to characteristic eco-environments.

Ethnobotany is the most recent branch of Botanical sciences, which is further very recent to India and trans Himalayas in particular. The regions are endowed with characteristic plant biodiversity which are closely associated with day-to-day life of local inhabitants. There are only scattered information on ethnobotany of Ladakh and Lahaul Spiti. Field Research Laboratory of Defence Research & Development Qrganization, Leh - Ladakh has carried out extensive ethnobotanical survey covering far-flung areas of Leh, Nubra, Changthang, Suru, Zanskar and Lahaul-Spiti valleys. The cold desert ethnobotanical wealth has been classified into medicinal, aromatic, edible, fodder, fuel, magico-religious and forest plants enlisting their characteristic growth features and ethnobotanical uses.

It gives me immense pleasure to compliment the tireless effort of the authors Dr. OP Chaurasia, Scientist 'D' and Dr. Zakwan Ahmed, Director and Dr. Basant Ballabh, STA of Field Research Laboratory (DRDO), Leh-Ladakh to compile this most valuable and multifaceted information over decades. The work will be of immense uses to cold desert development agencies, research organizations, educational institutes and to all those interested in development of cold desert trans Himalayas.

(Dr. W. Selvamurthy)

डॉ. नरेन्द्र कुमार
निदेशक एवं सचिव
जैव विज्ञान अनुसंधान बोर्ड
Dr. Narendra Kumar
Director and Secretary
Life Science Res. Board

दूरभाष / Tele : 011-23017752
फैक्स / Fax : 011-23017753, 23007347
भारत सरकार, रक्षा मंत्रालय
Government of India, Ministry of Defence
रक्षा अनुसंधान तथा विकास संगठन
Defence Research & Dev. Org.
जैव विज्ञान निदेशालय
Directorate of Life Sciences
347 'बी' ब्लाक, डी आर डी ओ भवन, राजाजी मार्ग, नई दिल्ली–110011
347 'B' Block, DRDO Bhawan, Rajaji Marg, New Delhi–110011

INTRODUCTION

Plant Biodiversity has always been the source of food, medicine and other necessities of life since the beginning of human civilization. Our country has a very rich plant biodiversity due to prevailing very highly divergent ecosystem and significant altitudinal variations ranging from sea level to the highest ranges of the Himalayas and allied factors. The Himalayas ranges are one of the best examples, which abound approximately 18,440 species of Angiosperms, Gymnosperms, Pteridophytes, Bryophytes, Lichens and Fungi, of which about 25 percent species are endemic to the Himalayas.

Cold deserts are usually confined to high altitudes and circumpolar regions of the Universe. The major portions under cold desert trans Himalayas belong to Ladakh (Jammu & Kashmir) and Lahaul -Spiti (Himachal Pradesh) followed by very small portions in Garhwal (Niti and Mana) beyond Badrinath and Nelang in Uttarkashi district of Uttranchal state. A very harsh and characteristic climate prevails in trans Himalayan cold deserts where the summer temperature touches upto 40°C while the winter temperature drops below minus 35°C and are characterized by barren mountains, nail biting winters, low humidity and sparse vegetation. The majority of local tribes of these regions follows Budhism followed by good number of Muslim population in Kargil and Leh districts. The dominant tribes are Boto, Balti, Drogpa, Argons etc. The cold arid regions of Jammu & Kashmir and Himachal Pradesh may be divided into seven major geographical valleys for exploration and enlisting the plant biodiversity viz. Indus, Nubra, Changthang, Suru, Zanskar, Lahaul and Spiti.

The flora of Ladakh and Lahaul-Spiti of trans Himalayan belts comes under alpine and high alpine zones and is dominated by annual and perennial herbs, followed by few stunted shrubs and bushes. The trans Himalayan cold desert has a characteristic vegetation differing from other parts of Himalayas due to prevailing unique climatic conditions and physiography. The vegetative growth starts at the commencement of summer when the melting snow provides abundant moisture. The alpine mountain slopes, meadows and alpine pasture lands abound in varieties of alpine and high alpine flowers making trans Himalayan barren mountains lush green for brief summer period (June -September). In broad sense, the vegetation of trans Himalayan cold desert may be studied under three categories - Alpine mesophytes, Oasitic vegetation and Desert vegetation.

Most parts of trans Himalayan cold desert of Ladakh and Lahaul-Spiti are inaccessible and remain cut off from the rest of the world due to rugged and difficult terrain and prolonged subzero temperatures which have forced the local people to be self sustained for most parts of the history. Like other parts of the Himalayas, these regions are considered treasure of medicinal, aromatic and other important plants which have been in use by the local tribes for day to day needs and this long dependence on plant wealth had enriched their knowledge about multifarious uses of plant biodiversity. Much of such valuable information are confined to tribes and ethnic groups. The science of ethnobotany plays an important role in collection and documentation of ethnobotanical wealth of such regions.

Field Research Laboratory (FRL), Leh of Defence Research & Development Organisation has carried out detailed ethnobotanical exploration covering far flung areas and higher passes of Ladakh and Lahaul - Spiti and documented in medicinal, aromatic, edible, aesthetic, fodder, etc based on ethnobotanical studies. The collected plant specimens were identified with the help of available literatures and Botanical survey of India, Dehradun. The ethnobotanical information were verified with local amchi physicians, senior citizens and available literatures.

Since the dawn of history man has been in search of ways to find cure and relief from mental and physical illnesses. India has one of the oldest, richest and the most diverse cultural traditions associated with the uses of medicinal plants. In India, over 7500 plant species are estimated to be used by approximately 4635 ethnic communities for human and veterinary health care across the various ecosystems ranging from Ladakh in the trans Himalayan cold desert to the southern coastal tip of Andaman & Nicobar Islands and from the deserts of Rajastan & Kutch to the Hills of the north east. The people of Ladakh and Lahul-Spiti of the trans Himalayan region has its own medical system called "Amchi system of medicine" (Tibetan Medicine) and the practitioners are called Amchis (Superior of all). Amchi medicine is based on principally on a skillful uses of plants, minerals and animal products. The important medicinal plant species of Ladakh and Lahaul - Spiti have been collected alongwith therapeutic uses during the botanical survey by the authors, some of them are *Aconitum sp, Colchicum sp, Corydalis sp, Dactylorhiza sp, Ephedra sp, Hippophae sp, Hyoscyamus sp, Hypericum sp, Inula sp, Gentiana sp, Podophyllum sp, Rhodiola sp, Saussurea sp*, etc.

Remoteness, cold desert high altitudes and prolonged frozen winters forced the local inhabitants for the search of edible wild plants. During the ethnobotanical exploration the authors have collected several such wild plants which are being consumed in the form of cooked as well as raw even now in far flung areas. Some of the potential plant species viz. *Lepidium sp, Rhodiola sp, Taraxacum sp, Urtica sp* have been analysed with respect to their nutritional profile.

Similarly there is a great scarcity of fodder plants and proper cultivation of fodder species is not in practice due to shortage of cultivable land and allied constraints. In addition to it, the natural pastures are also declining day by day due to environmental changes. However, the regions abound in several plant species which are being grazed by local livestock and are also being harvested by the locals for winter fodder. The important fodder plants collected on large scale are *Aconogonum tortuosum, Cicer microphyllum, Hippophae rhamnoides, Prangos pabularia* etc, the nutritional profile also support their fodder value.

The region is known for several such plants which have aesthetic importance. These plant species are being used as ornamental, cut flowers, dry decoration as well as for aesthetic value, by the locals even now in far flung areas. The important species are *Aquilegia fragrans, Epilobium sp, Meconopsis sp, Primula sp, Rosa sp,* etc. Similarly the region abounds with several plants which are being collected by local people as fuel plants, some of them are *Acantholimon lycopodioides, Artemisia sp, Berberis sp, Caragana sp, Hippophae rhamnoides* etc.

As we know the region is devoid of natural forest due to several climatic factors, FRL has surveyed and documented the indigenous plants which may be multiplied for afforestation in these cold arid zones. *Hippophae sp, Juniperus sp, Myricaria sp, Sophora sp, Tamarix sp, Ulmus sp* etc. are the potential forest species of these regions.

The rich diversity in flora and fauna, economically useful species, rare and endangered species, the genetic diversity, high degree of endemism, presence of alien elements in the ecosystems, usefulness of one of world' highest fresh water and brackish water lakes, high altitude passes and unique cultural heritage etc. make the trans Himalayan cold desert ecosystem unique and the most fragile in the world. During the ethno-botanical survey, the authors have observed that the frequency of some of important plants especially medicinal plants of Indian cold desert Himalayas with which it was earlier abound, have considerably declined due to their unscientific exploitation, natural calamities, road construction, uprooting for fuel, overgrazing and other activities. A list of rare, endangered and threatened plant species viz. *Aconitum heterophyllum, Bunium persicum, Dactylorhiza hatagirea, Ephedra gerardiana, Podophyllum hexandrum* and *Saussurea lappa* etc. has prepared based on botanical survey of these regions. FRL has actively engaged in collection, evaluation and maintenance of valuable germplasm of important plants. A field gene bank has been established with the objective to standardize, multiply and conserve rare endangered threatened plant species for scientific based sustainable utilization.

Lastly it gives me immense pleasure to appreciate the tireless efforts of the authors Dr OP Chaurasia, Scientist 'D', Dr Zakwan Ahmed, Director and Dr Basant Ballabh, STA of Field Research Laboratory, Leh for documenting and compiling this very valuable research findings based on ethnobotanical studies. The book will be of immense utility to scientists associated with ethnobotany, traditional medicinal systems, environments, horticulture, plant sciences and allied fields.

(Dr Narendra Kumar)

PREFACE

Plant biodiversity has been the source of food, medicine and other necessary things for human beings since times immemorials throughout the world. Due to characteristic climate, altitudinal variations and allied factors the Himalayas are endowed with rich biodiversity in flora and fauna which are closely associated to day to day life of local inhabitants of these regions.

Cold deserts are usually confined to high altitudes and circumpolar regions of the world. Indian cold deserts come under the trans Himalayan zone of Ladakh (Jammu & Kashmir) and Lahaul-Spiti (Himachal Pradesh) which are known for their characteristic environment. The region remains landlocked for six months due to heavy snowfall and prolonged subzero winters when the temperature drops below minus 35^{0}C while the temperature touches upto 40^{0}C in summer. The cold arid zones of Ladakh and Lahaul-Spiti of trans Himalayas may be classified into Indus, Nubra, Changthang, Zanskar, Suru and Lahaul-Spiti for systematic ethnobotanical studies.

The trans Himalayan belts are rich in ethnobotanical wealth for which several ethnobotanical explorations were conducted by the authors under various R&D projects to collect and identify plant specimens alongwith traditional knowledge with the objective to scientific based sustainable utilization of ethnobotanical wealth of trans Himalayan regions. The collected plant specimens and ethnobotanical information have been classified and documented in medicinal, aromatic, edible, fodder, ornamental, magico-religious, fuel and forest plants and the present compilation is the outcome of the same.

A total of 704 plant species of 298 genera, belonging to 63 families of ethnobotanical importance have been described alongwith their brief botany, families, habit, habitat and ethnobotanical utility under different chapters viz. medicinal plants, aromatic plants, edible wild plants, fodder plants, ornamental plants, magico-religious plants, fuel plants and forest plants. Besides some general chapters viz. physiography, plant survival in cold desert, ethnobotanical exploration, endangered plants, conservation & sustainable utilization and propagation & cultivation of important plants have also been discussed in the present compilation. In addition, all the listed plant species have been summarized in tabular form in category wise for quick go through the ethnobotanical diversity. The book also includes **438** photographs related to physiography, vegetation and different plant species of ethnobotanical importance, taken by the authors during the botanical exploration.

The present compilation will be extremely useful to researchers and scientists associated with ethnobotany, traditional medical systems, natural products, horticulture, environment and allied fields of plant sciences for scientific based sustainable and commercial utilization for the benefit of society in general and defence forces deployed in these sector in particular.

First of all the authors feel pleasure to express deep sense of gratitude to his excellency Dr A P J Abdul Kalam, Hon'ble President of India and Ex SA to RM, Secy Defence R&D & DG R&D for his earnest motivation and encouragement for ethnobotanical survey & documentation of trans Himalayan plant biodiversity.

We extend our sincere thanks to Shri M Natarajan SA to RM, Secy Defence R&D & DG R&D, Dr W Selwamurthy Chief Controller R&D (LS & HR), DRDO HQ, New Delhi and Dr Narendra Kumar Director Dte of Life Sciences & Secy Life Sciences Res. Board, DRDO HQ, New Delhi for technical guidance and financial support through various R&D projects. The authors greatly appreciate the support of Dr V K Aatre Ex SA to RM, Secy Defence R&D & DG R&D and Dr R V Swamy Ex Chief Controller R&D, DRDO HQ. We would like to express our special gratitude to Dr Brahma Singh Ex Director, FRL & Ex DALS, DRDO HQ for his immense guidance and constructive critics during ethnobotanical survey and its compilation.

It is our pleasure to acknowledge Col S K Sareen, Col D V Singh, Col (Dr) D P Attrey, Col B Raut the then Directors and all officers & staff members of FRL for their inspiration, valuable suggestions and assistance provided during every phase of research as well as during the compilation of the book.

We owe our deep regards to Sowa Rigpa (Amchi System of medicine) Research Centre (CCRAS), National Medicinal Plant Board, New Delhi, Botanical Survey of India Dehradun and Forest Research Institute Dehradun for providing herbarium facilities in identification of plant specimens, verification of ethnobotanical knowledge and allied facilities. During the compilation of the book, the authors consulted several publications, books, journals/periodicals, Ph D thesis regarding ethnobotanical information, we owe deep regards to them all.

We are also very much thankful to Amchis and local inhabitants of Ladakh and Lahaul-Spiti for sharing their valuable information to us during the ethnobotanical exploration. The authors extend our sincere thanks to Forest Department and Wild Life Department Leh, Kargil and Keylong for providing facilities as and when required during field studies. We also acknowledge Ladakh Autonomous Hill Development Council Leh & Kargil, local administration, 14 Corps, 3 Inf Div and 8 Mtn Div for providing necessary facilities during the field work.

(Authors)

CONTENTS

1

Physiography

Cold deserts are usually confined to high altitudes and circumpolar regions of the universe. Sixteen percent of total landmass are under cold arid zones. In northern hemisphere, these cold deserts include the perennial polar ice packs, ice deserts of Greenland, northern Canada, Alaska, the Arctic sea board of Eurasia covering Iceland, Spitzbergen island, northern Scandinavia, northern Siberia, Tundra desert and Tibetan plateau while in the Southern hemisphere, the continent of Antarctica is the land of ice. Gobi cold desert is one of the world's largest deserts and covers a vast area of land in the Mongolian plateau, covering parts of Mongolia, and China. Sometimes Gobi desert has been applied to all the desert and semi-desert lands east of the Pamirs and north of the plateau of Tibet. Similarly, Tundra cold desert plain which form the arctic lowlands of Europe and Asia, is also usually applied to a similar area in north America.

Indian cold deserts come under the trans-Himalayan zone. Major parts are confined to Ladakh in Jammu and Kashmir (J&K) and Lahaul & Spiti in Himachal Pradesh (HP). In J&K, the cold desert lies between 32° 15′- 36° N and 75° 15′ - 80° 15′ E. It covers approximately 68,321 sq. Km, besides 27,555 sq Km area which is under illegal occupation of Pakistan and China while in HP, the cold desert area lies between 31° 44′ 57″ - 32° 59′ 57″ N and 76° 46′ 29″ - 78° 41′ 34″ E and covers approximately 6488 sq Km. A very small pockets in Garhwal (Niti and Mana) beyond Badrinath and Nelang region in Uttarkashi district of Uttranchal state have similar climatic condition and terrain. Due to recent increase in humun activities, particularly in the form of tourist activities in this world's most fragile and threatened trans Himalayan cold desert, the region has acquired a special importance.

The trans Himalayan part of India stretches from the vicinity of Upper Nepal in the east to an area in north where Kashmir joins China and Afganistan. The two biggest ranges are the mid Himalayan range (Lesser Himalan range) and the inner Himalayan range (Great Himalayan range) which run parallel to each other and enclose between them an interminable land of plateau and valley, narrow, high and extremely arid where lie the cold desert lands of Ladakh and Lahaul-Spiti.

Ladakh, the largest, the loftiest, the land between earth & sky and the land of rock ruins, is the remotest part of India. The greatest extent is from northwest to southeast, being 360 km and widest distance is 335 km from Karakoram pass in the north to Rohtang pass in the south and the region is bounded in the north and north east with Tibet, north west with Balistan, in the west with Srinagar and in south with Himachal Pradesh. From the north to south the major ranges are Karakoram, Kailash, Ladakh and Zanskar range. More than 85 % of the area of Ladakh part lies above 5000 m or more above mean sea level. Physiographically the entire region may be classified in five major valleys : Indus, Nubra, Changthang, Suru and Zanskar which lie along the major rivers such as Indus, Shyok, Nubra, Zanskar and Suru rivers. The Indus river is the main river which follows the course from south east to north west through the greater part of the country. The region is famous for several lakes viz. Pangongtso (4,218 m), Tsomoriri (4,511 m), Tsokar (4,530m) etc. Similarly a number of hot springs abound in Ladakh, of which Puga and Panamik are well known.

Lahaul and Spiti : The area is separated in the north from Ladakh by Baralacha range and bounded in the east by Tibet and in the south east by Kulu and Chamba while Spiti is separated from Lahaul by a high mountain running towards the north from the mid Himalayan range. The both Lahaul and Spiti have a tenuous link over the Kunzam pass (4500 m). The average elevation of the peaks in Lahaul is between 4550 m and 5400 m. The lowest point is about 2750 m where the river Chenab makes its exit from the district. The region is really three valleys in one Chandra river valley, Bhaga river valley and Chandra - Bhaga river valley (Chenab valley). Spiti region in contrast to Lahaul, is higher in elevation with more rugged and difficult terrain.The average elevation is about 4600 m. Across a broad saddle between Lahaul, Spiti and the routes to Tibet, near the north eastern face of the arc, lies the Baralacha pass which gives both Lahaul and Spiti access into each other and into Ladakh. Closer to the middle of the arc, Lahaul and Spiti are separated from each other by a short off-shoot of the mid Himalayan range, the Kunzam ridge with Kunzam pass. Across the mid Himalaya are the two passes, the Hamptu pass and the Rohtang pass and both these passes lead into a backyard of Lahaul. Three major rivers Chandra, Bhaga and Spiti, with their tributaries, constitute the drainage system of Lahaul and Spiti belts.

Physical characteristics of cold desert

- Coarse, porous and immature sandy soils prone to wind and water erosions,
- Heavy influx of infra red and ultraviolet radiations
- Low air density - reduced oxygen levels
- Low concentration of atmospheric carbon dioxide

- Short agriculture (cropping) season 5 months (May-Sep)
- Very long freezing winters 5 to 7 months (Nov-April)
- Large variation in seasonal temperature : 40°C to -40°C
- Large diurnal variations in daily atmospheric temperature during cropping season ranging from 0°C to 35°C
- Fast blowing winds 40 to 60 km per hour mainly in afternoon hours,
- Precipitation mostly in the form of snow mainly during winters
- Irrigation water mostly from snow melts, major portion of which is unharvested
- Low relative humidity during growing season
- Soil moisture remains frozen during winters
- Very little to no rainfall.

Geology

Ladakh lies behind the line of the highest elevation and is a zone composed of a continuous series of highly fossiliferous marine sedimentary rocks, ranging in age from the earliest Paleozoic to the Eocene. As a tectonic unit, the great Himalayan range of which Ladakh forms a part, is made of the root of Kashmir nappe, the principal geanticline within the main Himalayan geocyncline, consisting of the Archean and Precambrian sedimentary rocks together with large bodies of intrusive granites and basic masses.

With the exception of a part of Ladakh, which consist of tertiary rocks and a basin of mesozoic sedimentary rocks on the northern flanks of Zanskar mountain, by far the larger part of the inner mountains is composed of igneous and metamorphosis rocks, granites, gneisses and achiest Spiti shales and Kioto limestone of Jurassic and Monotis shales of Triassic period are found in the central and southern parts of Ladakh, resting on upper Triassic limestone. The evidences for the existence of cretaceous series are obtained from *Gryphae vesiculosa* fossils on Leh-Yarkand road, consisting of calcareous sandstone. During the Tertiary period, retreating waters of Tethys Sea, which left a basin of lower Tertiary deposits from Rupshu to Kargil and Dras, occupied Ladakh. The territories of Ladakh rest over genesis and metamorphic rocks. The salt rock of Ladakh bear evidence of a progressive desiccation of the Changcheno, Dipsang and Lingzhitang at an elevation of produced by the desiccation and silting up of saline lake basin without any outlet. Limestone is the most prevalent rock found in the range separating Rukchu from Zanskar. All valuable minerals are found in Ladakh. The region has huge quantities of gold, silver, iron, mica, precious stones, sulphur and other minerals. Borax is found in huge quantity. Soda is also found in many places in Ladakh.

Similarly Lahaul-Spiti is characterized by the presence of a complete sequence of geological formations dating from the Precambrian to the cretaceous period with short breaks in carboniferous and Jurassic periods and by the occurrence of fossils of ancient vascular plants. The stratigraphic sequence of the formations consists of two groups viz. Schistose group and Calcareous group. The Schistose group comprises biotite-scists, schistose phyllites, phyllites, slates, paragneissic bands, quartzites and quartz-mica-schists. The crystalline limestones, flaggy and slaty limestones, calcareous phyllites, calcareous schists, dark grey phyllites interbedded with limestones.

Climate

The climate is characterised by great extremes of heat and cold, coupled with excessive dryness. The intensive solar radiation and rarification of the atmosphere cause this southerly wind. The air thus moves towards the North Pole in a southerly current. The highly rarified atmosphere offers so slight a check to the transmission of sun rays that the noon temperature often becomes 25°C to 40°C more than it is any part of India, while during night the temperature falls below freezing points. The temperature shows great fluctuation during the different months of the year. The summer months are short, while the winter months are long with the temperature going as low as -59°C (lowest recorded from Dras), -23°C at Kargil and -30°C at Leh. During prolonged winter the rivers freeze and form natural roads. (Table 1).

Table 1 : Meteorological data at Leh (JAN-DEC 1991-2005)

MONTHS	AV. TEMP (°C)		TEMP (°C) LIMITS	
	Max.	Min.	Max.	Min (%)
JAN	03.1	-11.1	08.4	-30.2
FEB	06.8	-09.1	12.4	-30.0
MAR	08.3	-05.3	18.4	-21.5
APR	14.2	-0.03	22.5	-7.8
MAY	17.6	03.5	27.6	-1.5
JUN	23.5	07.7	34.8	02.0
JUL	27.6	12.0	35.0	08.0
AUG	25.7	12.1	34.2	04.0
SEP	25.5	08.3	30.5	04.0
OCT	15.7	-01.7	25.4	-07.0
NOV	10.1	-07.8	17.4	-13.0
DEC	06.1	-14.8	19.1	-20.5

Source : Climatic data recorded at Field Research Laboratory, Leh.

Rain and Snow

Rainfall / snowfall pattern, amount and its distribution during the cropping season directly influence the plant biodiversity and agricultural productivity by affecting soil moisture, soil aeration, soil temperature, air temperature and relative humidity. It is surprising to note that the rain is very little in these areas. Except parts of Lahaul valley, in most of valleys of trans-Himalayas, it drizzles for an hour or two at the most three to four times in a year. Snow falls more often but the quantity is not much ever being more than six to eight inches deep. Even in the summer months light snow falls at night in the upper reaches of the mountains is common. Dras, however, which is one of the coldest places in the world, has a very heavy snowfall. Similarly average relative humidity remains around 40% – 50%, which makes the area totally arid. Absence of humidity helps to keep the atmosphere clear. (Table 2).

Table 2 : Precipitation (rainfall / snowfall) % RH& wind speed in Leh-Ladakh (1991-2005)

Months	Rain/Snowfall (in cms.)	Relative Humidity (%RH)	Max Wind Speed (knots)
January	1.06	65.0	54
February	1.60	56.8	50
March	0.86	55.2	70
April	0.61	48.1	83
May	0.59	49.3	88
June	0.48	43.9	89
July	1.27	51.4	82
August	1.65	51.9	75
September	0.91	50.0	75
October	0.31	50.8	63
November	0.18	52.8	60
December	0.96	59.9	62

Source : Regularly recorded data at FRL, Leh.

Wind

The constancy of the prevailing winds in different parts of Ladakh and Lahaul-Spiti is one of the most interesting phenomenon of the nature. During the warmer part of the day the early morning breeze soon transforms into a mighty wind or whirlwind, through the changes produced by the varying diffusion of solar heat through the atmosphere. Due to dry climatic conditions, wind blows with high velocity; sometimes it blows 50 to 60 km per hour mainly in the afternoon, in some lofty parts of the region, with the sandy plateaus.

Soil

The soils are predominantly sandy, derived from weathered debris of the rocks and are subject to the great diurnal as well as seasonal alteration of temperature which leads to a mechanical disintegration of the rocks, producing an abundance of loose debris. The soils in general have been classified as grey, light arid soils of low fertility status. In most parts the soil is poor in organic matter and nitrogen content. The plant nutrients however, are sufficiently available in the soils due to almost nil weathering of the rocks, which is due to extremely low precipitation and temperature. The soils are rich in potassium and its availability to plants is also high, since the weathered complex containing potash cannot be leached under low precipitation. Micronutrients have been found deficient in most of the villages of Ladakh and there is no practice of application of micronutrients by the farmers of the region. Hence, it becomes necessary to apply all the micronutrients regularly, especially in deficient areas. The soil of Nubra and Drass valleys is morainic, because of glaciers and Kargil soil is black due to grey shale while the soils of arid regions of Lahaul & Spiti are of almost similar nature, texture and quality.

History, People & Culture

In Ladakh

The earliest settlers in Ladakh were the Drogpa emigrants from Dardistan who occupied the lower reaches of the Indus valley, an area popularly known as Sham. These people brought with them the type of faith and beliefs that had prevailed in Dardistan at the time of their migration. The Mon-Aryan by race-who came later from the Karja or Kulu side and occupied the Rong area with Gya as its capital also brought with them a form of Buddhism which prevailed in their original home at the time. The last wave of immigrants rolled into Ladakh from Tibet side about the close of the 10th century. This wave which introduced the Mongolian strand into the racial mosaic of Ladakh eventually became the predominant element in its population and brought with it the form of Buddhism that had by this time been established in Tibet by Indian teachers (Kaul *et al.*, 1995).

Mongols : A Race of Tibetan Origin (Jina, 1996)

It is believed that many tribes from Central Asia : Meyak, (Me-Nyag), Do'ng, etc were among popular nomads who first came from the upper side of Indus valley and settled in Changthang sub-division of Leh district. However the Mongols of Tibetan origin migrated later from Central Tibet during 9th century A.D. when Langdarma came into power in Central Tibet. The descendants of Langdarma, their forecommers were able to establish the first political state in Ladakh and ruled upto 1834 (Marx, 1980; Kaul *et al.*, 1995; Koshal, 2000).

Dards: An Indo-Aryan Race

The Himalaya is the original home of many ancient tribes. Like other tribes Dards migrated to inner and outer Himalaya from various entrance points beginning from pre-historic to early Christian era. Ethnologically Dards are of the Indo-Aryan stock (Jina, 1996). There is no definite point and time when Dards, migrated towards Indus valley. However, they came in many batches and established their colonies at Skardu, Dras, Skabuchan and Kha-la-che villages in Ladakh. (Kaul *et al.,* 1995; Jina, 1996). Drogpa ancestors were nature worshippers. They worshipped the sun, the stars, the water and the mountains. They were also animists. Drogpa worships the God of beauty for good health, happiness and joy. (Kapur, 1987; Kaul *et al.,* 1995 and Jina, 1996).

Mons : An Ancient Race

Mons called the inhabitants of the country of Manus (Desi-si-manus). Manus was the country lying between Kulu and Lahaul. Abidin believes that Manus would be Monhali (Manali). In early days there was no king who ruled over whole 'Za-Zung' area. It was divided into many small independent republics, which were ruled by 'Magspons'. So, there is a possibility that some tribes of Aryan origin migrated from Kulu, Manali, Lahaul and Spiti into Ladakh under different Magspons and settled down at the upper reaches Kashmir. Later from these Magspons, their present race Mon developed (Jina, 1996).

Tibetan also believes that the aboriginal tribes were only known as 'Mons', who settled at the north of Tibet, where the whole area was covered with dense forest and high mountains. They had made special distinction with their own identity among the nomads and settled on the upper side of Kashmir in Zanskar (Kaul *et al.,* 1995). Mons are fond of music. It is therefore, certain that Mons were the first settlers and missionaries of Buddhism in lower Ladakh during 1st or 2nd century A.D. (or may be earlier). Afterwards Mongols came into Ladakh and established their political powers over Ladakh. They extended their Kingdoms upto Zanskar and then Mons became their citizens (Jina, 1996).

Baltis : An Ancient Race

The ancient history of Baltistan proves that Baltis were Buddhists and consisted the elements of Aryan and Tibetan cultures, they were braver than the average western Tibetan. Baltis came in Ladakh during 10th century for the first time. But they were then Buddhists, however, they were the first western Tibetans who latter became Mohammedans (Kaul *et al.,* 1995; Jina, 1996).

Buddhism

Buddhism was introduced in Ladakh probably around 3rd Century BC when Ashoka sent his missionaries in Ladakh. Many inscriptions of this period have been discovered in Zanskar. Besides, many 'Mon' castles have also been discovered in Zanskar, which reveals : (a) Buddhism was brought in Ladakh by Buddhist missionaries sent by King Ashoka (b) Mons had arrived in Zanskar earlier than King Ashoka's Buddhist missionaries (c) When the Buddhist missionaries arrived in Zanskar, Mons embraced Buddhism and later propagated it in Zanskar (Jina, 1996 & 2000).

The Buddhists of Ladakh have a number of peculiar customs and social institutions. The most striking of these is polyandry, which has been in vogue in the land almost since its history began. The custom was abolished by law more than four decades ago, but the momentum, acquired for centuries past is not yet exhausted, particularly in the country side.

Social Customs : People in Ladakh are simple, honest and very hospitable, particularly in villages. Local beer (Chhang) and Sattu are the main drink and dish of the people. Channg is an important item for the success of every feast. Exhibitions of female dancers frequently form a principal part of an entertainment; but the performers are more remarkable for their costume than their graceful movements (Jina, 1996).

In Lahaul & Spiti

The socio-culture aspects of Ladakh and Lahaul-Spiti are more or less very much similar due to similar topography. The name of Lahaul-Spiti derives from the names of its two divisions viz. Lahaul and Spiti. The term Lahaul has probably originated from the Tibetan word Loh-Yul means the southern country or Lhashi-Yul means the country of Gods. The name Spiti is locally pronounced as 'Ptit' meaning a middle province in Tibetan dialet, appears to have been given to territory on account of its location between the then British India, Kashmir, Tibet and Bashahr.

The people of Lahaul-Spiti are a mixed race. The study of their language, which possesses strong characteristic of the Mundari language, induces philologists to believe that times immemorial, (2000 BC), a Munda-speaking race spread from Bengal and Central India to the frointiers of Tibet (Francke, 1977). The Aryans and semi-Aryans also came from the west and the south. Swangla and Gaddi are the main tribes of Lahaul & Spiti areas.

The features of Lahaulis are a combination of Mongolian and Aryans traits. The men are possessing considerable business acumen while the women are very hard working. The Women are very much involved in agricultural and livestock activities. The people of Spiti area are of Mongoloids and by and large men are lazy and lethargic and like Lahaulis the Spitian women are hard working and are actively involved in most of works.

The traditional dress of the men consists of an undershirt, a long shirt (Cholu) and pajamas. A long piece of cloth (gachi) is often used as belt to keep the cholu tied to the body. Women wear a jacket over cholu in winter. A scarf or jogi is used as a head-wear. Brownish-red (marron) and dark-brown colours are popular among women. The common ornaments of women are ear-rings made of silver/ gold, 'kanthi' (a silver necklace) and 'Chully' or 'long' (a gold nose-ring or nose-pin).

The most ancient religions were probably 'phallus' and snake 'worship' the two representing the creative powers of the Sun and Water. At present, the most common religion followed is a combination of Hinduism and Buddism. The monasteries have preserved many types of paintings of saints, demons, evil looking beasts, devil masks, Yak's tail and sword, drums and other instruments. The people believe in super natural powers, worship God and Goddesses and obey the orders of haunted men/women.

The staple food of these regions are wheat, barley, buckwheat, potato and meat. Now-a-days rice is getting very much popular. Drinking is very common in these belts and the local beer (lugari) and local whisky (Chhang) are the popular drink. Saltis tea is taken by all aged group.

❑❑❑

2

High Altitude Vegetation

The flora of Ladakh and Lahaul-Spiti of trans Himalayan belts comes under alpine and high alpine zones and is dominated by annual and perennial herbs, followed by few stunted shrubs and bushes. The trans Himalayan cold desert has a characteristic vegetation differing from other parts of Himalayas due to prevailing unique climatic conditions and physiography.

The vegetative growth starts at the commencement of summer when the melting snow provides abundant moisture. Due to brief growing period, majority of plant species are in full bloom during July - August but start disappearing by the end of September. The alpine mountain slopes, meadows and alpine pasture lands abound in varieties of alpine and high alpine flowers making trans Himalayan barren mountains lush green for brief summer period. It is very interesting to note that plants that are in fruit at lower altitudes, may be found in flowering stage at higher altitude.

The vegetation of Ladakh and Lahaul-Spiti differs significantly from the rest of the Himalayas. In broad sense, the vegetation of trans Himalayan cold desert may be studied under three categories - Alpine mesophytes, Oasitic vegetation, Desert vegetation. At the same time the vegetation of Lahaul valley is slightly differing from other parts of Cold desert, because of high humidity and that is why the entire area may be studied under Temperate vegetation, Alpine mesophytes, Oasitic vegetation, Desert Vegetation.

a) Temperate vegetation : This vegetation is very much localised and is available between tract of Thirot and Udaipur of Lahaul valley. The vegetation is characterized by woody elements. The main composition of this vegetation in woody groups are *Pinus wallichiana* (Blue pine), *Picea smithiana* (Spruce), *Cedrus deodara* (Devdiar), *Betula utilis* (Birch), *Juniperus macropoda* (Juniper), *Juniperus recurva* (Juniper), *Juglans regia* (Walnut), *Prunus armeniaca* (Apricot) and different species of *Salix, Populus, Malus, Sorbus* etc.

The major shrub species of this zone are *Berberis pachyacantha, Cotoneaster sp, Ephedra gerardiana, Fraxinus xanthoxyloides, Hippophae rhamnoides var turkestanica, Hippophae salicifolia, Juniperus communis, Lonicera sp, Rosa macrophylla, Rosa webbiana* etc while *Aquilegia fragrans, Codonopsis clematidea, Galium aparine, Geranium pretense, Gaultheria tricophylla, Inula grandiflora, Lepidium sp, Origanum vulgare, Pedicularis sp, Podophyllum hexandrum, Primula denticulata, Ranunculus lateus, Rubus saxatilis, Rumex sp, Senecio sp, Silene sp,* etc. belong to herbaceous as forest undergrowth.

The marshy areas are covered with *Bromus sp, Eleocharis palustris, Halerpestis sarmentosa, Poa sp, Polygonum sp, Carex sp, Scirpus sp,* etc. The common species of meadows and slopes are *Achillea millefolium, Capsella bursa-pastoris, Heracleum lanatum, Origanum vulgare, Lotus corniculatus, Thymus serpyllum, Phlomis bracteosa, Trigonella sp,* etc. Similarly the common grasses are *Agrotis sp, Bromus sp, Carex sp, Elymus sp, Festuca sp, Poa sp, Stipa sp,* etc.

b) Alpine mesophytes : This zone is also characterized by high humidity and more rainfall. The flora of Suru valley and parts of Lahaul valley shows the characteristic of alpine mesophytes. Majority of plants growing in this zone, are also found in temperate regions. The common mesophytic plant species are *Podophylllum hexandrum, Lavetera kashmiriana, Lotus corniculatus, Astragalus rhizanthus, Verbascum thapsus, Oxyria digyna, Capsella bursa- pastoris* etc. Only about 10% of alpine mesophytes are typical cold desert plants like *Delphinium brunonianum, Potentilla sp, Leontopodium sp, Taraxacum sp* etc. which are widely distributed in high altitude areas.

Salix daphnoides, Morus alba, Myricaria elegans etc. are found along river belts and in marshy depressions. *Prangos pabularia, Ferula jaescheana, Juniperus macropoda* are dominant on mountain slopes and rocky areas. Among herbs, *Colchicum luteum, Tulipa stellata var chrysantha, Euphrasia officinalis* are very common.

c) Oasitic vegetation : Plants of this zone are generally found near habitations like Kargil , Leh, Khoksar, Keylong, Kaza, Trilokinath, Udaipur etc. and are cosmopolitan in nature. This type of vegetation is represented by a variety of exotic as well as indigenous species, growing near habitation, along water channels, streams, nullahs and in moist places. The common taxa of this vegetation is *Chenopodium foliosum, Dianthus anatolicus, Epilobium roseum, Mentha longifolia, Potentilla cuneata, Sedum ewersii, Rhodiola quadrifolia, Lancea tibetica, Trifolium pratense, Melilotus alba, Melilotus officinalis, Galinsoga parviflora, Stachys tibetica,* etc.

Except *Hippophae rhamnoides* and *Lonicera species,* there is no natural vegetation. The commonest species introduced in these areas, are species of *Salix, Populus, Juglans, Morus, Pyrus, Prunus* etc.

d) Desert vegetation : This zone is characterized by little rainfall, low humidity, extreme fluctuation of diurnal temperature and high velocity winds etc. The long stretches of Ladakh and Lahaul-Spiti regions viz. Central Ladakh,

Puga areas, Tangtse, Kaza, Pin valley are characterized by stunted, cushion type desert vegetation of tufted plants. Due to high velocity winds, which are constantly blowing in these areas, the plants tend to become prostrate, thick woolly, cushion forming, bushy, hardy, spongy and with long deep rooted roots and small leaves. The herbaceous species show a mixture of Tibetan, Siberian, Chinese and Afganistan flora. Majority of plant species are characterized by stunted, cushioned, prostrate, thick woody, spongy and long roots and succulent small leaves.

Common plant species which are growing around barren valleys of Indus, Changthang and Spiti, are *Polygonum avicule, Atriplex crassifolia, Corydalis flabellata, Caragana pygmea, Echinops cornigerus, Peganum harmala, Capparis spinosa, Berberis ulicina, Tanacetum sp,* etc. Similarly desertic elements found growing around high passes like Khardungla (5,520 m.), Changla (5,286 m.), Penzila (4,204 m.), Tanglangla (5,255 m.) and Kunzum pass (4500 m) are *Thylacospermum caespitosum, Acantholimon lycopodioides, Arnebia euchroma, Caragana pygmea, Corydallis crassima, Dracocephallum heterophyllum, Krascheninnikovia ceratoides, Meconopsis aculeata, Primula rorea, Rhodiola imbricata, Rosularia alpestris, Waldheimia tomentosa, Euphorbia tibetica, Arenaria bryophylla Lindelofia stylosa, Nepeta longibracteata, Christolea crassifolia, Draba setosa, Saussurea bracteata, Saxifraga sp, Waldheimia sp etc.*

Although there are no natural forest in Ladakh and parts of Lahaul-Spiti, there are frequently thickets along streams and along river belts. These have a typical association of *Myricaria sp, Hippophae sp, Rosa sp, Lonicera sp,* etc. The introduced forest species are *Populus, Salix, Prunus, Malus,* etc.

Cultivated vegetation

A number of cultivated plants of various categories are grown in Ladakh and Lahaul-Spiti which are the major source of economy, besides their own consumption. Besides indigenous cultivated crops, a number of cultivated crops of various utility have been introdued time to time.

Cereal crops : The main food crops are *Triticum aestivum* (Wheat), *Hordeum vulgare* (Barley), *Zea mays* (Maize).

Oil crops : Brassica juncea (Lahi) *Carthamus tinctorius* (Safflower), *Glycine max* (Soyabean), *Helianthus annuus* (Sun flower), *Linum usitatissimum* (Lentil),

Pulse crops : Cicer arietinum (Black gram), Lathyrus sp (Bakla). *Lens esculenta* (Lentil), *Phaseolus aureus* (Moong), *P. mungo* (Urd) *P. vulgaris* (Lobia), *Pisium sativum var. arvense* (Local pea), etc.

Fruit plant : Fragaria grandiflora (Strawberry), *Juglans regia* (Walnut), *Malus pumila* (Apple), *Morus alba* (Mulberry), *Prunus armenica* (Apricot), *Vitis vinifera* (Grape).

Vegetable crops : Alliun cepa (Onion), *A. porum* (Leek), *A. sativum* (Garlic), *Brassica caulorapa* (Knol khol), *Brassica chinensis* (Chinese cabbage), *B. oleracea*

var. botrytis (Cauliflower), *B. oleracea var. capitata* (Cabbage), *B. oleracea var. gemmifera* (Brussels sprouts), *B. oleracea var. italica* (Broccoli), *Beta vulgaris* (Beet root), *Brassica rapa* (Turnip), *Daucus carota* (Carrot), *Raphanus sativus* (Radish), *Helianthus tuberosus* (Sunflower), *Solanum tuberosum* (Potato), *Phaseolus vulgaris* (French bean), *Pisum sativum* (Peas), *Vicia faba* (Broad bean), *Citrulus lanatus* (Watermelon), *Cucumis melo* (Muskmelon), *C. melo var. utilissimus* (Longmelon), *C. sativus* (Cucumber), *Cucurbita moschata* (Pumpkin), *C. pepo* (Summer squash), *Lagenaria siceraria* (Bottle gourd), *Momordica charantia* (Bitter gourd), *Amaranthus tricolour* (Amaranth), *Beta vulgaris var. benghalensis* (Spinach beet), *Brassica juncea* (Vegetable mustard), *Coriandrum sativum* (Coriander), *Mentha piperita* (Mint), *Spinacea oleracea* (Spinach), *Trigonella foenum-graceum* (Fenugreek), *Apium graveolens* (Celery), *Lactuca sativa* (Lettuce), *Petroselinum crispum* (Parsley), *Capsicum annuum* (Capsicum), *Lycopersicon esculentum* (Tomato), *Solanum melongena* (Brinjal), *Abelmoschus esculentus* (Okra), etc.

Fodder Crops : Avena sativa (Oats), *Brassica juncea* (Mustard), *Medicago falcata* (Lucerne), *M. media* (Lucerne), *M. sativa* (Lucerne), *Pisum sativum* (Peas), *Sorghum vulgare* (Sorghum), *Zea mays* (Maize).

Ornamental plants : Acroclinum roseum (Paper flower), *Althaea rosea* (Hollyhock), *Antirrhinum majus* (Snapdragon), *Bellis spp* (Daisy). *Calendula officinalis* (Calendula), *Callistephus chinensis* (Aster), *Celosia argentea* (Cock's comb), *Centaurea cyanus* (Corn flower), *C. moschata* (Sweet Sultan), *Chrysenthemum coronarium* (Chrysanthemum), *Clarkia elegans* (Clarkia), *Coreopsis tinctoria* (Coreopsis), *Cosmos bipinnatas* (Cosmos), *Dahlia variabilis* (Dahlia), *Delphinium hybridum* (Larkspur), *Dianthus barbatus* (Sweet Willium), *D. caryophyllus* (Carnation), *Dimorphotheca sinuate* (Dimorphotheca), *Eschscholzi californica* (California poppy), *Gaillardia pulchella* (Gaillardia), *Gladiolus spp* (Gladioli), *Helianthus annuus* (Sunflower), *Helichrysum bracteatum* (Straw flower), *Iberis amara* (Candytuft), *Impatens balsamina* (Balsam), *Kochia scoparia var. trichophylla* (Kochia), *Lathyrus odoratus* (Sweet peas), *Linaria maroccana* (Linaria), *Linum grandiflorum var. rubrum* (Linum), *Matthiola incana* (Stock), *Mesembryanthemum criniflorum* (Ice flower), *Molucella laevis* (Bells of Ireland), *Papaver rhoeas* (Poppy), *Petunia hybrida* (Petunia), *Pimpinella monoica* (Lady's lace), *Phlox drummondii* (Phlox), *Portulaca grandiflora* (Portulaca), *Rosa indica* (Rose), *Spanoria calabrica* (Saponaria), *Syringa spp.* (Lilac) , *Tagetes erecta* (Marigold), *T. patula* (French marigold), *Thuja spp.* (Thuja), *Tropaeolum majus* (Nasturtium), *Verbena hybrida* (Verbena), *Viola tricolor* (Pansy), *Zinnia elegans* (Zinnia).

Forest plants : Betula utilis, Elaeagnus angustifolia, Hippophae rhamnoides spp. turkistanica (Seabuckthorn), *Hippophae salicifolia* (Seabuckthorn), *Myricaria squamosa, Populus caspica, P. ciliata, P. deltoides, P. euphratica, P. nigra, P. sylvestris, Robinia pseudacacia, Salix alba, S. angustifolia, S. babylonica, S. caesia, S. daphnoides, S. denticulata, S. excelsa, S. hastata, S. sclerophylla, S. tetrasperma etc.*

❑❑❑

3

Plant Survival in High Altitudes

The cold desert trans-Himalayan belt, characterized by physical factors such as sub zero temperature, low atmospheric pressure, low partial pressure of metabolic gases (oxygen and carbon dioxide), high solar radiation, turbulent winds and steepness of slopes has resulted in dramatic and drastic changes in the ecosystem.

Plant species of cold desert zones of India are mostly xerophytes followed by mesophytes. The flora is dominated by annual and perennial herbs followed by a few dwarf bushes or shrubs. The vegetative growth starts with the onset of summer. The melting snow provides abundant moisture to the plants to run their full course of development with great rapidity. The mountain slopes, alpine meadows, moraines and pasture lands give a spectacular display of varieties of flowers in otherwise barren mountain during July and August. The remarkable feature of this zone is continuous and successive changes in the floristic pattern during growth period. After melting of snow and onset of summer the cruciferous plants such and *Draba, Christolea, Choriospora, Mathiola, Arabis* are the first to bloom followed by species of *Corydalis, Gentiana, Primula, Saxifraga* and many others. The late summer (Aug - Sept) of this zone is dominated by Asteraceous taxa viz. *Saussurea, Anaphalis, Leontopodium, Jurinea, Waldheima,* and others.

Adaptation

Adaptation to the environment by the potential flora of a region leads to certain changes in underground and aerial parts of plants for their survival. The vegetation of the cold desert trans-Himalayas consists of a highly specialized group of plants with metabolic and reproductive strategies suited for maximizing their activity in specialized harsh climatic conditions. The very little rainfall, extreme diurnal fluctuations of temperatures, low humidity, high intensity of UV radiations and soil textures are some of characteristics features, the impact of which are clearly observed in high altitude plants. The plants of high altitude cold desert exhibit a number of ecological, morphological and physiological adaptations which help them to counteract the impact of harsh climate prevailing in this zone. The following peculiarities in plant parts have been observed in cold desert region:

(i) Root System

The root system in most of the plant species is very deep and extensive in order to absorb water from depth as this zone is known for aridity. Such root system also helps to withstand strong winds, snow blizzards and is escaping damage from subzero ambient winter temperature. In some cases roots become swollen to retain moisture and food storage to meet plant requirement. *Capparis spinosa, Hyoscyamus niger, Inula racemosa, Prangos pabularia, Rheum webbianum, Saussurea lappa* etc. belong to this category.

(ii) Growth Peculiarities

The vegetation is dominated by herbs followed by a few stunted and much-branched shrubs or bushes such as *Caragana sp, Hippophae sp, Lonicera sp, Myricaria sp, Tamarix sp*. The dwarf condition results from suppression of internodes due to slow growth and brief growing period (May- October) which may be correlated with extreme cold, high solar radiation and retardation of plant sap flow. In some cases like *Lancea tibetica, Arabis sp, Draba sp, Gentiana sp, Corydalis sp* etc. the plants become so small, that one may not even notice them in fields.

Another adaptation is the underground modified stem viz. rhizome and bulbs by which the perennial herbs survive in the winter such as *Bunium persicum, Carum carvi, Podophyllum hexandrum, Rhodiola imbricata* etc. Poplar and Salix, the main forest tree species become dormant during winter but with the onset of summer they break dormancy and the dormant buds of these plants unfold and start growing.

Similarly leaves show high degree of adaptation ranging from the most delicate to the conspicuously coriaceous which in turn provide characteristic shapes such as spinescent cushion, soft cushion, solid mat and woolly appearance etc. Spinescent cushion is the simplest adaptation in which the stem is reduced and the leaves are either much dissected or linear with pointed apex as noticed in *Acantholiman lycopodiodies, Astragalus rhizanthus* and *Caragana versicolor etc.*

The rocky plants which are exposed to the fury of icy winds form soft cushion in which all the aerial parts particularly leaves are densely hairy, pubescent and villous. This is prominent in the members of Asteraceae e.g. *Saussurea gossypiphora, Leontopodium leontopodium, Waldheimia tomentosa* and *Jurinea sp*. These features may be correlated with thermal insulation and to control rate of transpiration. In some species the stem is much reduced and followed by closely rosette arrangement of tiny leaves which give the appearance of solid mat as in case of *Arenaria bryophylla, Stellaria sp* and *Androsace sp, Rhodiola imbricata. Thylacospermum caespitosum* of Caryophyllaceae, a very striking species of extreme high altitude regions, forms a large hermispheric mound on the exposed rocky slopes.

(iii) Physiological Specialization

Transpiration, the loss of water in the form of vapour through stomata, indicates the nature of plants. Though the area is arid, most of the plants have developed adaptative features to minimize the loss of water. Species of *Rhodiola, Sedum* and *Bergenia* possess thick, fleshy, leathery and succulent leaves whereas the leaves of *Acantholimon* and *Astragalus* become needle like thus minimizing the transpiration rate. In the species of *Saussurea, Leontopodium* and *Waldheimia* the surface of leaves is hairy with waxy coating which checks the transpiration rate.

In high altitude cold desert zone the temperature plays an important role. The high altitude plants are able to survive in low temperature as they have a higher rate of metabolic activity at low temperature. But at the same time this ability has a great disadvantage to them because the metabolic activity increases with the rise in temperature which leads to the depletion of carbohydrate at a very fast rate. This rapid utilization rate of carbohydrate reserve due to higher respiration rate in warm climate has been reported to be a limiting factor in the extension of the cold desert species to warmer region.

Another physiological specialization in the high altitude plants are frost resistance. The freezing injury is mainly caused by the ice crystal formation within the cell and less either in the form of inhibition or reduction in the ice crystal formation or a sort of increased resistance of protoplasm against the deformations caused by the formation of ice crystals. The starch grains are accumulated in the chloroplast through photosynthesis which is converted into sugar as night falls or often they even produce soluble sugars directly through photosynthesis instead of starch. This sugar increases the concentration of cell sap and therefore the increased cell sap concentration lowers the freezing point by several degrees. In this way the cold desert plants are able to survive in low temperature (-15^0C to -20^0C).

(iv) Reproductive Strategies

The successful survival of a species in a given environment depends upon its effective mode of reproduction which help the species for both multiplication and perennation during unfavorable periods. The cold desert plants have to complete the entire lifecycle starting from seedling growth to sprouting of leaves and flowers, fruiting and dispersal of seeds during the short growing season. Thus the reproduction and seed dispersal among the cold desert plants enable them to survive in sparse vegetation. Reproduction is carried out both by sexual and vegetative methods.

Majority of cold desert plants multiply vegetatively. The vegetative propagating organs like runners and stolons, sub-aerial modified form of stem as found in *Ranunculus, Saxifraga stenophylla, Geranium pratense, Nepeta longibracteata* help in proliferation while on other hand modification of stem such as rhizomes, tubers and bulbs perform the function of both multiplication and perennation as observed in *Colchicum luteum, Rhodiola heterodonta, Primula rosea, Podophyllum hexandrum* and *Allium carolianum* etc.

In cold desert the sexual reproduction is uncommon due to low temperature and short growing period. In order to make sexual reproduction easy and effective these plants have developed certain physiological adaptation which help them during the various stages of reproduction like development of pollens and ovules and attracting the pollinating agents etc. for successful pollination and seed formation. During day time the flowers as in *Anemone, Ranunculus, Saxifraga* etc., are usually heliotropic and have highly reflective inner surface of petals which act as small disc antenna focusing the reflected light and heat on the centre of flower through which reproductive parts such as stamens and carpals gain extra energy. This accumulated energy helps in pollination. It has been reported that visiting insects on a flower for nector feeding also produce energy. In this way the heat so required enhances their metabolic activities including reproduction. Therefore a sort of symbiotic relationship is formed between the insects and plants in cold desert areas.

The colour of flowers also helps in sexual reproduction. Majority of cold desert plants such as *Delphinium brunonianum, Aconitum heterophyllum, Corydalis flabellata, Geranium pratense, Meconopsis aculeata, Saxifraga stenophylla, Primula macrophylla, Taraxacum officinale,* have attractive colour like yellow, orange, red, pink and purple etc. which help in pollination.

Family Identification

The adaptative peculiarities of trans Himalayan cold desert plants may help in identification roughly upto families on the field spot. Commonly found families of this zone are Brassicaceae, Asteraceae, Ranunculaceae, Caryophyllaceae, Lamiaceae, Primulaceae, Crassulaceae, Saxifragaceae, Gentianaceae, Scrophulariaceae, Poaceae and Roasaceae.

Plant families which are easily identifiable with their characteristic plant features have been deseribed in brief below :

Ranunculaceae

Frequent in moist places. Plants usually perennial herbs, dwarf, much branched and tufted with radical leaves. Plants belong to genus such as *Aconitum sp, Anemone sp, Delphinium sp* and *Ranunculus sp.*

Brassicaceae

Common in semi-arid localities. Plants usually hairy, tufted, much branched and dwarf. *Capsella sp, Christolea sp, Draba sp, Matthiola sp* and *Parrya sp* are common.

Caryophyllaceae

Common on open rocky slopes. Herbs densely tufted, with small and linear leaves which usually form cushion to hard mat. *Arenaria sp, Stellaria sp* and *Thylacospermum sp* dominate in this family.

Fabaceae

Frequent in alpine grassy meadows. Plants usually herbs or shrublets with much branched and silky haired leaves . *Astragalus sp, Caragana sp* and *Oxytropis sp* are not uncommon.

Saxifragaceae

Common at high altitude passes. Plants usually small and densely tufted with long slender often red runners. *Saxifraga sp* is prominent.

Crassulaceae

Frequent at the elevation of 3,604 m - 5,105 m. Plants usually succulent perennial herbs with simple and fleshy leaves in rosette arrangement. Root stock is usually thick and fleshy with persistent scales *Rhodiola sp* and *Sedum sp* are important genera having economic importance.

Asteraceae

Frequent in rocky places. Plants usually small herbs with characteristic development of profusely wooly growth on vegetative and floral parts. Plants belonging to this family are *Saussurea sp, Jurinea sp, Leontopodium sp* and *Waldheimia sp.*

Lamiaceae

Common on semi-arid steppes at the elevation of to 4,955 m. Plants usually aromatic herbs or shrublets. *Dracocephalum sp, Nepeta sp, Scutellaria sp* and *Stachys sp* dominate the flora under this family.

❑❑❑

4

Ethnobotany and trans Himalayan Plants

Elementary needs of pre-historic man were fulfilled by the plants of his surroundings. He was totally dependant on surrounding plants for food, medicine and other domestic purposes. This interaction between man and plants in day to day life continued over long period and a systematic study of its nature came to be known as a science of ethnobotany in recent past. A good majority of the human population in different parts of the world, more particularly in the third world countries, still continue to live in and around the forest or far flung areas and their lifestyles and tradition did not change for centuries and remained almost unchanged for hundreds of years. It is often said that such tribes or other such less evolved (primitive) communities are the living archaeological museums of ancient traditions and cultural heritage and these traditions add to the texture and luster of humankind.

The history of ethnobotany reveals that this particular branch of knowledge was included in Economic Botany before John W. Harshberger's publication (1895), who coined the term **Ethnobotany** used by current researchers. The term refers to two components viz. ethnic and botany, implying study of ethnic group from the point of view of their interaction with vegetational environment. Since then the definition of ethnobotany has changed and evolved along with the formation and evolution of the field (Ford 1978, Jain 1989, Jones 1941, Schultes & Reis 1995). It was Faulks (1958) who wrote the first 'Introduction to Ethnobotany' about four decades ago. According to Schultes, R. E. (1962), the well known authority on the subject, ethnobotany means "the relationship which exists between people of primitive societies and their plant environment. Sometimes, it is understood as anthropological or ethnographical approach to botany". According to Martin (1995), the term Ethnobiology implies an explanation on local's perspectives on cultural and scientific knowledge. Ethnobotany in general context denotes the entire realm of useful and close interaction between plants and man. It includes everything from interaction, interrelation of human communities with plants. Human - plant relations fall under two broad categories:

concrete (material) and abstract (socio-cultural or religious and spiritual). Pushpangadan (1990) treated ethnobotany as a study of the knowledge system pertaining to the multidimensional perspective of life culture, traditions as well as interaction of traditional or less advanced human communities like tribals with their local flora. Ethnobotany in general, thus refers to the study about the multifarious uses of plants for human needs such as food, medicine, fodder, fibre and goods required for his material culture and amenities.

India is a country with a spectrum of ethnocultural groups residing in diversified socio-geographic locations, which have origin in different racial stocks. Schultes (1962) writes in his forward note in Glimpses of Indian Ethnobotany (1981) and Contribution of Ethnobotany (1991) that Indian sub-continent represents without a doubt one of the greatest emporia of ethnobotanical wealth. Many living groups of people, still more or less isolated from the influence of the modern world and who continue to live in close association and vital dependant on their ambient vegetation, provide scientist with unparalleled opportunities for profound research (Jain, 1991). Jain has also pointed in his prefatory note, "India is very rich in ethnobotanical information. India has a population of about 102 million out of which about 50 million people belong to 550 tribal communities. There are 325 languages spoken by different communities with a large number of dialects. About 80 percent of India's population lives in villages and a considerable proportion comprises tribals living in remote forest areas. The population of several district in Central and Eastern India is predominantly tribal. The different traditions, beliefs, needs and cultures of the various tribes and the diversity of flora in India richly contribute to plant folklore.

In the past two decades a considerable work in the field of ethnobotany has been done in India. Janki Amal has undertaken the first ethnobotanical study in Indian settings through his study of subsistence food plant of certain tribals of South India. However, the credit to really establish ethnobotanical research in Indian Institutes and Universities goes to Dr SK Jain, former Director, B.S.I. Researchers of Faulks (1958), Schultes (1962-63), Barrau (1959, 1966) and Conklin (1962) have promoted Indian researchers to undertake ethnobotanical studies. Jain and Mitra traced retrospects and prospects of ethnobotany in India and reviewed about 87 studies undertaken by different investigators in varying localities, communities and in varying localities and tribal groups. They complied major works on the subject in books entitled, "Contribution to Indian Ethnobotrany", "Glimpses of Indian Ethnobotany" and "Methods and Approaches in Ethnobotany", which provide significant insight into the nature methods and findings of ethnobotanical research in Indian sub-continent and paved the way for ethnobotanical investivgations all over India. Thereafter a number of sporadic studies were undertaken. In the first work, uses of medicinal plants by Santals reported earlier by Bodding (1925 & 1926) were presented and in second work, ethnobotany of Kumaon Himalayas was reported with various uses of plants such as medicinal, food, condiments, incense, etc along with the etymology and philology of the vernacular names and a brief of introduction of tribal communities. After establishment of the "**Society of Ethnobotanists**" during 1980, ethnobotany in India was streamlined and a common platform was made available. During 1982-1983, an "All India Coordinated Research Project on Ethnobiology" was initiated at Regional Research Laboratory, Jammu, under the leadership of Dr P Pushpangadan. It was a multidisciplinary, multi-institutional and action oriented research programme for documenting the multidimensional perspectives of the tribal culture, traditions and knowledge. The entire work on ethnobotany in India was summarized (Binu *etal.* 1992), Maheswari (1970) and Radahakrishnan *etal.* (2000). However, a graphic view of ethnobotany from 1982 to 2000 has been presented by Jain & Srivastava (2000) and recently, its status and its relationship with several other discipline has been narrated by Jain (2001). Presently there are a number of institutions and universities where ethnobotanical studies are being carried out actively in India.

Ethnographical Aspects

The main aspects of ethnography with their importance are listed as under :

i) *Tribals or ethnic communities* : After identifying a tribal community for ethnobotanical studies one should include its history, origin and geographical limits, note their habit, wheather pastoral/stationery and also the means of their livehood in brief. Before visiting the field, all such information should be made ready from various anthropological or ethnological literature.

ii) *Phonetics* : Vernacular or tribal names must be recorded with correct pronunciation by listening carefully to the voice of tribals and local guide for proper transcribing in phonetic transliteration. Phonetics is the most important part of study, but rarely proper attention is paid by researchers to this aspect.

iii) *Etymology and philology* of vernacular names convey much more about their relationship of the plants with the tribal communities. De Candolle (1886) explained in detail the importance of philology in relation to origin of cultivated plants. Usually the native plants names may be derived from folk speech based on structure, morphological appearance, specific uses, nuisance as weeds, presence of thorns, spines, acrid juice, odour, taste, poisonous properties, blossoming time, location or natural habitat, mythical and technical aspects of plants (Shah 1987, Patil 1998).

iv) *Social and cultural association* : Plants are usually associated with various festivals, religious ceremonies, rituals, rites etc. throughout the world. Women folk are the custodians of culture and are very particular in observing these festivals, ceremonies and rituals as a part of their daily life. A number of plants are closely associated with social and cultural association.

v) *Multifarious uses of plants* : The multifarious uses of plants around surrounding areas are closely associated with man in their day to day life as under : Food (cereals, vegetables, fruits, beverages, spices, condiments, seasoning, etc; edible dye or as dye colours); ethno-medicine and ethnoveterinary medicine, aromatic for fragrance & incense; flowers of aesthetic value; timber value of plants; fibre yielding plants; fodder plants; fuel plants; plants of magico-religious importance, etc.

vi) *Indigenous Knowledge* : This knowledge encompasses a wide spectrum of areas which is not only technical in nature, but also incorporates integrative insights, wisdom, ideas, perceptions and innovative capabilities which pertain to ecological, biological, geographical and other physical phenomena. The indigenous knowledge particularly concerned with the utilization of their local genetic resources for the treatment of various diseases has provided leads for development of biologically active molecules by the technologies enriched countries. The intrinsic potential of ethnobotanical knowledge as a key resource for developing new kinds of foods, cosmetics, drugs and pharmaceuticals has been increasingly realized. As a result, the potential and actual economic values of genetic materials are increasing day by day and have stimulated international trade in genetic resources commonly known as biotrade which soon led to biopiracy or gene piracy.

Ethnobotanical Knowledge & Intellectual property rights

Knowledge and technologies in the historical past have been exchanged freely without any legal barriers. The intellectual property rights as the term suggests are certain legally enforceable rights granted to individuals or corporate bodies to have monopolistic ownership or control over things emanating from the exercise of human brain, such as thoughts, ideas and information. Especially regarding new inventions, innovation and process (Pushpangadan 2002). It is through IPR, and particularly patents, that control and ownership over traditional knowledge are being usurped by commercial interests. IPR's emerged strongly during the industrial revolution, and have been an important driving force behind rapid industrial growth and prosperity of the western countries during last three centuries.

The convention on Biological Diversity (CBD), which has come into force from December 1993, is a unique international legal instrument to protect the sovereign rights of the state and people over its/their biological resources. The main objectives of the CBD concerned are conservation of biological diversity and sustainable use of its components and fair and equitable sharing of benefits arising out of the utilization of genetic resources against the background of sustainable development (CBD Art 1). Prior to CBD, there was no legal basis for the traditional societies, individuals or countries to claim any right or share of financial benefits accruing from the products developed by developed countries from biological resources and or traditional knowledge systems of the developing nations. Thus CBD provides a fundamental framework for the parties to develop appropriate national legislations and policy guidelines to facilitate fair and transparent means and mechanisms for access to and equitable sharing of benefits from the use of genetic resources and traditional knowledge.

After CBD, there came another international agreement, the WTO section of this agreement on Trade Related Aspects of Intellectual Property Rights. While CBD dwells upon the sovereign rights of nations over their biological diversity, the TRIPs provides for effective and adequate protection of IPRs of individuals or corporates or new inventions, technologies and products developed by them. TRIPS which came into force in 1994, stipulates that " the protection and enforcement of intellectual property rights should contribute to the promotion of technological innovation and to the mutual advantage of producers and users of technological knowledge and in a manner conducive to social and economic welfare, and to a balance of rights and obligations" (Article 7, TRIPS).

India has the distinction of being the first country in the world that experimented a benefit sharing model that implemented Article 8(j) of CBD. The Tropical Botanic Garden and Research Institute (TBGRI) Kerala, has demonstrated that indigenous knowledge system merits support, recognition and fair and adequate compensation. The model that "TBGRI Model" or "Kani Model" or "Pushpangadan Model" of benefit sharing, it was indeed an outcome of an "All India Coordinated Project on Ethnobiology" sponsored by Ministry of Environment & Forests, Govt of India (1983-1998). Dr Pushpangadan and his team carried out scientific investigation on a medicinal plant (*Trichopus zeylanicus* from Kani tribe) and finally developed a herbal formulation, namely 'Jeevani' and the same was transferred to a pharmaceutical company against payment of a licence fee and royality. TBGRI at this point of time resolved to share the benefits (licence fee and royalty) with Kani tribe on 1 : 1 basis. This TBGRI Model of benefit sharing received wide acclaim, acceptance and popularity throughout the world, being the first of its that recognized the resource rights and IPR of a traditional community by a way of sharing equitably the benefits derived out of the uses of traditional knowledge that has been developed , preserved and maintained by that community for many generations (Anand 1998, Mashelkar 2001).

Ladakh region comprising two districts namely, Leh and Kargil including valleys, snow capped mountains and glaciers, is very rich in ethnobotanical lore. In Ladakh there are many tribal groups exist. These tribal groups are represented by majority of Boto (Buddhist), Phaaris, Gujjars, Bakarwals, Gaddhis, Dards, Balti, Brokpas Changpas and Khache and represent specific socio-culture. Similarly the people of Lahaul-Spiti of Himachal Pradesh are a mixed race and the most common religion followed is a combination of Hinduism and Buddism. These tribes have attracted the attention of historians, anthropologists, sociologists and lately of ethnobotanist.

A perusal of ethnobotanical studies of the region reveals that Stewart (1917) was the first to compile a checklest of flora of Ladakh, Western Tibet. Thereafter, many other workers have attempted to gather information on utilization of plants by different tribal groups. The flora of Ladakh was compiled by Kachroo, Dhar and Sapru in 1978. Similarly the fascinating floristic wealth of Lahaul-Spiti has attracted several botanists and ethnobotanists and the major plant explorers were JD Cunningham, T Thomson, H. A. Jaeschke, D. Brandis, Stoliczka, R. N. Parker, R. R. Stewart, S. R. Kashyap, M. A. Rau, N. C. Nair, Y. K. Sarin, R. S. Negi. The account of the flora and vegetable products of Lahaul-Spiti was published by Aitchison then followed by several workers. M. R. Uniyal, Y. K. Sarin, Arora etal and many others published several papers on ethnobotany of this region. The detailed flora of Lahaul-Spiti was published by Aswal and Mehrotra in 1994.

5

Ethnobotanical Exploration

As we know, the Himalaya is considered as the treasure house of medicinal, aromatic and other important plants since time immemorial and the Indian cold desert of Ladakh and Lahaul-Spiti (trans-Himalaya) is not the exception to it. The local inhabitants of these areas are using surroundings flora for various utility in their day to day life.

Keeping in view the rich diversity of ethnobotanical wealth of cold arid zones of the Himalayas, Field Research Laboratory, Leh-Ladakh, an establishment of Defence Research and Development Organisation has carried out detailed extensive ethnobotanical surveys of all the seven valleys - Indus, Nubra, Changthang, Zanskar, Suru of Ladakh region and Lahaul and Sipiti of Himachal Pradesh, covering far-flung and higher passes over the decade as shown in map (Fig. 1). The ethnobotanical explorations were usually conducted from April to October with a view to collect plant specimens, field notes and ethnobotanical information. Camps were also established during summer months for studying the plants of various uses. The ethnobotanical information were gathered with the help of local people, shepherds, nomads and the same were also verified through various sources like Amchis (herbal doctors), old aged people and other available literatures.

During the stay in fields, data on ethnobotanical uses and application of plants were gathered from tribal chief, reliable persons and other informants. Furthermore, whenever opportunities arose, festivals and ceremonial occasions were directly observed by the authors for uses of plants amongst them and in other cases the information were collected indirectly through interviews with local people. Articles of various plant materials were observed. Allied literatures are also consulted during the investigation. The local and vernacular names were also noted by tape-recording and listening carefully the pronounciation and then transcribing these in phonetic transliteration.

Maximum plant species of ethnobotanical importance were collected in triplicate from the area under study. Field observations (habit, habitat, altitude and locality etc.) were also recorded on the spot. All necessary precautions were taken and normal procedure prescribed has been followed for preserving the specimens. These specimens are preserved in the Herbarium section of Field Research Laboratory, Leh. All the specimens were critically examined and identified with the help of various floras, monographs, revisions and research papers. The identification of these plants was further confirmed with the help of Botanical survey.

The brief information about valleys, important ethnobotanically plants and their allied information, collected from a particular valley are as under:

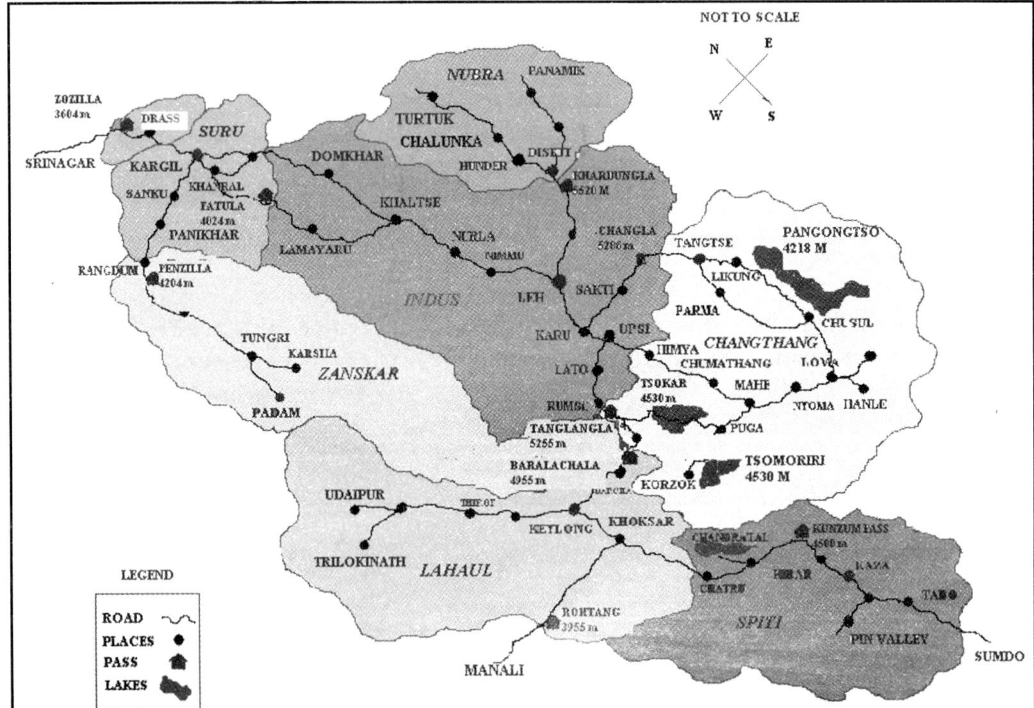

Map 1. Area surveyed : Ladakh & Lahaul-Spiti

Indus Valley

The Indus valley is the most populated and the central valley of Leh district of Ladakh region. The majority of settlement of the valley is under 180 km long and 55 km wide. The important places of this valley are Dha-Hanu, Lamayaru, Khalsi, Saspol, Likir, Nimmu, Phyang, Spituk, Leh, Saboo, Choglamsar, Stok, Thiksay, Karu, Hemis, Upshi etc. It is surrounded by Khardungla on the north-east, by Changla on south-east and by Fatula on the west.

Though the Indus is the main river of Ladakh region, but it has been used for irrigation in only few villages of Ladakh, rest of the water enters Pakistan from Batalik side. Beside Ladakh Indus River has great significance in Indian history and socio cultural and religious context. The cultivated area lies between 2,853 m to 3,904 m. above mean sea level. Wheat, barley and peas are the major crops, in vegetables potato, cabbage, cauliflower, radish, carrot, knol khol, spinach, beans etc. are grown in good quantity and apricot, apple and walnut are dominant fruits. The dominant plant species of the valley are as under:

Species	Common Name	Status
Delphinium brunonianum	Himalayan Larkspur	Vulnerable
Ephedra gerardiana	Somlata	Endangered
Hippophae rhamnoides var. turkestanica	Seabuckthorn	Common
Peganum harmala	Wild Rue	Frequent
Rhodiola imbricata	Rose root	Frequent

Nubra Valley

This is the northern most valley of Ladakh. Khardungla pass (5,520 m.), the world's highest motarable pass, is the gate way of the valley and it includes Khardung, Khalsar, Partapur, Thoise, Chalunka, Turtuk, Tegar, Panamic and Siachen glacier area. The altitude of this valley is less in comparison to Leh valley. Population is very sparse and the major concentration of population is on the confluence of Nubra and Shyok river. The majority of population of Nubra valley is Boto and they follow Buddhism, followed by few Muslim community; settled in Turtuk and Bogdang area. Nubra valley experiences very high temperatures during summer due to lower altitudes. High velocity wind, sand storm and dusty atmosphere are not uncommon. During winter months soil and water sources etc. get frozen. Nubra and Shyok rivers are main source of water, besides several glaciers flow.

Nubra valley is known for its rich flora. Infact Nubra means green valley or garden of flowers. Barley, wheat, buckwheat, peas, potato, onion, mustard are major crops in Nubra valley. Natural forests are noticed on both the sides of Nubra river. A number of shrubs, trees and herbaceous plants are found in the forest. Alpine grasses are on mountain peaks which maintain sheep and goats of the valley during summer months. Unique feature of this area is double hump camels which are used as means of transport, providing wool, and getting milk and manure. Panamic area is having hot water spring which is known to cure various diseases. The dominant plant species of the valley are as under:

Species	Common Name	Status
Aconitum heterophyllum	Atees	Endangered
Berberis ulicina	---	Common
Dactylorhiza hatagirea	Salam Panja	Endangered
Perovskia abrotanoides	Wild lavender	Common
Rheum webbianum	Himalayan Rhubarb	Endangered

Changthang Valley

The maximum area of Changthang plateau is situated between an elevation of 3,604 to 4,505 m. above sea level. Changla pass (5,286 m.) and Taglangla pass (5,255 m.) are the gateways to Changthang valley from Ladakh region. This is one of the most elevated valleys of Ladakh comprising large number of villages and hamlets like Lintsey, Chumathang, Karey, Mahe, Nyoma, Changla, Tangtse, Chusul, Hanle, Kyul, Demzok, Puga, Rupsu and Kharnak etc. Changthang valley is famous for its borax, sulphur and salt deposits, hot springs and high quality Pashmina wool production etc. It has some of the world's highest lakes like Pangongtso (4,218 m.) Tsokar (4,530 m.) and Tsomoriri (4,511 m) sheltering large number of beautiful and endangered birds and wildlife around its rich green wetlands. From floristic point of view Changthang valley has scattered vegetation. Agriculturally this valley is underdeveloped in comparison to other valleys due to its harsh and cold climatic conditions. The cultivated area lies between 3,303 m. to 4,354 m. where only barley crops are grown in small quantity.

The local inhabitants of Changthang valley are known as Changpas and they are Buddhist by religion. The majority of Changpas are Nomads whose main occupation is livestock rearing especially famous pashmina goats and sheeps. They move with their livestock's with complete luggage, baggage, tent and food from one pasture to another, setting apart for winter quarters a particular pasture in a less snowfall area and where the flocks are not allowed to graze during summer. Being always close and traveling with nature Changpas are largely dependent on wild plants for vegetables, medicine for both human and animal and other needs. The dominant plant species of the valley are as under:

Species	**Common Name**	**Status**
Arnebia euchroma	Rattanjot	Endangered
Cicer microphyllum	Wild gram	Frequent
Dracocephallum heterophyllum	--	Frequent
Nepeta longibracteata	--	Frequent
Rheum spiciforme	Himalayan Rhubarb	Endangered

Zanskar Valley

Parkachik is the last Balti village of Suru valley and after few kilometers from Parkachik; Shagma is the first hamlet of Rangdum area of Zanskar valley. Rangdum is a small green valley with rich herbal wealth before crossing Penzila (4,204 m.) to enter main Zanskar valley. Zanskar is one of the most remote, underdeveloped and least populated valleys of Ladakh. The important places are Lungnak, Stongday, Padum, Karsha, Sani, Phey, Aksho, Abran and Rangdum etc. This beautiful valley of Kargil district is cut off from rest of the world during most of the time, it is reachable by road during summer season for only four to five months and during winter time it can be only reached by harsh Chader track (seven days track on the frozen Zanskar River).

Situated at an altitude between 3,303 m. to 4,204 m. Zanskar valley is quite rich in diversity of flora as well as fauna. The inhabitant of Zanskar valley is mainly Buddhist followed by very few Muslim families settled in Padum town. The Zanskar River is the main river of this valley and joins the Indus River at Nimmu village of Indus valley. The cultivated area lies at an altitude between 3,183 m. and 3,604 m. Economy of the region is mainly dependent on agriculture and animals. The dominant plant species of the valley are as under:

Species	Common Name	Status
Artemisia dracunculus	Tarragon	Frequent
Bergenia stracheyi	Pasanbhed	Endangered
Delphinium cashmerianum	Larkspur	Vulnerable
Meconopsis aculeata	Blue Poppy	Endangered
Podophyllum hexandrum	Himalayan Mayapple	Endangered

Suru Valley

After Drass we enter Suru valley from village Kharul where Drass River meets Suru River. The valley is also named after the Suru River. The maximum geographical area of this valley is situated on the bank of Suru River, the average altitude of the valley is 2727 m. Summers are warm and longer in Suru as compare to other valleys of Ladakh region, soils are most fertile. It includes Kargil, Sanku, Panikhar, Batalic, Paskum, Shakar-Chiktan and Bothkharbu areas etc. The cultivated area lies between 2,553 m. & 3,183 m. Most of the area is double cropped.

Suru valley is famous for quality production of apricot. In the month of May the entire valley becomes awash with fragrant white apricot blossom while in August the ripening fruits lead it an orange hue. The majority of inhabitants of this valley are known as Baltis and they are follower of Shia sect of Islam followed by few Buddhist populations in Bodh-Kharbu and Darchik areas. The valley is comparatively greener in comparison to the rest of valleys. The flora is also a mixture of temperate as well as of alpine desert vegetation. The dominant plant species of the valley are as under:

Species	Common Name	Status
Aconitum heterophyllum	Atis	Endangered
Bunium persicum	Kala Zeera	Endangered
Physalis alkekengi	Winter cherry	Vulnerable
Podophyllum hexandrum	Himalayan Mayapple	Endangered
Rubia cordifolia	Manjistha	Vulnerable

Lahaul Valley

The Lahaul valley of Himachal Pradesh, is surrounded by the main Himalayan ranges on the north, mid Himalayan range or Pir Panjal on the south, the Kunzum range which separates Lahaul from Spiti on the east and off-shoot of the Pir Panjal range on the west. The Rohtang pass (3,955 m) the gateway of Lahaul valley, connects this valley from Kullu district. As soon as one crosses Rohtang pass towards Lahaul valley the entire topography changes; a view of bleak, sunny, gigantic ranges, snow clad peaks and massive glaciers. The valley is drained with two main rivers Chandra and Bhaga and the both rivers unite near a place Tandi and thereafter called as Chandrabhaga. The Chandarbhaga parts are the most fertile belt and thickly populated of Lahaul valley.

The term Lahaul has probably derived from the Tibetan word 'Loh - Yul means the southern country or Lhashi - Yul' means the country of Gods. The people of this valley are industrious and educated. The feature of Lahaulis are a combination of Mongolian and Aryan traits. The staple foods are wheat, barley, buckwheat, potato and meat. The important places are Kelong, Tandi, Khoksar, Udaipur, Trilokinath etc. The region is famous for quality potato and pea production. Like other valleys of Cold desert, Lahaul Valley is known for its characteristic flora and fauna. The dominant plant species of the valley are as under:

Species	Common Name	Status
Hippophae salicifolia	Seabuckthorn	Common
Inula racemosa	Pushkarmool	Endangered
Jurenea macrocephala	Dhoop	Endangered
Picrorhiza kurrooa	Kutki	Endangered
Saussuea lappa	Kuth	Endangered

Spiti valley

The Spiti valley is situated beyond the Kunzum pass and lies amid the north-eastern ranges of the Lahaul-Spiti district and in some adjoining part of the central Himalaya. The mystic valley of Spiti unravel another world. The craggy cliffs, narrow gorges, rushing streams and snowy peaks strike beautiful vistas, a world apart, a wild and unspoilt landscape exists with little villages perched on rocky slopes. The spiti river, the lifeline of this valley originates from Kunzum pass. Spiti valley is a broad and flat valley intersected by high vertical cliffs of blade, grey and brown rocky terrain.

Unlike Lahaul, Spiti valley is dry and barren valley. The people are mongoloids and they follow Mahayana Budhism. On contrary the people of Spiti are backward and poorer due to their indolence, general poverty, apathy, lack of resources, barrenness of land and limited working season. The headquater of Spiti valeyy is at Kaza and the other important places are Lossar, Rangrig, Kee, Kiber,Taboo, Pin valley etc. The agriculture and livestock rearing are the main source of economy. Unlike other valleys, Spiti valley has its own characteristic flora and fauna. The dominant plant species of the valley are as under:

Species	Common Name	Status
Allium carollianum	Wild garlic	Frequent
Carum carvi var gracile	Caraway	Common
Eremurus himalaicus	Desert candle	Rare
Hippophae tibetana	Seabuckthorn	Frequent
Hyssopus officinalis	Hyssop	Rare

6

Medicinal Plants

Since the dawn of history man has been in search of ways to find cure and relief from mental and physical illnesses. The history of medical botany dates back to the origin of civilization. The Sumerian ideograms depicting various medicines of plant origin date back approximately to 2000 B.C. The Assyrians and Mesopotamian pharmacopoeias had approximately 250 herbal plant species listed in them. These included poppy, belladonna, hemp, saffron, thyme, garlic, onion, cassia and asafoetida.

The earliest written records of herbal plants are preserved in scrolls of Papyrus, a precursor of paper made from pith of paper Nile of bulrush. The most famous scroll, the Ebers papyrus, which is a compilation of earlier works, contains 877 prescriptions and recipes of *Cannabis, Opium, Aloe, Juniper* and *Henna* dating back to the sixteen century B.C.

India has one of the oldest, richest and the most diverse cultural traditions associated with the uses of medicinal plants. It represents a striking example of the intimate linkage between biodiversity and cultural diversity. The history of medicinal plants can be traced back to the Vedic period (4500 B.C. to 1500 B.C.). The identity of several plants like Semal, Pithuan and Pipal referred to in the *suktas* of *Rig Veda* can be fixed with reasonable certainty. *Atharveda* contains detailed information on approximately 2000 medicinal plants and their uses. After the Vedic era, the works of Charak and Susruta namely *Charak Samhita* and *Susruta Samhita* deal with 700 drugs of daily and specific uses. The period of the 16th and 17th century witnessed a spurt in the development of medical botany and most of the world famous medical drugs came into light during this period. For this reason, the period is also referred "The Age of Herbal Medicines".

In India, over 7500 plants species are estimated to be used by approximately 4635 ethnic communities for human and veterinary health care across the various ecosystems ranging from Ladakh in the trans Himalayan cold desert to the southern coastal tip of Andaman & Nicobar Islands and from the deserts of Rajasthan & Kutch to the Hills of the north east.

S. No.	SYSTEMS	USED (approx) NO. OF SPECIES
1	Ayurveda	1769
2	Folk	4671
3	Homeopathy	482
4	Siddha	1121
5	Tibetan / Amchi	279
6	Unani	751

In the codified medical texts of Ayurveda, a recent study enumerates around 1,800 species of plants that are fully documented in terms of their biological properties, actions and drug formulations for a range of health conditions from a common cold to raising of the body's general immunity. According to WHO report, over 3/4th of the five billion world population cannot afford the products of modern medicines and have to rely upon the uses of traditional medicines of plant origin. WHO currently encourages, recommends and promotes the inclusion of plant derived drugs in national health care programme because of low cost and within the reach of a common man and as such are time-tested and considered to be much safer.

The people of Ladakh and Lahul-Spiti of the trans Himalayan region has its own medical system called "Amchi system of medicine" (Tibetan Medicine) and the practitioners are called Amchis (Superior of all). Amchi medicine is based on skillful uses of plants, minerals and animal products. The origin of Amchi medicine is traced back to India, where more than 2500 years ago Lord Budha delivered a medical practice called Bzi (Chatus tantre) which was translated in Tibetan language during 8th centrury AD by Acharya Chandrananda and Varochana. In order to popularise these texts in Tibet, the senior Yusthog Youthan Gonbo had made some changes in it to make it harmonious with the culture and environment of Tibet. Since then this system is very popular in Tibet and in most parts of the Himalayan belt. This system was introduced in cold arid zones during 10th/11th century AD approximately.

The Amchi system of medicine has been in vogue in Ladakh and other parts of trans Himalaya for past many centuries. Amchis are looking after 60% of public health. Amchis have not only theoritical texts but also practical experiance handed over from generation to generation. They have also a reputation of having high ethical standard in social system. Generally single herbal preparation and combinations of two or more herbs are prescribed and administered.

Bioactive constituents & Medicinal plants

Bioactive constituents of medicinal plants may be classified into two major groups :

a)Toxic : causing poisonous effect on organisms.

b)Non toxic : those do not cause poisoning but are therapeutically active.

The major groups of bioactive compounds occurring in medicinal plants are described in brief here :

a) *Vitamins* : Vitamins play major role in keeping good health of human. Some of plants are natural source of vitamins and they contain significant amounts of vitamins viz. Vit 'A' (Carrot, Tomato), Vit 'B' (Maize, Carrot, Peas), Vit 'C' (Fruits of orange, anola, seabuckthorn), Vit 'E' (Apricot, Rose, Seabuckthorn), Vit 'K' (Lucerne, Spinach, Strawberry) etc.

b) *Alkaloids* : Alkaloids are the major bioactive constituents available in plants and are characterized by the presence of nitrogen which make them pharmacologically active. They usually occur in all plant parts but their constituents vary and appear to optimum at or around flowering stage.The first alkaloid isolated from opium was morphine in 1803 and since then several alkaloids have been discovered. The important medicinal plants and their alkaloids are :

	Botanical Name	**Alkaloids**	**Therapeutic uses**
1.	*Aconitum heterophyllum*	Aconitine	Antipyretic
2.	*Atropa belladona*	Atropine	Anti-spasmodic
3.	*Colchicum luteum*	Colchicine	Gout
4.	*Ephedra gerardiana*	Ephedrine	Asthma
5.	*Podophyllum hexandrum*	Podophyllotoxin	Anti-cancer

c) Cynogenic Glycosides : These groups are characterized by presence of a sugar portion attached by a special chemical bond to a non sugar portion. They have a helpful sedative and relaxant on the heart and muscles in small doses viz. *Prunus serotina, Sambucus nigra.*

d) Cardiac Glycosides : These compounds have a strong, direct action on heart, helping to support its strength and rate of contraction. These compounds are of diuretic properties. The common plants are *Digitalis purpurea* etc.

e) Saponins : There are two kinds of saponins triperpenoid and steroidal saponins. The later get their name from their similarity to the human body's own naturally steroid hormones. They are characterized by the fact that they produce a soapy lather when shaken with water. Saponins have a range of systemic effects such as anti-inflammatory, antiallergic, immunomodulatory, antistress, sedative etc. *Glycyrrhiza glabra, Panax ginseng, Primula sp* are the best example.

f) Glucosilinates : These are pungent tasting compounds found mainly in the Brassicaceae family. These compounds have an irritant effect on the skin, causing inflammation and blistering. Usually applied as poultice of painful or aching joints to increase blood flow, helping to remove waste products. They also help to reduce thyroid function e.g. *Raphanus sativus, Sinapis alba* etc.

g) Anthaquinones : These are yellow-brown pigments, most commonly occurring as O-glycosides or C-glycosides. These pigments are the main bioactive constituents found in several plants having laxative properties. The common plants are *Aloe barbadensis, Cassia senna, Rheum species,* etc.

h) Anthocyanins : These pigments which flowers and fruits a blue, purple or red, help to keep the blood vessels healthy. *Rubus sp, Vitis sp* are the best example.

i) Flavonoids : Flavonoids are the very important bioactive groups having a wide range of actions. They are antiflammatory and are especially useful in maintaining healthy body. Seabuckthorn, Buckwheat, *Citrus* sp are the best example.

j) Tannins : These are non nitrogenous plant products, soluble in water and alcohol and when exposed to air they turn black. They are widely distributed in the bark of trees, insect galls, leaves, stems and fruits. Tannins are usually astringent, promote rapid healing, used in inflammation of gums, binding effect in gut, etc.e.g. *Quercus sp, Acacia, Hippophae sp, Geranium sp.*

k) Coumarins : These compounds are having divergent actions and are widely distributed in several plants. Coumarins have antimicrobial and fungicidal activity. The common plants are *Melilotus officinalis, Apium graveolens, Prangos pabularia* etc.

l) Bitter principles are non poisonous, non-nitrogenous substances of varied chemical composition with characteristic bitterness but not unpleasant taste. The bitterness itself stimulates secreations by the salivary glands and digestive organs which improve the appetite and strengthen the overall function of the digestive system. The common plants are *Artemisia absinthium, Capparis spinosa, Saussurea lappa, Swertia chiretta* etc.

m) Phenols : Phenols are one of the largest groups of secondary plant constituents. These compounds are antiseptic in nature and reduce inflammation. Salicylic acid is found in many plants viz. *Gaultheria procumbens, Populus sp, Salix alba, Thymus vulgaris* etc.

n) Mucilages : Mucilages are the products of metabolic activity in plants and are polysaccharides in nature that soak up water, producing a sticky jelly like mass. These compounds line the mucous membranes of the digestive tract, protecting against irritation, acidity and inflammation. Some of plants are *Lavetera sp, Malva sp, Plantago sp, Ulmus sp.*

o) Fixed oils : Fixed oils are lipids, classified as primary metabolites, and therefore essential for life. Unlike secondary metabolites, fixed oils are universally present in plants. These are of saturated fatty acids, monosaturated fatty acids, polysaturated acids. Some of main fatty acids are linoleic acid, oleic acid, palmitic acid, omega series. Seabuckthorn, walnut, apricot, almond, etc. are the best example.

o) Essential oils : These oils are aromatic compounds of complex substances known terpenes and their compounds or derivatives. The oils are liquid components of plant cells generally volatile having multiple effects such as antiseptic, germicidal, scabies, anthelmintic, anti-inflammatory etc. The common plant families are Apiaceae, Asteraceae, Lamiaceae, Myrtaceae, Pinaceae. The important aromatic plants and its major volatile compounds are as follows.

S. No	Plant species	Volatile compounds	Therapeutic uses
1.	*Achillea millefolium*	Azulenes	Anti-inflammatory
2.	*Artemisia annua*	Artemisin	Anti-malarial
3.	*Carum carvi*	Carvone	Carminative
4.	*Origanum vulgare*	Thymol	Anti-spasmodic
5.	*Salvia sclarea*	Linalool	Nervousness

Stages of collection of plant parts

Potency and efficacy of medicinal plants depend upon collection / harvesting of different plant parts at their right stage.

a) *Roots, rhizomes, tubers and bulbs :* The underground parts are the storage organs of the plant and accumulate active bioconstituents during summer months. Roots and rhizomes of the perennials are usually harvested after two to five years of growth either at the end of growing season or in the beginning of growing season. The underground parts of annuals are not generally collected.

b) *Aerial parts :* These include all the parts of the plant growing above ground - stems, leaves, flowers. Stems are usually cut just above ground at the beginning of flowers while the perennials may be cut higher above ground to encourage further crops. Leaves are harvested throughout the growing season. Usually tender or young leaves are considered to have maximum content of bioactive constituents. Similarly flowers or the entire inflorescences are collected at the beginning of flowering period.

c) *Fruits & berries :* These plant parts should be collected when they are fully ripen. They should be picked individually or in bunches.

d) *Seeds :* Seeds are usually collected when fruits / berries fully muture and seeds should be gathered before scattering.

e) *Bark :* This plant part is collected either in spring when the trees or shrubs start to sprout or in autumn when they shed their leaves because during these time the flow of sap is considered to be maximum and barks readily detach from the wood.

Storage of raw materials

Processing and proper storage of medicinal plants parts are the most important steps for retaining maximum potency and quality. Underground parts/ leaves / stems / bark should be thoroughly washed, cleaned and dried in shed or ovens at proper temperature. Similarly fruits and seeds should be properly dried.

Storage of raw materials is very important for maintaining the quality of crude drugs. Leaves, roots, flowers and other plant parts should be stored in sterilized, dark glass containers with airtight lids or they may also be stored in brown paper / cotton bags and should be kept in dry and cool places to protect from insect damage and moulds.

Field studies

The extensive ethno-medico-botanical explorations reveal that Ladakh and Lahaul-Spiti of cold desert Himalayas abound in medicinal plants and approximately 35 percent of total flora are of medicinal value. The widely used medicinal plants in traditional systems of Indian medicines including Sowa Rigpa system (Amchi system) are *Aconitum heterophyllum, Colchicum luteum, Dactylorhiza hatagirea, Ephedra gerardiana, Hippophae rhamnoides, Inula racemosa, Rheum webbianum, Saussurea lappa etc.* .

The present chapter deals with brief description of the plant species, their families, english (E), hindi (H) & local names (L), distribution, habitat, their traditional (T) and other uses mentioned in literatures (Li). (Table 3).

Enumeration of Medicinal Flora

1. *Acantholimon lycopodioides* (Girard) Boiss. *Statice lycopodioides* Girard.

Family : Plumbaginaceae

Name : Longze (L)

Distribution : Central Asia & Kashmir to Kumaun.

Habitat: Rocky dry slopes, Leh, Zanskar and Changthang valleys around 4600 m.

Plant Characters : A small prickly tufted shrub. Leaves linear-lanceolate, with needle sharp apex, crowded. Bracts with papery margins. Flowers pale pink in dense clusters. Petals notched, spreading, fused below.

Flowering & fruiting : July to August.

Uses (T) : Usually plant ash is used with milk in cardiac disorders.

2. *Achillea millefolium* Linn.

Family : Asteraceae

Name : Gandana (H), Milfoil / Yarrow (E), Chuang (L)

Distribution : Widely distributed from Kashmir to Sikkim, temperate Eurasia.

Habitat : Along road sides & dry places, Zanskar, Suru, Changthang and Lahaul valleys around 3,680 m..

Plant Characters : A perennial herb. Stem branched, furrowed reaching upto 50 cm tall, pubescent, glabrous. Leaves narrowly oblong lanceolate. Flower heads terminal, corymbose, white short pedicel. Involucral bracts obtuse. Achenes small, 1-2 mm long.

Flowering & fruiting : July - September.

Uses (T) : The extract of leaves is useful in urinary problems, toothache and gum inflammation. The leaves are used for stomach complaints. Flowers are used in tonic.

(Li) : The whole herb is credited with astringent, stimulant, tonic, diaphoretic, cold, colic, heartburn, hysteria, epilepsy and rheumatism The herb contains alkaloids and quartenary bases like achilleine, achilletin, betaine, betonicin and choline. Major sterols and triterpinoids are stgmasterol, campesterol and B-amyrin. It is one of the constituents of Liv-52.

3. *Aconitum heterophyllum* Wall ex. Royle

Family : Ranunculaceae

Name : Attees (H), Aconite (E), Bona-karpo (L)

*Distribution :*Pakistan, Central Nepal to Kashmir.

Habitat : Moist & damp places, Zanskar, Suru and Lahaul valleys around 2,865 m.

Plant Characters : An erect, perennial herb. Roots tuberous. Stem branched, 30-100 cm. Leaves ovate-heart-shaped to rounded blades, the upper clasping the stem; lower leaves deeply lobed, long stalked. Flowers large, greenish purple usually in lax spike like clusters. Hood rounded, broader than long. Seeds small and black.

Flowering & fruiting : June - mid September.

Uses (T) : A very important plant of Amchi system and other traditional systems of Indian medicine. The roots are used against toothache, high fever, stomach complaint, gastric trouble and headache. The roots are collected at the time of maturity in October and dried properly in shade. These are made into a powder and used in different medicines.

(Li) The roots are source of alkaloids heterophyllisine heterophylline, atidine, hetisine, atisine. The roots are considered as astringent, anti-pyretic, anti-diabetic and for dyspepsia, abdominal pains, vomiting and cough.

4. *Aconitum laeve* Royle

Family : Ranunculaceae

Name : Aconite (E)

Distribution : Kashmir to Sikkim & Western Nepal.

Habitat : Meadows, Lahaul valley around 3,890 m.

Plant Characters : An erect, perennial herb upto 1.0-2.5 m tall. Stem much branched, glabrescent or pubescent in the upper part. Leaves orbicular, 15-25 cm across, deeply 5-9 lobed nearly to the base; segments lobed and sharply toothed; lower leaves long petioled; upper nearly sessile. Flowers in racemes, pale-yellow to dull purple. Helmet pointed infront and produced upwards in a high obtuse beak. Lateral sepals much shorter than the helmet. Follicle 3.

Flowering & fruiting : July to September.

Uses (T) : An important medicinal plant used in various indigenous systems of medicine as a substitute of *Aconitum heterophyllum*.

5. *Aconitum rotundifolium* Kar. & Kir.

Family : Ranunculaceae

Name : Aconite (E)

Distribution : Kashmir to Himachal Pradesh.

Habitat : On alpine slopes, Lahaul valley around 3,970 m.

Plant Characters : An erect or decumbent, tufted, perennial herb. Stem 15-30 cm. Leaves few lower long petioled, 5-partite or lobed; lobes further divided into narrow, linear segments. Flowers in short racemes or corymbs, greenish-blue. Helmet shallow. Follicles 5, hairy.

Flowering & fruiting : Mid July to September.

Uses (T) : The roots are used to cure fever and cold.

6. *Aconitum violaceum* Jacq. ex Stapf.

Family : Ranunculaceae

Name : Pattees (H), Aconite (E), Bona-nagpo (L)

Distribution : Kashmir to Himachal Pradesh & Kumaun and Pakistan to Central Nepal.

Habitat : Along water stream & moist places, Nubra, Zanskar and Spiti valleys between 2,960 – 3,600 m.

Plant Characters : A perennial herb. Stem 15-30 cm. Leaves few 4-8 cm across, palmately cut to the base into narrow segments. Flowers in dense spike, dark blue, 2-2.5 cm, bracts small, linear. Follicles 3-5, hairy.

Flowering & fruiting : Mid June - August.

Uses (T) : The roots are sometimes used as a substitute of *Aconitum heterophyllum*. These are used against cold, cough, asthma, fever, and gastric problems.

(Li) Air-dried roots of the plant are reported to contain 1 per cent indaconitine. The roots are used to cure cough, fever and as antiperiodic.

7. ***Aconogonum tortuosum*** (D. Don.) Hara. (*Polygonum tortuosum* D. Don.)

See Sr. No. 2, Chapter IX.

Uses (T) : The whole plant has got medicinal properties and play important role in Amchi system of medicine.

8. ***Actinocarya tibetica*** Benth.

Family : Boraginaceae

Distribution : Kashmir to Sikkim in Temperate Himalayas.

Habitat : Found around dry areas, Leh valley 3,700 to 4,250 m.

Plant Characters : A prostrate, perennial, herb. Stem branched. Leaves alternate, obovate-oblong, margin entire. Flowers minute. Calyx deeply 5-lobed, corolla tubes slightly longer than the sepals. Fruits small with spiny outgrowth.

Flowering & fruiting : June - August.

Uses (T) : A decoction is prepared from the roots, which is said to be beneficial for fever.

9. ***Ajuga bracteosa*** Wall. ex. Benth.

Family : Lamiaceae

Name : Neelkanthi (H), Ratpacho (L)

Distribution : Afganistan to Bhutan & Lahaul to Kashmir.

Habitat : Occasional on dry areas, Lahaul valley around 3,640 m.

Plant Characters : An erect-ascending, pubescent-villous, perennial herb with a woody base. Leaves lower long petioled, upper sessile, all leaves oblanceolate-obovate with a narrow base, crenate-dentate. Flowers subsessile, in axillary clusters. Calyx glandular-villous outside. Corolla blue or pink, hairy outside. Stamens exerted.

Flowering & fruiting : Mid May to July.

Uses (T): The leaves are having stimulant, diuretic, tonic and apparient properties.

(Li) : The plant contains glycoside and tannin having medicinal value. Mostly consumed in traditional medicines and in Ayurvedic formulations.

10. *Allium carolinianum* DC.

See Sr. No. 2, Chapter VIII.

Uses (T) : The leaves are used against indigestion.

11. *Anaphalis busua* (Buch. ex D. Don.) DC.

Family : Asteraceae

Name : Telgang (L)

Distribution : South West China to North Western Himalayas.

Habitat : On alpine slopes and grazing grounds, Leh, Suru and Spiti valleys around 3,400-4,000 m.

Plant Characters : An erect perennial plant reaching upto 30 to 100 cm. Stem usually branched from the base. Leaves numerous, acute with enrolled margins, green above, white woolly beneath. Flower-heads white with blunt involucral bracts which are erect in flowers and spreading in fruits.

Flowering & fruiting : Mid July to September.

Uses (T) : The flower heads are used to check the bleeding. A decoction is made from leaves and flowers, which are used against cold and cough.

12. *Anaphalis cuneifolia* (DC.) Hk. f.

Family : Asteraceae

Name : Simula (L)

Distribution : From Kashmir to Ladakh & Tibet.

Habitat : Rocky places & dry areas, Leh, Nubra and Lahaul valleys between 3,410-3,680 m.

Plant Characters : A small perennial herb. Stem 15-30 cm, branched, slender, cottony. Leaves small, linear-oblong. Flower-heads 5-10 cm, corymbose. Involucral bracts white-papery and lanceolate.

Flowering & fruiting : July to August.

Uses (T) : The extract of plant is used in treatment of skin diseases.

13. *Anaphalis triplinervis* (Sims) C.B. Clarke var. ***monocephala*** (DC) Airy Shaw (*A. nubigena* DC .)

Family : Asteraceae

Name : Phulumentock or sPra-rgod (L)

Distribution : South West China to Western Himalayas

Habitat : Alpine rocks & dry places, Leh and Nubra valleys around 4,230-4,575 m.

Plant Characters : A dwarf tufted alpine herb. Stem leafy, 4-9 cm. Leaves many, narrow lanceolate, white woolly, entire or toothed. Flower-heads small, white, solitary or few in terminal clusters.

Flowering & fruiting : Mid July - August.

Uses (T) : The paste of flower heads is used against wounds. The stem, leaves, flowers and fruits are beneficial in chronic diseases and genital problems.

14. *Anaphalis triplinervis* (Sims) Clarke var. ***intermedia*** (DC) Airy Shaw *A. nepalensis* (Spreng) Hand. Mazz.

Family : Asteraceae

Name : Yaktso (L)

Distribution : South West China to Kashmir to Kumaun.

Habitat : Moist grounds & alpine slopes, Leh & Nubra valleys around 2,990-3,285 m.

Plant Characters : A small tufted herb. Stem 10-25 cm, leafy. Leaves elliptic-lanceolate, awned tip, hairy. Flower-heads 1-3 cm diameter. Involucral bracts subacute, 7-9 mm long. Achene oblong.

Flowering & fruiting : June to September.

Uses (T) : A paste of leaves and flower is used against skin diseases.

15. *Anaphalis virgata* T. Thoms ex. Clarke.

Family : Asteraceae

Name : Simula (L)

Distribution : From Kashmir to Sikkim.

Habitat : On moist rocky slopes, Suru, Zanskar and Lahaul valleys around 3,120-3,640 m.

Plant Characters : An annual tufted herb. Stem many, slender, branched above, reaching upto 40 cm. Leaves spreading, linear with recurved margins. Flower-heads many in campanulate. Involucral bracts linear-oblong, white or yellow.

Flowering & fruiting : July - August.

Uses (T) : A decoction of flowers is used in treatment of fever.

16. *Androsace mucronifolia* Watt. *A. microphylla* Hk. f.

Family : Primulaceae

Name : Zigsolo or Ga-tig smug-po (L)

Distribution : Afganistan to Kumaun to Kashmir.

Habitat : On open alpine slopes, Nubra and Spiti valleys between 4,650-4,750 m.

Plant Characters : A rather lax or dense chusion or mat forming plant. Stem densely tufted. Leaves 2-4 mm, obtuse, mucronate. Flowers borne in umbel, fragrant and bracts bristly haired.

Flowering & fruiting : August - mid September.

Uses (T) : An infusion of whole herb and flowers is used against ringworm infections, excess serous fluid, fever, wounds in the body and in other stomach complaints. An infusion of the plant is used to treat abdominal pains.

17. *Androsace rotundifolia* Hardw.

Family : Primulaceae

Distribution : Afganistan to Bhutan & Himachal Pradesh to Kashmir.

Habitat : On alpine meadows & dry slopes, Lahaul valley 3,980 m.

Plant Characters : A hairy or glandular-pubescent, perennial herb. Leaves long-petioled, orbicular to rounded-cordate, lobes toothed, crenate or entire. Flowers light to deep pink or pinkish-white, in 2.5 to 10.0 cm long flowering scapes.

Flowering & fruiting : July to mid August.

Uses (T) : The whole plant is used as tonic.

18. *Androsace villosa* Linn.

Family : Primulaceae

Name : Fujung (L)

Distribution : Himalayas in Kashmir to Ladakh.

Habitat : Dry areas & open slopes, Leh valley around 4,205 m.

Plant characters : A prostrate perennial herb. Stem tufted. Leaves small linear. Scape short. Umbel sessile. Corolla with a prominent ring at the mouth.

Flowering & fruiting : June to September.

Uses (T) : The leaves are used for making health tonic mainly given weak person by local Amchis.

19. *Anemone rivularis* Buch. Ham. ex DC.

Family : Ranunculaceae

Name : Kikisorng or Srub-ka (L)

Distribution : Himalayas from Kashmir to Sikkim & South West China.

Habitat : On moist places, Changthang and Lahaul valleys around 3,760 m.

Plant Characters : A perennial herb. Stem branched from the base. Radical leaves long petioled, lobbed. Flowers white-bluish in compound cyme, bracteate. Sepals 5-8, silky outside. Achenes large, oblong with hooked style.

Flowering & fruiting : July to mid September.

Uses (T) : The fruits are used to restore digestive heat of the stomach and beneficial in treatment of indigestion, pathogenic disease and poisoning caused by snake bite, heals infected wounds and reduces excess serous fluid. The fruits are commonly collected by the local tribes during summer.

(Li) The fruits are beneficial in treatment of indigestion, pathogenic disease and poisoning. The plant contains saponin glycoside rivularinin and betulinic acid.

20. *Anemone rupicola* Camb.

Family : Ranunculaceae

Name : Simaso (L)

Distribution : Afganistan to South West China, Himalayas from Himachal Pradesh to Kashmir.

Habitat : Near grazing grounds and moist places, Zanskar and Suru valleys around 3,710 m.

Plant Characters : A small perennial herb. Radical leaves long-petioled, segments more or less petioled. Scapes 15-20 cm long, silky. Flowers 1-2, large showy. Sepals broadly oval, downy outside. Fruits achene, ovate.

Flowering & fruiting : Mid June to August.

Uses (T) : The whole plant is considered useful against stomach complaints.

21. *Aquilegia fragrans* Benth.

See Sr. No. 4, Chapter X.

Uses (T) : The herb is diuretic, diaphoretic, antiscorbutic and tranquillizing. It is employed for hysteria and chronic skin troubles. Roots are used for removal of kidney stones.

(Li) Seeds contain fatty acid and resin. It also contains water soluble proteins of low-molecular weight and an alkaloid, aquilegine.

22. *Arabidopsis multiflorum* Hk. f. & T.

Family : Brassicaceae

Name : Imatso (L)

Distribution : From Kashmir to Kumaun.

Habitat : Stony slopes & along roadsides, Nubra valley around 3,120 m.

Plant Characters : A small herb. Stem 10-20 cm, branched. Leaves small, more or less toothed. Flowers in terminal raceme. Pod long slender with ovoid seeds.

Flowering & fruiting : July to August.

Uses (T) : The whole plant is considered useful in diarrhoea.

23. *Arabis glandulosa* Kar. & Kir. *Dontostemon glandulosus* (Kar & Kir.) O. E. S.

Family : Brassicaceae

Name : Umakno (L)

Distribution : From Kashmir to Kumaun & Western Tibet.

Habitat : Rocky places & roadsides, Nubra, Changthang and Spiti valleys around 3,625 m.

Plant Characters : A prostrate perennial plant. Stem many from the base and leafy. Radical leaves stalked; cauline leaves sessile. Flowers rose-purple in terminal clusters. Pod linear, 1-2 cm long, hairless.

Flowering & fruiting : July to September.

Uses (T) : The extract of leaves is used for treatment of abdominal pains and considered digestive. The plant infusion is used against stomach upset.

24. *Arctium lappa* Linn.

Family : Asteraceae

Name : Burdock (E), Jangli kuth (H), Pizums (L)

Distribution : Throughout Himalayas.

Habitat : Common on dry and wet areas, Zanskar and Suru valleys between 3,450-3,560 m.

Plant Characters : A tall coarse herb reaching upto 1 m. Stem much branched. Leaves petioled, ovate-cordate, cottony. Flower-heads terminal, hooked. Corolla and stamens usually purple. Achenes oblong, glabrous.

Flowering & fruiting : Mid June to August.

Uses (T) : The roots of the plant are used as diuretic, urinary and kidney problems.

(Li) The drug possesses stomachache, depurgative, and anti-phologistic properties and is used in chronic gastritis. The tincture prepared from the fresh roots is used in pulmonary catarrh, rheumatism and gout. The plant contains butyric acid, caffeic acid, chlorogenic acid, lauric acid, linoleic acid, oleic acid, etc.

25. *Arenaria bryophylla* Fern; *A. musciformis* Wall. ex Edgew. & Hk. f. & T.

Family : Caryophyllaceae

Name : Lekhum (L)

Distribution : From Kashmir to Kumaun, Tibetan borderlands.

Habitat : Stony slopes and high passes, Leh & Changthang valleys between 4,980-5,320 m.

Plant Characters : A densely tufted perennial plant forming tight round cushion. Leaves linear, minutely ciliate, acute apex, persistent on old stem. Flowers small, solitary, white. Petals linear-lanceolate.

Flowering & fruiting : August to mid September.

Uses (T) : The plant, dried in shade, is made into powder and used against kidney and urinary troubles in Ladakh.

26. *Arenaria griffithii* Boiss.

Family : Caryophyllaceae

Name : Sokhtam (L)

Distribution : Himalayas from Kashmir to Sikkim & Afganistan.

Habitat : Stony slopes, Leh valley around 3,450 m.

Plant Characters : A loose tufted perennial herb. Leaves slender-pointed, ribbed, densely tufted. Flowers few, white borne on nearly leafless stem. Petals spathulate. Sepals broadly elliptic, sharp pointed.

Flowering & fruiting : Mid July to August.

Uses : The whole plant is made into an extract, which is considered useful in menstrual regulation by local Amchis.

27. *Arnebia benthamii* (Wall. ex G. Don.) Johnston

Family : Boraginaceae

Name : Ratanjot (H); Balchar (L)

Distribution : Pakistan to West Nepal, Himalayas from Kashmir to Himachal Pradesh & Kumaun.

Habitat : On moist alpine slopes, Lahaul valley between 4,690-5,120 m.

Plant Characters : An erect, perennial hairy herb with thick purple roots. Stem solitary, fistular, unbranched, densely covered with white tricomes. Basal leaves many, lanceolate, entire, acute; middle and upper cauline leaves gradually reduced in size upwards. Flowers purple to pink in dense cylindrical inflorescence. Bracts linear-lanceolate, acuminate, exceeding the calyx. Calyx divided to the base. Corolla usually shorter than the calyx. Nutlets broadest near the middle, narrowed towards the base.

Flowering & fruiting : June to August.

Uses (T) : The powdered root mixed with the hair oil and is considered as a hair tonic. The root has also got medicinal properties.

28. *Arnebia euchroma* (Royle.) John. I. M Johnston.

Family : Boraginaceae

Name : Ratanjot (H); Demok (Dre-mog) (L)

Distribution : Himalayas from Kashmir to Kumaun & Afganistan.

Habitat : On dry rocky slopes, Leh, Nubra, Suru, Zanskar, Changthang and Spiti valleys between 4,650-4,790 m.

Plant Characters : A tufted very bristly haired perennial with stout rootstock. Stem many, arising from the rootstock, 25-40 cm. Leaves linear to narrow lanceolate, bristly haired. Flowers many in rounded cluster, pale pink-purple turning blackish-purple.

Flowering & fruiting : June to September.

Uses (T) : In Amchi system of medicine the root is used as blood purifier, blood disease, cold, cough, lung and pulmonary problems. The roots are also used as hair tonic by the locals.

(Li) : The roots contain Ac-Shikonin and four related compounds - alkamin-B, B-di-Me-acrylate. It has been reported to be an antipyretic, anti-microbial and anti-inflammatory.

29. *Arnebia guttata* Bunge. *A. tibetana* Kurtz.

Family : Boraginaceae

Name : Ratanjot (H), Demok (L)

Distribution : Throughout Himalayas from Kashmir to Sikkim & Afganistan.

Habitat : On dry sandy places, Changthang and Spiti valleys around 3,870m.

Plant characters : A low tufted very bristly-haired perennial herb. Stem branched covered with the base. Leaves strap shaped, blunt covered with bristly white hairs. Flowers yellow tubular, borne in terminal coiled inflorescence. Nutlets conical, hairless.

Flowering & fruiting : June to September.

Uses (T) : The root has got same application as in *Arnebia euchorma.*

30. *Artemisia annua* Linn.

Family : Asteraceae

Name : Sweet wormwood (E)

Distribution : Afganistan, Kumaun to Kashmir.

Habitat : Introduced and cultivated, Leh valley around 3,410 m.

Plant Characters : An annual, erect, glabrous plant. Radical leaves petiolate, bipinnatisect. Heads very small, globose, yellow, pedicellate, and herbaceous.

Flowering & fruiting : Mid June to September.

Uses (Li) : It is cultivated for an essential oil, which contains a drug named 'Artemisinin', used against malaria throughout the world.

31. *Artemisia biennis* Willd.

Family : Asteraceae

Name : Margrass (L)

Distribution : North Asia, North America, Himalayas from Kashmir to Himachal Pradesh.

Habitat : Dry and sandy areas, Changthang, Suru and Spiti valleys between 3,015 -3,380 m.

Plant Characters : A subsucculent annual or biennial herb. Stem 20-40 cm, stout and soft. Leaves radical, several, petioled, cauline leaves sessile, 3-5 cm, linear-oblong. Flower-heads 1 cm diameter in panicled racemes. Involucral bracts pubescent with scarious margins.

Flowering & fruiting : July to mid August.

Uses (T) : The extract of leaves and flowers is useful in stomach complaints, digestion, seeds are considered useful against obesity they also reduce fat deposition in stomach.

32. *Artemisia brevifolia* Wall. ex DC. *Artimisia maritima* Linn.

See Sr. No. 05, Chapter VII.

Uses (T) : Leaf extract is also used against stomach problems and internal worms of intestine mainly in children by Amchis.

(Li) The plant is a source of a drug 'Santinin' which is effective against round worms and threadworm or *Taenia*.

33. *Artemisia desertorum* Spreng.

Family : Asteraceae

Name : Khamlol or Tshar-bong smug-po (L)

Distribution : Western Himalayan region, Turkestan.

Habitat : Wastelands & dry places, Zanskar valley around 3,605 m.

Plant Characters : A suberect, foetid, perennial herb. Stem much branched; branches grooved and ribbed. Leaves green, pinnatified. Segments usually acute. Flower-heads usually pedicelled, dark, gray, solitary or in clusters. Outer involucral bracts green.

Flowering & fruiting : Mid June to mid September.

Uses (T) : The young aromatic leaves are supposed to be effective in intestinal complaints.

34. *Artemisia dracunculus* Linn.; *A. glauca* Pallas ex Willd.

See Sr No. 07, Chapter VII.

Uses (T) : It is observed that the leaf or flower extract is mainly used against toothache, urinary problems and sometimes against intestinal problems. The roots are useful in treating pharyngitis, lung diseases and swelling due to hot disorders.

(Li) Two compounds namely capillene and 3(2-butynyl) isocumarin reported to be isolated from roots. It is treated as diuretic, control menstrual cycle, against intestinal worms, also relieves toothache.

35. *Artemisia gmelinii* Web. ex Stechm.

Family : Asteraceae

Name : Russian wormwood (E), Nurcha or Khampa shridi (L)

Distribution : Afganistan to Central Nepal, Kashmir to Kinnaur and Kumaun.

Habitat : Grows abundantly on dry, rocky slopes, Leh, Changthang and Spiti valleys between 3,330-3,590 m.

Plant Characters : An erect, strong aromatic, perennial plant (40-140 cm). Leaves long petioled, ovate, pinnatified, green or white. Flower-heads 15-20 florets, short stalked, green in spike-like clusters.

Flowering & fruiting : Mid July to mid September.

Uses (T) : Leaves and flowers extract are recommended for headache, cold, cough and in abdominal upsets. The leaves are used internally for abdominal pains. A decoction of leaves is given in case of cold and cough.

(Li) The chief constituents are flavonoids, scopoletin and monoterpenes & sesquiterpenes, used in hepatitis.

36. *Artemisia laciniata* Willd.

Family : Asteraceae

Name : Bintso (L)

Distribution : From Kashmir to Kumaun.

Habitat : On open places, Changthang valley around 3,870 m.

Plant Characters : A tall perennial herb. Stem hoary, grooved, reaching upto 1-1.5 m. Leaves petioled, pinnatisect, segments pectinatifid, puberulous on both surfaces. Flower-heads 15-20 flowered, hemi-spheric, 2-4 mm diam, hoary. Involucral bracts hoary with green disc and scarious margins.

Flowering & fruiting : July to September.

Uses (T) : The powdered leaves are used in toothache.

37. *Artemisia nilagirica* (Clarke) Pamp.

Family : Asteraceae

Name : Chirmara (L)

Distribution : From Kashmir to Himachal Pradesh.

Habitat : Open slopes & wastelands, Lahaul valley around 3,650 m.

Plant Characters : An erect or ascending, aromatic, perennial herb or undershrub. Stem many from the woody base. Leaves sessile, pinnatipartite to 2-pinnate; segments lanceolate-oblong, entire, glabrate or thin hairy above, densely white-lanate beneath. Heads sessile in spiciform racemes, combined into panicles. Achenes glabrous.

Flowering & fruiting : July to September.

Uses (T) : The leaves are used as blood purifier.

(Li) : The plant contains camphor, 1:8 - cineole and borneol and is having larvicidal, insecticidal, antimicrobial, antifungal properties.

38. *Artemisia parviflora* Buch. - Ham ex D. Don.

Family : Asteraceae

Name : Khamang (L)

Distribution : Himalayas from Kashmir to China and Nepal.

Habitat : On dry open areas, Changthang and Spiti valleys around 4,380 m.

Plant Characters : A shrubby, aromatic, perennial herb. Leaves variable, sessile, palmately segmented. Flower-heads pedicelled, greenish in panicle raceme. Fruits achenes, ellipsoidal, smooth.

Flowering & fruiting : July to mid September.

Uses (T) : The plant parts are used to prepare local drug against menstrual problems and also believed as diuretic by local Amchis.

(Li) The plant is reported to possess anti-viral properties.

39. *Artemisia persica* Boiss.

Family : Asteraceae

Name : Burnak (L)

Distribution : Western Tibet to Afganistan.

Habitat : Open slopes & wastelands, Suru valley around 2,980 m.

Plant Characters : An erect, strongly aromatic perennial plant. Stem 1-1.5 m. tall, grooved and ribbed, much branched. Leaves sessile or petioled, very finely cut, segments minute, linear or obovate. Flower-heads many, yellow in long racemes. Involucral bracts tomentose, outer linear, inner orbicular. Corolla very short, pubescent or glabrous. Achenes small.

Flowering & fruiting : July to September.

Uses (T) : The flowers are used as tonic. An extract of flowers is also used against stomachache. The plant is also used to prepare tonic, febrifuge and vermifuge.

(Li) The leaves and stems of the plant yield scopoletin.

40. *Artemisia salsoloides* Willd.

Family : Asteraceae

Name : Amango (L)

Distribution : Western Himalayas from Kashmir to Kumaun & Pakistan, Mongolia.

Habitat : On Sandy & dry slopes, Zanskar & Suru valleys between 3,650-3,710 m.

Plant Characters : An aromatic, perennial herb. Stem branched from the base, angled, glabrous. Leaves, glabrous, 2-3-pinnatisect; segment slender, obtuse or acute. Flower-heads subsessile or rarely pedicelled, ovoid. Involucral bracts subulate, hooked. Corolla and stamens usually purple. Achenes glabrous, oblong.

Flowering & fruiting : June to August.

Uses (T) : The plant is useful against intestinal complaints.

41. *Artemisia scoparia* Waldst. & Kit.

Family : Asteraceae

Name : Khamtspa (L)

Distribution : Central Europe, Afganistan, Kashmir to Garhwal, Punjab and the upper Gangotric plains.

Habitat : Open areas & along roadside, Changthang valley around 3,890 m.

Plant Characters : A faintly scented, branched, annual, herb. Stem prostrate, branches often capillary. Leaves radical, long petioled, 1-3 pinnatisect, ovate, segments linear. Flower-heads minute in slender racemes; achene small.

Flowering & fruiting : July to September.

Uses (T) : The plant parts are used in preparation of medicine against burns and intestinal problems by the local Amchis of Ladakh.

(Li) The essential oil is reported to be used in therapy of urolithiasis. It is found to possess some sedative and anti-convulsant properties.

42. *Artemisia sieversiana* Ehrh ex Willd.

Family : Asteraceae

Name : Khamchu (L)

Distribution : China, Western Himalayas from Kashmir to Lahaul.

Habitat : Rocky areas and dry slopes, Leh, Suru and Spiti valleys between 2,890 - 3,320 m.

Plant Characters : An aromatic, annual or biennial plant. Stem tall (50-90 cm) ribbed, branched above. Leaves petioled, broadly ovate, pinnatisect, segments obtuse grey- hoary on both surface. Flower-heads yellow in terminal branched spike like clusters.

Flowering & fruiting : July to September.

Uses (T) : A decoction is prepared from dried leaves and flowers, which is helpful to cure intestinal worms in childs.

(Li) : The aerial parts contain sieversinin and siversin. Sieversinin is having antimicrobial properties.

43. *Artemisia stracheyi* Hk. f. & T. ex Clarke.

Family : Asteraceae

Name : Rumonlo (L)

Distribution : Northern Himalayas from Kashmir to Lahaul & Garhwal.

Habitat : Common on stony slopes, Changthang valley around 4,560 m.

Plant Characters : An aromatic, dwarf, annual herb. Stem woody branched, tomentose. Leaves long, linear, subacute, oblong; cauline sessile. Flower-heads 4-6 mm broad, hemispheric on stout leafless raceme. Involucral bracts numerous, ovate, tomentose. Achenes small, cuneiform

Flowering & fruiting : Mid June to August.

Uses (T) : Useful in toothache.

44. ***Artemisia tournefortiana*** Reichb.

See Sr No. 17, Chapter VII.

Uses (T) : A decoction of leaves and flowering buds is used against intestinal worms.

45. ***Asperugo procumbens*** Linn.

Family : Boraginaceae

Name : Zingsha (L)

Distribution : From Kashmir to Western Tibet and North Asia.

Habitat : Cultivated fields and wastelads, Suru and Lahaul valleys around 2,930 m.

Plant Characters : A procumbent annual herb. Stem branched, 30-50 cm high.

Leaves alternate, subopposite, spathulate-oblong. Flowers small, purplish-blue, subsessile, solitary or paired. Calyx-lobes oblong. Fruits nutlets. Persistent calyx reticulately nerved in fruits.

Flowering & fruiting : July to mid September.

Uses (T) : The extract of roots is useful in stomach complaints.

46. ***Asperula oppositifolia*** Reg. & Schmalh ssp. *pseudo-cynanchica* Ehrend.

Family : Rubiaceae

Name : Zingsha (L)

Distribution : From Himachal Pradesh to Western Tibet and Kashmir.

Habitat : Rare on rocky crevices, Lahaul valley around 3,560 m.

Plant Characters : An erect or ascending, glabrous, perennial herbs. Stems many from the woody base. Leaves linear-oblong, acute or obtuse, flat, thick, nerveless, lower whorled upper opposite. Flowers pink in terminal umbel-like fascicles.

Flowering & fruiting : July to September.

Uses (T) : The whole herb is used as tonic.

47. ***Aster flaccidus*** Bunge

See Sr No. 7, Chapter X.

Uses (T) : The flowers are useful in treating infectious, common cold and poisoning. It is also beneficial in treating psoriasis and spasms of the tendons and the ligaments.

48. ***Astragalus multiceps*** Wall.

Family : Fabaceae

Distribution : From Kashmir to Himachal Pradesh, Kumaun and Garhwal.

Habitat : Dry places, Suru valley around 3,610 m.

Plant Characters : A tufted, hairy, diffused, perennial herb. Tufts densely congested, armed. Stipules lanceolate, adnate. Branchelets with densely crowded nodes, flowers 1-2 together in leaf axils sessile. Pod sessile, oblong.

Flowering & fruiting : June to September.

Uses (T) : Seeds are sometimes used for relief in colic.

49. ***Astragalus rhizanthus*** Royle ex Benth.

Family : Fabaceae

Name : Sarma (L)

Distribution : From Pakistan to J&K, Himachal Pradesh and Uttranchal.

Habitat : On open stony slopes, Leh, Zanskar, Suru, Changthang & Nubra valleys between 3,120-3,705 m.

Plant Characters : A stemless perennial herb, densely covered with silky hairs. Leaves pinnate, pressed to the ground and radiating from the centre; imparipinnately compound, oblong, pilose beneath. Flowers bright yellow in dense stalkless cluster at the centre, aggregated together at the base. Bracts linear. Calyx with unequal lobes. Corolla yellow. Pods 1.5-2 cm, oblong, densely silky and bilocular.

Flowering & fruiting : June to September.

Uses (T) : A decoction of plant is useful in skin diseases.

50. ***Astragalus strictus*** Graham ex. Benth.

Family : Fabaceae

Name : Serpang (L)

Distribution : From Pakistan to Kashmir, Kumaun and Tibet.

Habitat : On rocky slopes, Leh & Changthang valleys between 4,780-4,830m.

Plant Characters : A semi prostrate perennial herb with tufted spreading branches. Stem short slender, ascending clothed. Leaves 2-4 cm, petioled, oblong-blunt with adpressed hairs; leaflets close, oblong, obtuse, numerous, pinnate. Flowers small, pale purple to pinkish in cluster. Calyx with black hair. Pod white or black hairs.

Flowering & fruiting : July to September.

Uses (T) : The whole plant is commonly used as diuretic.

51. *Astragalus tribulifolius* Benth.

Family : Fabaceae

Name : Yanglo (L)

Distribution : Throughout J & K to Himachal Pradesh, Pakistan and Bhutan.

Habitat : Cultivated fields, Nubra valley around 3,510 m.

Plant Characters : A prostrate perennial herb. Stem glabrous, densely tomentose. Rootstock, long, woody, fusiform Leaves 1-2 inches long; leaflets many, finely, haired, crowded, oblong. Stipule minute. Flowers few, in a dense heads. Fruits ovoid, black. Pod finely haired.

Flowering & fruiting : July to September.

Uses (T) : A decoction of the whole plant is useful as diuretic and against kidney complaints.

52. *Atelanthera perpusilla* Hk. f. & T.

Family : Brassicaceae

Distribution : From Kashmir to Western Tibet and Zanskar.

Habitat : Sandy places, Zanskar valley around 4,210 m.

Plant Characters : A slender, annual herb. Stem hairy, simple, subsessile. Leaves few, linear, entire. Flowers small, subsessile. Sepals erect, equal at the base. Petals narrowly spathulate, emarginated. Pod few, erect, linear, compressed. Seed 1-serrate, flattened.

Flowering & fruiting : July to August.

Uses (T) : The decoction of leaves is used against fever.

53. *Berberis jaeschkeana* Schneid.

Family : Berberidaceae

Distribution : Himalayas from Pakistan to Kashmir & Nepal.

Habitat : Dry slopes, Lahaul valley around 3,140 m.

Plant Characters : A deciduous shrub upto 1.0 m tall. Stem stout, angular at first slightly reddish, yellow-brown at maturity; spines 3-fid. Leaves oblong-elliptic, sessile or subsessile. Flowers in umbels or subumbellete, 3-5 (-8) flowered. Fruit oblong-ovoid.

Flowering & Fruiting : June to October.

Uses (T) : The roots are used as remedy of eye troubles.

54. *Berberis kunawarensis* Royle.

Family : Berberidaceae

Distribution : Himalayas from Kashmir to Nepal.

Habitat : Open slopes, Lahaul valley around 3,670 m.

Plant Characters : A shrub upto 1.5 cm tall. Young shoots red, minutely puberulous; mature stems glabrous. Leaves narrowly obovate or elliptic, subsessile, spinose serrate at the margins. Panicles about as long as leaves, densely 15-40 flowered. Pedicels red-brown.

Flowering & Fruiting : June to August.

Uses (T) : The roots are used as remedy of eye troubles.

55. *Berberis pachyacantha* Koehne.

Family : Berberidaceae

Name : Karpa (L)

Distribution : Himalayas from Himachal Pradesh to Pakistan & Nepal.

Habitat : Moist slopes, Lahaul valley around 3,670 m.

Plant Characters : A deciduous, glabrous shrub, 2.0-3.0 m tall. Stem dark-red to pale-brownish, subterete; spines often absent. Leaves 3-6 cm long, 1-2 cm broad, oblong-obovate. Flowers yellow, in 15-20 flowered racemes. Fruits oblong ovoid.

Flowering & Fruiting : May to September.

Uses (T) : The ripe fruits are used in stomach problems.

56. *Berberis ulicina* Hk. f. & T.

Family : Berberidaceae

Name : Berberry (E), Khizer or Kiraring (L)

Distribution : Western Himalayas from Pakistan to Central Nepal.

Habitat : Common along river belts on sandy & drier areas, Nubra valley around 3,120 m.

Plant Characters : A much branched, shrub reaching upto 1 m. Branches short, stout, spiny. Leaves 1-2, linear, toothed. Flowers orange-yellow, drooping, petals bifid. Fruits ovoid, black.

Flowering & Fruiting : June to August.

Uses (T) : Roots and bark are a source of tonic.

57. *Berberis zabeliana* C. K. Schn.

Family : Berberidaceae

Name : Berberry (E)

Distribution : From Kashmir to Kumaun in the temperate Western Himalayas.

Habitat : Along sandy belts, Nubra valley around 3,015 m.

Plant Characters : A stiff, stout-stemmed, perennial shrub. Stem branched, woody, yellow, and spiny. Leaves deciduous, elliptic to obovate with mucronate apex. Flowers many in raceme. Berries ovoid, red.

Flowering & fruiting : July to September.

Uses (T) : A decoction of roots is used in fever by locals of Ladakh.

58. *Bergenia ligulata* Yeo *Bergenia ciliata* Sternb.

Family : Saxifragaceae

Name : Pashanbhed (H), Dhoklambu (L)

Distribution : From South East Tibet to Himalayas & Kashmir to Himachal Pradesh.

Habitat : Moist slopes, Lahaul valley around 3,780 m.

Plant Characters : A perennial herb. Rootstocks very stout, creeping. Stem short, thick & fleshy. Leaves stalked, sheathing at the base, ovate or orbicular, margin fringed with short & stiff hairs. Flowers many, pink or white or lilac, in terminal corymbs. Fruits capsule.

Flowering & fruiting : June to September.

Uses (T) : The roots are used against kidney stones.

(Li) : The rhizome contains bergenin, gallic acid, tannic acid, mucilage, glucoside, albumens, starch and mineral salts. The rhizomes are astringent, diuretic, tonic properties. The presence of beta-sitosterol and flavonoids is also reported.

59. *Bergenia stracheyi* (Hk. f. & T.) Engl. *Saxifraga stracheyi* Hk. f. & T

Family : Saxifragaceae

Name : Pashanbhed (H); Tiang or Gaotlis or Gatikpa (L)

Distribution : From Afganistan to Uttranchal, Western Himalaya and Kashmir to Himachal Pradesh.

Habitat : Open slopes, Zanskar and Lahaul valleys around 3,610 m.

Plant Characters : A perennial herb. Rootstocks very stout, creeping. Stem simple. Leaves with ovate or rounded, blade, 20 cm, glabrous with crenate margins. Flowers many, pink or white or lilac, in a drooping cluster. Calyx hairy with marginal bristles. Fruits ovate-lanceolate.

Flowering & fruiting : Mid June to September.

Uses (T) : In Zanskar areas roots are used against stomach complaints. The roots are also used as febrifuge, anti-inflammatory, against contagious diseases, influenza, inflammation of lungs and nerve, swelling of limbs etc.

(Li) The rhizome contains gallic acid, tannic acid, mucilage, glucoside, albumens, starch and mineral salts. The rhizomes and roots are bitter, astringent, diuretic, tonic properties. The roots are reported to be used in menorrhagia and in urinary diseases.

60. *Biebersteinia odora* Stephan ex Fischer.

See Sr No. 18, Chapter VII.

Uses (T) : The paste of whole plant is used to cure of cuts, ulcers, skin sores and wounds. In Amchi (Tibetan) system of medicine it is used in healing of wounds, cuts and in peptic ulcers.

61. *Blumea bifoliata* DC.; *Conyza bifoliata* DC. Prodr.

Family : Asteraceae

Name : Leuman (L)

Distribution : From Pakistan to Bengal, Malaya, Burma and Ceylon.

Habitat : Open slopes, Changthang valley around 3,595 m.

Plant Characters : An annual or perennial, prostrate, herb. Stem branched, pubescent, ascending. Leaves 3-5 cm, sessile, obovate or oblong, serrate, obtuse or acute. Flower- heads 3-10, solitary, yellow, pedicelled. Bracts acuminate, receptacle glabrous. Corolla yellow. Achenes oblong with white pappus.

Flowering & fruiting : June to August.

Uses (T) : The dried roots are used by the locals to control fever.

62. *Caltha palustris* Linn.

Family : Ranunculaceae

Name : Marsh marigold (E), Chhomlap (L)

Distribution : Kashmir to Nepal, temperate Europe, Asia & North America.

Habitat : Marshy places, Lahaul valley 3,760 m.

Plant Characters : An erect branched herb 15-50 cm tall. Radical leaves surrounded by membranous sheath, long petioled, reniform, finely or coarsely toothed; cauline leaves reniform or deltoid, subsessile to sessile, obtuse, dentate or crenate-dentate. Flowers yellow, 4-5 cm across, on long peduncles. Follicle sessile, flattened, asymetrically oblong, gradually or abruptly passing into a short or long straight or curved beak.

Flowering & fruiting : June to September.

Uses (T) : The powdered leaves are used to keep maggots out of cattle wounds. The roots are used as cure for gonorrhoea, the tribals sometimes used the green leaves to clean their hands.

63. *Capparis spinosa* Linn.

See Sr No. 37, Chapter VIII.

Uses (T) : Bark and roots are used in affections of liver, rheumatism, tubercular gland and paralysis. The plant extract is used in treatment of viral hepatitis, cirrhosis of liver and old age ailments. The fruit paste is sometimes applied externally as pain reliever.the juice and volatile fraction from leaves showed anticystic, fungicidal, bactericidal and phytoncidal activities.

(Li) The bark is, aperient, diuretic and expectorant. An extract of the plant is one of the constituents of the Ayurvedic preparation 'Liv 52'. The root bark contains stachydrine, rutic acid and a volatile substance of garlic odour. The plant contains glocosinolates, glucocapparin and glucocapangalin.

64. *Capsella bursa-pastoris* (Linn.) Medik.

See Sr No. 38, Chapter VIII.

Uses (T) : The whole plant is used against swelling, pus and pain, stop vomiting, beneficial in treating kidneys, lungs, nerve disorders and fluid retention in the body by Amchis. The leaves and dried parts yield tea, which is used for stopping haemorrhages of stomach, lungs, uterus and kidney.

(Li) The fruits are used in dried pus and diseased blood, analagesic. The plant contains luteolin, quercetin, B-sitosterol, 3-rutinoside.

65. *Carum carvi* Linn. var. ***gracile***

See Sr No. 42, Chapter VIII.

Uses (T) : The fruits and seeds are used as febrifuge, improve eye vision, digestive. Useful for hot disorders, weak eye sight, indigestion and poisoning.

(Li) : The seeds contain a mixture of ketone, carvone, terpene and traces of carvacrol. The plant is reported to be abortifacient, anthelmintic, cardiac, carminative, stimulant, stomachic and tonic.

66. *Cassiope fastigiata* (Wall.) D. Don

Family : Ericaceae

Name : Sulu (L)

Distribution : Himalayas from Pakistan to Nepal.

Habitat : A common associates of *Rhododendron anthopogon.*, Lahaul valley around 3,360 m.

Plant Characters : A dwarf, tufted, decumbent herb or shrub. Stems much branched. Leaves sessile, thick, ovate-oblong, acute, closely mbricate. Flowers white, axillary, solitary or 2-3 together. Corolla bell-shaped. Capsule erect, globose, 5-valved.

Flowering & fruiting : Mid June to mid September.

Uses (T) : The plant is of medicinal value as tonic.

67. *Catabrosa aquatica* (Linn.) Beauv.

Family : Poaceae

Name : Leema (L)

Distribution : North temperate regions to Kashmir, Tibet & Asia.

Habitat : Moist areas, Suru valley around 2,895 m.

Plant Characters : A soft, perennial, grass. Stem erect, from the creeping, floating base upto 20-40 cm high. Leaves linear, obtuse, flat, smooth. Ligule ovate, hyaline. Panicle upto 20 cm open, loose. Spikelets pale-green or yellowish 1-3 flowered. Grain oblong subacute at the both ends.

Flowering & fruiting : June to August.

Uses (T) : The plant is believed to be effective in reducing high fever.

68. *Centaurea depressa* M Bieb.

Family : Asteraceae

Name : Vashakha (L)

Distribution : From Afganistan to Asia, Australia and Ceylon.

Habitat : Drier areas, Suru and Leh valleys around 3,890 m.

Plant Characters : An annual, adpressed tomentose herb. Stem simple or branched. Lower leaves petioled, entire or pinnatifid. Cauline leaves linear-lanceolate. Flower-heads ovoid, blue. Ray flowers large spreading blue. Involucral bracts with scarious toothed margins. Achene shining.

Flowering & fruiting : June to August.

Uses (T) : The whole plant is useful in fever, headache and chest pain.

69. *Cerastium cerastioides* (Linn.) Britton

Family : Caryophyllaceae

Name : Shimnar or Tyndong (L)

Distribution : Temperate North Europe, Afganistan to Bhutan, Kashmir and Himachal Pradesh.

Habitat : Open and stony slopes, Suru and Lahaul valleys between 3,010-3,420 m.

Plant Characters : A prostrate, tufted, dwarf, annual or perennial herb. Stem slender, branched. Leaves linear acute, sessile. Flowers small, white to greenish-white in lax branched cluster. Petals conspicuously shallowly bi-lobed. Capsules longer than calyx.

Flowering & fruiting : July to September.

Uses (T) : The plant extract is recommended in headache.

70. *Chaerophyllum reflexum* Lindl.

Family : Apiaceae

Name : Lcha-wa or Young (L)

Distribution : Himalayas from Himachal Pradesh to West Bengal and Kashmir.

Habitat : On grazing grounds, Suru valley around 3,125m.

Plant Characters : An erect, slender, glabrous, perennial herb. Leaves 2-3 pinnate, reflexed, segments small. Flowers white, in small umbels and fruits oblong, narrowed into a thick cylindrical beak.

Flowering & fruiting : July to September.

Uses (T) : The roots are used to promote heat. Used against anemia, swelling of limbs, stomach problem and wind disorder.

71. *Chenopodium botrys* Linn.

See Sr. No. 46, Chapter VIII.

Uses (T) : The plant is used in catarrh and asthma, also used as anthelmintic. The whole herb is used in liver and stomach complaints by local tribes of Ladakh.

(Li) The herb is reported to possess stimulant, laxative, anti-spasmodic, diuretic, pectrol, carminative, emmenagogue and anthelmintic properties. The plant contains flavanoids and alkaloids.

72. *Chrysanthemum griffithi* Clarke.

Family : Asteraceae

Name : Serpan (L)

Distribution : Pakistan to Kashmir, Bhutan and Ladakh.

Habitat : Drier areas, Changthang and Suru valleys between 4,430-4,980 m.

Plant Characters : A perennial shrub. Stem branched from the base. Leaves pinnatifid, segments linear narrow. Involucral bracts viscid. Flower- head long, peduncled. Achenes ribbed, pappus membranous.

Flowering & fruiting : July to August.

Uses (T) : The powdered flower-heads are said to be quite useful to correct menstruation.

73. *Chrysanthemum pyrethroides* (Kar. & Kir.) B. Fedtsch. *Chrysanthemum richteria* Benth.

See Sr. No. 22, Chapter VII.

Uses (T) : The flowers are used against rheumatism, fever and healing of external injury of boils, wounds and excess serous fluid. The powdered flowers are used to control high fever.

74. *Chrysanthemum tibeticum* Hk. f. & T.

Family : Asteraceae

Name : Phemantso (L)

Distribution : From Afganistan to Pakistan, Kashmir, Himachal Pradesh and Ladakh.

Habitat : Alpine & rocky slopes, Zanskar valley around 3,610 m.

Plant Characters : An aromatic, branched, perennial plant. Stem branched reaching upto 30 cm tall. Leaves pinnatifid, segments linear acute or obtuse. Flower-heads long, peduncled, small. Involucral bracts pubescent, vicid with scarious margins. Ligules large, numerous, purple.

Flowering & fruiting : Mid Junt to August.

Uses (T) : The plant paste is used against septic wounds in Ladakh.

75. *Cicer microphyllum* Benth. *C. soongaricum* auct. non Steph ex. DC.

See Sr. No. 49, Chapter IX.

Uses (T) : Seeds are considered as an adaptogen.

(Li) : Seeds are a good source of lecithin.

76. *Cirsium wallichii* DC. Prodr. 6: 643. 1838; *Cnicus wallichii* (DC.), Hk. f.

Family : Asteraceae

Name : Kakar or Vangtsar (L)

Distribution : Afganistan to South West China, Palni & Nilgiri hills in South India.

Habitat : Forest areas, Leh and Lahaul valleys around 3,640 m.

Plant Characters : An erect, thistle like annual-perennial herb. Stem robust, simple or branched, ribbed, glabrescent or cottony. Leaves stalkless, pinnately lobed, with long spines, hairless, above, cottony or woolly beneath. Flower-heads whitish purple in cluster. Involucre with lanceolate bracts ending in recurved spines.

Flowering & fruiting : July to September.

Uses (T) : A decoction or fresh extract prepared from leaves is used in severe stomach pains by local Amchis of Ladakh.

77. *Clematis tibetana* Kuntze.

Family : Ranunculaceae

Name : Clematis (E), Zakgic or dByi-mong nag-po (L)

Distribution : Western Asia, Pakistan to Uttranchal.

Habitat : Cultivated areas, Leh and Spiti valleys around 3,020 m.

Plant Characters : A woody perennial climber. Leaves pinnate, lanceolate-ovate; leaflets lobed, long stalked. Flowers yellow spotted with brown nodding, long stalked, borne on axillary branches. Stamens filaments hiary. Fruits achene, silky haired.

Flowering & fruiting : June to September.

Uses (T) : A water extract of root and stem is used to cure itch and skin problems. Root powder is applied to kill lice in cattle.

78. *Codonopsis clematidea* (Schrenk) Clarke.

Family : Campanulaceae

Name : Burkutang or Mokhting (L)

Distribution : From Afganistan to Kashmir.

Habitat : Cultivated fields & alpine slopes, Suru and Spiti valleys around 3,020 m.

Plant Characters : A strong smelling, perennial herbs. Stem slender, branched from the base, 30-70 cm high. Leaves alternate, hairy, ovate and acute at apex, margins irregularly crenate, base cuneate or rounded. Flowers white, solitary, axillary and terminal peduncles. Capsule cylindric, often curved.

Flowering & fruiting : July to mid September.

Uses (T) : The roots are considered useful in mild stomachache, digestion and stimulant. The whole plant is used against pain, swelling of joints due to gout, arthritis, rheumatism, stiffening of ligaments, nerves, paralysis and leprosy.

(Li) : The plant contains flavonoids, Vit A and C.

79. *Codonopsis ovata* Benth.

Family : Campanulaceae

Name : Ludut or Chamelia (L)

Distribution : From Pakistan to Kashmir.

Habitat : Cultivated fields, moist places & alpine slopes, Suru and Spiti valleys around 3,010 m.

Plant Characters : A sweet-scented, sub-erect or decumbent, hairy, perennial herbs. Stem 40-80 cm, much branched, covered with glandular hairs. Leaves compound, glaucous above. Flowers cream-coloured, large 3-5 cm with mild fragrance and hooked spurs. Spur 15-20 cm. follicles 6-8, densely hairy with long styles.

Flowering & fruiting : July to September.

Uses (T) : Same as in *C. clematidea.*

(Li) : The plant contains taraxerol acetate, apigenin and luteolin.

80. *Codonopsis rotundifolia* Benth.

Family : Campanulaceae

Distribution : From Kashmir to Himachal Pradesh.

Habitat : Cultivated fields, Lahaul valley around 3,565 m.

Plant Characters : A twining, hairy perennial herb. Leaves alternate, ovate, glabrous to sparsely pilose with rounded base and crenate margins. Flowers purple, in solitary, axillary and terminal long peduncles. Calyx teeth elliptic, lobes often crenate or toothed. Corolla widely bell-shaped. Capsule hemispheric.

Flowering & fruiting : July to September.

Uses (T) : Sometimes it is also used as substitute of *C. clematidea.*

81. *Colchicum luteum* Baker.

Family : Liliaceae

Name : Hirantutiya (H), Tukapa (L)

Distribution : Western temperate Himalayas & Pakistan, Turkistan to Himachal Pradesh.

Habitat : Open snow cleared slopes, Suru valley around 3,325 m.

Plant Characters : A bulbous, small perennial herb. Corm cylindrical, brownish in colour, almost conical in shape with narrow-oblong, blunt, increasing in size as the plant approaches fruiting stage. Flowers solitary or 2, large, with golden-yellow petals. Leaves narrow-oblong blunt. Fruits 2.5-4 cm with curved beaks.

Flowering & fruiting : Earliest in mid April to May.

Uses (T) : The bulbs are used against gout.

(Li) : Corms and seeds are source of drug ‚Colchicine' which is useful in pains and inflammation of gout.

82. *Conringia planisiliqua* Fisch.

Family : Brassicaceae

Name : Khaskar (L)

Distribution : From Pakistan to Kashmir, Tibet and Persia.

Habitat : On alpine slopes, Zanskar valley 4,310 m.

Plant Characters : An erect, annual, herb. Stem many from the base, reaching upto 1 m tall. Radical leaves few, obtuse, petioled; cauline sessile, cordate with amplexicaule base. Flowers small, yellow, pedicelled. Pod erect, narrow, and linear.

Flowering & fruiting : July to mid September.

Uses (T) : The decoction of dried roots is used as diuretic.

83. *Corispermum hyssopifolium* Linn.

Family : Chenopodiaceae

Name : Seimso (L)

Distribution : From Kashmir to Lahaul, Kumaun, Western Tibet and China.

Habitat : Dry rocky slopes, Zanskar valley around 3,505 m.

Plant Characters : An annual, stiff, herb. Stem branched from the base, glabrous; branches 10-20 cm long, rigid. Leaves alternate, linear-ovate, and obtuse. Flowers in axillary terminal clusters like spikes, dense 7-15 cm long. Bracts imbricate, oblong-lanceolate, green. Utricle oblong, shining, winged all round.

Flowering & fruiting : June to August.

Uses (T) : The decoction of dried roots is used against kidney problems.

84. *Corydalis cashmeriana* Royle.

Family : Papaveraceae

Name : Ralchatnakpo (L)

Distribution : From Kashmir to Pakistan & Himachal Pradesh.

Habitat : Alpine meadows, Lahaul valley around 4,560 m.

Plant Characters : A perennial herb 10-25 cm tall. Stem simple with 1-2 cauline and few redical leaves. Radical leaves long petioled, palmately 3-5 pinnate;

cauline leaves subsessile, deeply lobed or pinnatisect. Racemes simple 2-3 cm long, subumbellate, 3-8 flowered. Flowers blue to sky-blue with darker tips, turning pale when dried. Capsule linear-oblong.

Flowering & fruiting : July to August.

Uses (T) : The roots are diuretic, antiperiodic and employed in eye troubles.

85. *Corydalis crassissima* Camb. *C. crassifolia* Royle.

Family : Papaveraceae

Name : sKyes-pa dong-btags (L)

Distribution : Western Tibet, Pakistan to Himachal Pradesh.

Habitat : Open rocky slopes, Suru valley around 3,020 m.

Plant Characters : A very, stinking, perennial herb. Root-stock tuberous. Radical leaves few, petioled, cauline leaves 1-3 sessile. Flowers pinkish-white borne in dense cluster. Bracts green spathulate to lanceolate. Fruit inflated, balloon-like easily dispersed by the wind. Capsule ellipsoid-orbicular, many seeded.

Flowering & fruiting : July to September.

Uses (T) : The leaves, stems, flowers and fruits are used for drying up excess pus formation in the lung and visceral organs. It is also used to heal ruptured wounds fractured cranial bones and others.

86. *Corydalis flabellata* Edgew.

Family : Papaveraceae

Distribution : Afganistan, Kashmir to Himachal Pradesh.

Habitat : Common along roadsides and stony slopes, Suru and Spiti valley around 3,540 m.

Plant Characters : A much branched, perennial herb reaching upto 60 cm tall. Stem branched with cauline leaves and many radical leaves bi-or subtri-pinnatisect; pinnae ovate to suborbicular; cauline leaves pinnatisect with narrow lobules. Flowers yellow in dense spike like cluster. Capsule linear, recurved at apex.

Flowering & fruiting : June to August.

Uses (T) : The roots are useful against skin diseases.

87. *Corydalis govaniana* Wall.

Family : Papaveraceae

Name : Stong-Zil (L)

Distribution : From Kumaun to Kashmir.

Habitat : Forest areas & moist places, Suru and Lahaul valleys around 3,604m.

Plant Characters : A stout, branched, annual herb. Rootstock woody often branched. Leaves many, oblong, 2-pinnatisect. Flower many, dense, yellow in dense cluster.

Flowering & fruiting : June to August.

Uses (T) : The roots are used as antipyretic, diuretic, gastric pain and muscular pain. The leaves are used as febrifuge; useful against contagious fever, eye diseases and infections as well as swellings and burns.

(Li) : The plant contains govanine, protopine, stylopine.

88. *Corydalis meifolia* Wall.

Family : Papaveraceae

Name : Nyetapu or Tongzil (L)

Distribution : From Himachal Pradesh to South East Tibet and China.

Habitat : Alpine rocks, Changthang and Lahaul valleys around 4,760 m.

Plant Characters : An erect, perennial herb. Stem stout, branched, leafy. Radical leaves 3-pinnatisect, cauline leaves similar but smaller upwards. Flowers yellow-purple or orange in racemes. Capsule oblong-ellipsoid.

Flowering & fruiting : June to August.

Uses (T) : The whole plant is considered as useful in liver, stomach pain, tonic and diuretic.

(Li) : The stems and leaves contain various alkaloids, mainly of protoberberine and isoquinoline type.

89. *Corydalis thyrsiflora* Prain

Family : Papaveraceae

Name : Nyetapu (L)

Distribution : From Himachal Pradesh to South East Tibet and Bhutan.

Habitat : Alpine rocks, Changthang & Nubra valleys between 3,020-4,520 m.

Plant Characters : A small, delicate, perennial herb. Stem branched, 15-30 cm. Leaves hairless, somewhat glaucous, three times cut into oblong to linear usually pointed segments. Flowers yellow, borne in widely branched inflorescence.

Flowering & fruiting : June to August.

Uses (T) : The plant extract is used in skin diseases of man as well as in cattle as an ethno-veterinarian practice.

90. *Cousinia falconeri* Hk. f.

Family : Asteraceae

Name : Kreatising (L)

Distribution : From Afganistan to Western Nepal, Tibet and Kashmir.

Habitat : Open drier grounds, Suru valley around 3,104 m.

Plant Characters : An erect, spinescent, perennial herb. Stem cottony, branched. Radical leaves petioled, pinnatified, densely white, tomentose beneath. Segments triangular or lanceolate, lobes spinescet. Flower heads sessile to subsessile, solitary or fascicled, involucral bracts many, woolly ending in a stiff erect or spreading spine.

Flowering & fruiting : Mid June to August.

Uses (T) : The juice of plant is found to be effective as diuretic and in body pain.

91. *Cousinia thomsonii* Clarke.

Family : Asteraceae

Name : Megtham (L)

Distribution : From Afganistan to Western Nepal, Tibet and Kashmir.

Habitat : Open dry grounds, Suru and Spiti valleys around 3,120 m.

Plant Characters : An erect, thistle like perennial plant. Stem cottony reaching upto 50 cm high. Leaves pinnatifid, radical petioled, linear, white tomentose beneath, lobes ending in long rigid spines; cauline sessile. Flower-heads pink to purple with globular, spiny, woolly-haired involucres, solitary. Achenes glabrous. Involucral bracts woolly, numerous, ending in a stiff spreading or recurved spine.

Flowering & fruiting : Mid June to September.

Uses (T) : Use as substitute of *C. faconeri.*

92. *Cremanthodium ellisii* (Hk. f.) Kitam. var. *C. plantagineum* Maxim.

Family : Asteraceae

Name : Lukumentok (Ming-chen-ser-po) (L)

Distribution : Kashmir to South East Tibet.

Habitat : Open slopes & rocks, Leh, Changthang and Lahaul valleys between 4,790-5,230 m.

Plant Characters : An erect, perennial herb. Stem 15-30 cm. Basal leaves with an oblong blunt blade, 5-12 cm, widely spaced teeth, petioled. Flower-heads solitary, terminal. Rays florets yellow, elliptic; disk florets greenish to brownish. Involucral bracts narrow, lanceolate, woolly haired.

Flowering & fruiting : July to August.

Uses (T) : The whole plant is analgesic, useful in fever from contagious disease, swelling in the throat, reduce pain in upper body regions.

93. *Cremanthodium reniforme* Benth.

See Sr. No. 25, Chapter X.

Uses : (T) : Same as in *Cremanthodium ellisii.*

94. *Crepis flexuosa* (DC.) Benth.

Family : Asteraceae

Name : Hawksbread (E), Remang or Samboo (L)

Distribution : From Pakistan to Central Nepal, Central Asia and Tibet.

Habitat : Dry rocky slopes, Leh and Spiti valleys around 3,375 m.

Plant Characters : A much-branched, erect, perennial herb. Leaves few, linear, entire or toothed, radical pinnatifid; cauline leaves sessile. Flower-heads small, yellow, in terminal clusters; outer bracts minute, the inner ones linear and hairless.

Flowering & fruiting : Mid June to August.

Uses (T) : The leaf paste is used in muscle pain.

95. *Cuscuta approximata* Bieb.

Family : Cuscutaceae

Name : Amarlata (H), Large dodder (E)

Distribution : Pakistan to Nepal, J&K, Himachal Pradesh and Kumaun.

Habitat : Common as a parasite on many species of herb, shrub and plants, Nubra valley between 3,550-3,890 m.

Plant Characters : A small delicate, leafless, perennial parasitic plant. Leaves usually divided into oblong leaflets. Flowers few, pink to purple, globose on cluster. Calyx ovate-oblong. Stigmas as long as styles. Capsule ovoid.

Flowering & fruiting : July to August.

Uses (T) : The plant extract is said to be useful in skin diseases.

96. *Cuscuta capitata* Roxb.

Family : Cuscutaceae

Name : Amarlata (H), Large dodder (E)

Distribution : Afganistan to Central Nepal, J&K, Himachal Pradesh.

Habitat : A parasite on many species of herb, shrub and other plants, Zanskar and Lahaul valley between 3,389-3,620 m.

Plant Characters : A leafless, twining, parasite. Flowers rose coloured, capitate. Sepals ovate. Corolla ovoid, much contracted, lobes 5, triangular, covered with minute papillae. Capsule thin, fragile. Seeds 4-ellipsoid.

Flowering & fruiting : Mid June to August.

Uses (T) : The plant extract is mixed with ginger and given in kidney problems.

97. *Cuscuta europaea* Linn.

Family : Cuscutaceae

Name : Amarlata (H), Large Dodder (E)

Distribution : Pakistan to Central Nepal, J&K, Himachal Pradesh and Kumaun.

Habitat: A parasite on many species of herb, shrub & plants such as *Geranium, Galium, Phlomis, Potentilla;* Leh & Zanskar valleys around 3,350 m.

Plant Characters : A leafless, twining, pinkish parasite, annual or perennial plant. Flowers pinkish white in stakless glabrous, sessile, in clusters. Calyx lobes ovate, triangular; corolla lobes ovate. Capsule fragile, seeds 4 and brown.

Flowering & fruiting : July to September.

Uses (T) : A water extract of the plant is used against urinary, kidney complaints by Amchis of Ladakh. It is believed to be beneficial for patients suffering from backache and abdominal disorders.

98. *Cuscuta gigantea* Griff.

Family : Cuscutaceae

Name : Hande-thapa (L)

Distribution : From Pakistan to Central Nepal, China, Tibet, Bhutan.

Habitat : Common as a parasite on many species of herb, shrub and plants, Suru valley between 2,780-3,020 m.

Plant Characters : A leafless, much slender, twining parasitic herb. Flowers irregularly fascicled or shortly pedicelled. Bracts ovate. Sepals ovate, obtuse. Corolla-tube subcylindric with reflexed lobes. Capsule sub-quadrate, seeds black.

Flowering & fruiting : Mid June to September.

Uses (T) : The whole plant is considered antiseptic and used against septic wounds.

99. *Cuscuta reflexa* Roxb.

Family : Cuscutaceae

Name : Amarlata (H), Dodder (E)

Distribution : Afganistan to South West China and South East Asia.

Habitat : A parasite on herb, shrub & plants as a climber, Leh and Spiti valleys between 3,540-3,670.

Plant Characters : A herbaceous, parasitic, climber. Stem yellow to pinkish. Leaves reduced to scales. Flowers fragrant, white to pink, sessile in lax cluster (2.5-10 cm). Fruits fleshy, four seeded.

Flowering & fruiting : August to September.

Uses (T) : The paste of climber is used against warts and sores.

(Li) : The plant contains amerbelin, cuscutin and wax. The plant is purgative in nature.

100. *Cyananthus lobatus* Wall ex Benth.

Family : Campanulaceae

Name : Ralchatnakpo or sKyes-pa dong-btags (L)

Distribution : From Kashmir to Pakistan.

Habitat : Alpine meadows, Lahaul valley around 3,650 m.

Plant Characters : An ascending branched herb with woody rootstock. Stem 10-30 cm long glabrous below, hairy upwards. Leaves obovate-oblong, lobed sparsely hairy. Flowers blue in solitary, terminal heads. Calyx ovate with black hairs. Corolla funnel-shaped, lobes hairy in the throat.

Flowering & fruiting : July to September.

Uses (T) : The plant is useful against skin diseases.

101. *Cynoglossum wallichii* G. Don.

Family : Boraginaceae

Distribution : From Kashmir to Kumaun.

Habitat : Dry alpine slopes & along sandy riverbeds, Suru and Lahaul valleys between 3,010-3,450 m.

Plant Characters : An erect, biennial herb. Stem solitary to many. Basal leaves petiolate, shorter, lanceolate to obovate; lower cauline leaves usually longer and broader than basal ones. Flowers pale to deep blue in terminal to axillary, diviricate, short, scorpoid cymes.

Flowering & fruiting : June to August.

Uses (T) : The whole plant is useful against vomiting.

102. *Dactylorhiza hatagirea* (D. Don.) Soo.; *O. latifolia* Linn.

Family : Orchidaceae

Name : Salampanja (H), Spotted Heart Orchid (E), Ambolakpa (L)

Distribution : From Pakistan to South East Tibet.

Habitat : On damp & moist places, Nubra, Suru and Lahaul valleys around 3,280 m.

Plant Characters : A tuberous, perennial herb. Roots palmately divided into 2 to 5 finger like processes. Stem tall upto 70cm. leaves several, oblong-lanceolate upto 25 cm. Flowers pink or purple in dense flowering spikes. Sepals and petals nearly equal. Lower bracts longer than the flowers.

Flowering & fruiting : June to August.

Uses (T) : Local Amchis use the tuber roots for health tonic. An extract of roots is also used against urine and kidney problems. The tubers are considered energetic, and health improving, which are recommended for weak people also.

(Li) Tubers are astringent, expectorant, used as nervine tonic, aphrodisiac, useful in diarrhoea, bone fracture and chronic fevers. The tuberous roots contain starch, mucilage, sugar, albumen and ash of potassium.

103. *Delphinium brunonianum* Royle.

Family : Ranunculaceae

Name : Himalayan Larkspur (E), Lunde-kaown (L)

Distribution : Pakistan to Eastern Nepal, South East Tibet & Kashmir.

Habitat : Moist stony places, Changthang, Leh and Spiti valleys between 4,760-5,130 m.

Plant Characters : An erect, glabrous, long, perennial herb. Stem 15-20 cm long, leafy below. Leaves lobed, cuneate-ovate, obovate, long petioled. Flowers large, royle blue to pale blue, mostly in terminal corymbs or sometimes in compound corymbs. Seeds winged or angled. Follicle 5-6, hairy.

Flowering & fruiting : July to September.

Uses (T) : The flower juice and infusion of the whole plant are used for colic in Ladakh.

(Li) : The seeds and vegetative parts of the plant contain an alkaloid "Delphinine" which is used as an insecticide.

104. *Delphinium cashmirianum* Royle.

Family : Ranunculaceae

Name : Himalayan Larkspur (E), Lunde-kaown (L)

Distribution : West Tibet, Pakistan to Kashmir & Uttranchal.

Habitat : Snow melted alpine slopes, Suru and Lahaul valleys between 4,790-5,150 m.

Plant Characters : An erect, glabrous, perennial herb. Stem hispidaly hairy, branched, 20-45 cm, few leaved. Leaves deeply lobed, 3-5 partite, reniform, lobes cuneate-ovate, acute. Flowers bluish-purple with short spur in terminal racemes, long stalked. Fruits follicle, 3-7 and hairy.

Flowering & fruiting : July to September.

Uses (T) : The flower juice and infusion of the whole plant is used for colic in Ladakh.

(Li) : The seeds and vegetative parts of the plant contain an alkaloid "Delphinine" which is used as an insecticide.

105. *Delphinium vestitum* Wall. ex Royle.

Family : Ranunculaceae

Distribution : Nepal, Pakistan to Kashmir & Uttranchal.

Habitat : Alpine meadows & pastures, Lahaul valley between 4,690-5,120 m.

Plant Characters : An erect or decumbent, perennial herb 30-75 cm tall. Stem simple or branched from below or above, often reddish, densely hairy. Leaves mostly basal, with hairy petioles, dilated at base, the 2 or 3 cauline leaves with much shorter petioles, reniform in outline, stiff, hairy, palmately 5-7 lobed. Flowers purplish-blue to dull-violet in few or many-flowered, peduncled racemes. Follicles 3, hairy. Seeds narrowly winged.

Flowering & fruiting : July to October.

Uses (Li) : A source of drug 'delphinine' which is used as insecticide.

106. *Dianthus anatolicus* Boiss.

Family : Caryophyllaceae

Distribution : Pakistan to Himachal Pradesh and Kashmir.

Habitat : Dry places, Suru and Lahaul valleys around 3,050 m.

Plant Characters : A small, slender, densely tufted, perennial herb. Stem 6-10 inches. Very slender. Leaves small, slender, very narrow, midrib and margin thick. Flowers rosy. Calyx toothed, blade small, broad, crenate-toothed.

Flowering & fruiting : June to August.

Uses (T) : The leaves are used against stomach problems.

107. *Dianthus deltoides* Linn.

Family : Caryophyllaceae

Name : Khenchpa (L)

Distribution : Kashmir to Pakistan, West Tibet and Western Europe.

Habitat : Rocky slopes, Suru valley around 3,025 m.

Plant Characters : A dwarf, glabrous, branched, perennial herb. Stem branched, from the base. Leaves linear, obscurely serrulate. Bracts-2. Flower many in penciled cymes. Calyx tubular. Corolla with a long claw. Fruit a capsule.

Flowering & fruiting : July to August.

Uses (T) : The leaves are used as a substitute of *D. anatolicus*.

108. *Dianthus jacquemontii* Edgew. ex Hk. f. & T.

Family : Caryophyllaceae

Name : Khenchpa (L)

Distribution : Kashmir to Tibet and Western Himalayas.

Habitat : Rocky slopes, Zanskar valley 4,250 m.

Plant Characters : An erect, perennial herb. Stem branched from the base, slender, glabrous, 10-20 cm tall, leafy. Leaves sessile, coriaceous, acutely pointed. Flowers usually solitary, pedicelled. Bracts pointed. Calyx tip rigid. Petals long, pale with broad tips. Capsule rigid, valvulur, smooth. Pedicel very short. Nutlets ovate.

Flowering & fruiting : July to September.

Uses (T) : The leaves are used as a substitute of *D. anatolicus*.

109. *Dioscorea deltoidea* Wall. ex Kunth, Enum.

Family : Dioscoreaceae

Name : Singni-Mingni (L)

Distribution : From India to Pakistan & Nepal.

Habitat : Frequent in Juniperus forest, Lahaul valley 4,125 m.

Plant Characters : A glabrous, twining herb, with tuberous roots. Leaves alternate, long-petioled, variable, usually ovate-lanceolate, long-pointed, widely cordate. Spikes solitary male very slender, 7-35 cm long, female minute, in small distant clusters, 10-15 cm long. Capsule broadly winged.

Flowering & fruiting : July to September.

Uses (Li) : It is one of the most important plants for the pharmaceutical industry. The tubers are dug-out by the local people for sale, which is an excellent source of disogenin, a steroidal sapogenins, used as a starting material for oral contraceptive pills.

110. ***Draba tibetica*** Hk. f. & T.

See Sr. No. 59, Chapter VIII.

Uses (T) : The plant is used as health tonic.

111. ***Dracocephalum heterophyllum*** Benth.

See Sr. No. 23, Chapter VII.

Uses (T) : A decoction of dried flowers and leaves is used for cough and headache in Ladakh.

112. ***Dracocephalum moldavicum*** Linn.

See Sr. No. 24, Chapter VII.

Uses (T) : A decoction of the whole plant is used for intestinal problems.

113. ***Dracocephalum stamineum*** Kar. & Kir.

See Sr. No. 25, Chapter VII.

Uses (T) : It is used against cold and cough.

114. ***Echinops cornigerus*** DC.

Family : Asteraceae

Name : Globe Thistle (E); Aczema (L)

Distribution : Afganistan to Central Nepal and Kashmir.

Habitat : Wasteland areas, Leh & Nubra valleys between 3,305-3,890 m.

Plant Characters : An erect, spinescent, perennial herb. Stem very leafy, 30-60 cm long. Leaves 10-20 cm, lanceolate with spiny toothed lobes. Spherical inflorescence of numerous densely packed pale blue flower-heads. Achenes villous, obconic.

Flowering & fruiting : July to September.

Uses (T) : Leaves are made into paste and applied to septic wounds. Seeds are used as tonic. The powdered leaves are used to cure yellowish of eye (Jaundice).

115. ***Echinops niveus*** Wall. ex DC.

Family : Asteraceae

Name : Aczema (L)

Distribution : Afganistan to Central Nepal, Tibet and Kashmir.

Habitat : Wasteland areas, Changthang valley around 4,650 m.

Plant Characters : An erect, simple, spiny, perennial herb. Stem erect, branched, cottony. Leaves bipinnatifid, pubescent, beneath densely cottony, segment narrow with long spine. Balls 6-8 cm diam. Involucral bracts spinescent. Achenes villous.

Flowering & fruiting : July to mid September.

Uses (T) : A paste of leaves and aerial parts is applied to cure wounds.

116. *Ephedra gerardiana* Wall. ex Stapf.

Family : Ephedraceae

Name : Asmania / Somlata (H), Ephedra (E), Chhapat (L)

Distribution : From Afganistan to Bhutan.

Habitat : Stony dry areas, Leh, Nubra, Zanskar and Spiti valleys around 3,960 m.

Plant Characters : A low growing tufted shrub. Stem tall 30-60 cm, erect, green, branched. Branches scaly on joints. Male cones ovate, solitary. Female cones usually solitary. Fruits ovoid, red, fleshy bracts. Seeds 2, black.

Flowering & fruiting : July to September.

Uses (T) : The roots, stems and branches are used as a tonic, against fever, stop bleeding, blood purification, snuffing, hepatic diseases, rheumatism, toothbrush and bronchial asthma. It is also used against haemorrhage, hot disorders of gall bladder, liver, spleen, fever due to bile, fevers, inflammation of spleen, wounds, dissolve tumors. It is also used as an important fuel and fodder during winters.

(Li) : The fruit and dried stem constitutes a valuable drug "Ephedrine" which is used to cure Asthma. It is very effective cardiac stimulant, asthma, heart failure and rheumatism. The stems and branches are used as a tonic, against fever, stop bleeding and bronchial asthma. Stops bleeding, febrifuge and tonic.

117. *Ephedra intermedia* Schrenk *et* Mey.

Family : Ephedraceae

Name : Asmania (H), Ephedra (E), Chhapat (L)

Distribution : From Afganistan to Bhutan & Kashmir.

Habitat : Wastelands, Suru valley around 2,960 m.

Plant Characters : A low growing, erect, tufted shrub. Stem with green branches arising from a branched base. Male cones ovate, solitary or 2-3. Female cones usually solitary. Fruits ovoid with fleshy red, succulent bracts.

Flowering & fruiting : July to September.

Uses (T) : The dried stems are used to cure Asthma.

118. ***Epipactis helleborine*** (Linn.) Crantz.; *Serapisas helleborine* Linn.

Family : Orchidaceae

Name : Broad leaved helleborine, Penginlo (L)

Distribution : From Pakistan to South East Tibet, temperate Eurasia & Kashmir.

Habitat : Wastelands, Changthang and Lahaul valleys around 4,850 m.

Plant Characters : A small terrestrial orchid. Stem dwarf, slender. Leaves lanceolate, with recurved margin, glabrous. Flowers in lax raceme, glabrous, yellow.

Flowering & fruiting : July to August.

Uses (T) : An infusion of the plant is used as a blood purifier.

119. ***Eragrostis poaeoides*** P. Beauv.

See Sr. No. 69, Chapter IX.

Uses (T) : The paste is prepared from the roots, which is helpful against stomach complaints.

120. ***Erigeron alpinus*** Linn.

Family : Asteraceae

Name : Repan (L)

Distribution : Pakistan to Central Nepal, Bhutan, and Tibet.

Habitat : Open rocky areas, Zanskar and Spiti valley around 3,670 m.

Plant Characters : An erect, glabrous, perennial herb. Stem many from the base. Radical leaves petioled, spreading obovate or oblong-obovate; cauline, sessile, obovate or oblong. Flower-heads pinkish-white or pale, solitary or corymbose. Achenes silky. Pappus reddish.

Flowering & fruiting : June to August.

Uses (T) : The plant extract is used against cold and cough.

121. ***Ermania albiflora*** (T. Ander) O. E. S.; *Cheiranthus albiflorus* T. Ander.

Family : Brassicaceae

Name : Chunga (L)

Distribution : From Kashmir to Himachal Pradesh and Uttar Pradesh.

Habitat : Sandy places, Suru valley around 3,029 m.

Plant Characters : A small hoary, tufted, perennial herb. Stem slender, branched, covered below by withered petioles. Lower leaves loosely tufted; upper leaves linear-lanceolate, rarely toothed. Sepals linear, tomentose. Flowers in raceme. Petals white, spathulate, obtuse. Pods linear, tomentose.

Flowering & fruiting : July to September.

Uses (T) : The plant extract is useful in skin diseases.

122. ***Ermania lanuginosa*** (Hk. f. & T.) O. E. S.; *Parrya lanuginosa* Hk. f. & T.

Family : Brassicaceae

Name : Measlo (L)

Distribution : Afganistan to Central China and North India.

Habitat : Moist rocky places, Nubra valley around 3,020 m.

Plant Characters : A small gray haired, woolly, perennial herb. Stem several from the base. Leaves spathulate, and coarsely serrated. Flowers lilac in spike like clusters. Fruits finely haired, oblong, and linear. Pod acute.

Flowering & fruiting : Mid July to August.

Uses (T) : The leaves are diuretic and purgative to cure cuts and kidney complaints.

123. ***Euphorbia hispida*** Boiss.

Family : Euphorbiaceae

Name : Therno (L)

Distribution : Pakistan to China, India and South Asia.

Habitat : On moist & drier places, Suru and Lahaul valleys around 3,320 m.

Plant Characters :A stout, sub glabrous to hairy, annual herb. Stem branched, branches diverticate. Leaves opposite, sessile, pale-green in distant pairs, base almost auricled on one side. Involucres sub-solitary, sub-sessile; glands purple, white or pink rounded limb. Capsule glabrous, shortly pedicelled.

Flowering & fruiting : July to September.

Uses (T) : The latex of the plant is used against skin eruptions.

124. ***Euphorbia kanaorica*** Boiss.

Family : Euphorbiaceae

Name : Kianglo (L)

Distribution : Kashmir to Sikkim, China and Central Asia.

Habitat : Alpine slopes, Zanskar valley around 4,201 m.

Plant Characters : A flaccid perennial herb. Stem much branched from the rootstock, decumbent. Rootstock slender. Stems 6-10 inches, flaccid flexous. Leaves alternate, coriaceous, obovate-spathulate. Involucres in axils of 2-opposite orbicular leaves. Style quite free. Capsule smooth, seeds oblong.

Flowering & fruiting : June to September.

Uses (T) : The latex of the plant is used against skin diseases.

125. *Euphorbia stracheyi* Boiss.

Family : Euphorbiaceae

Name : Tingling or Khron-bu (L)

Distribution : Kashmir to China and Central Asia.

Habitat : Drier places, Leh and Suru valleys between 3,560-4,167 m.

Plant Characters : A small, milky, perennial herb. Stem spreading. Leaves small, elliptic-ovate, overlapping. Flower-heads dark red. Involucre 5-lobed. Glands kidney shaped, dark purple. Underground stems long; rootstock stout. Bracts ovate. Fruits smooth, hairless. Capsule globose.

Flowering & fruiting : July to September.

Uses (T) : The plant latex is used against boils.

126. *Euphrasia laxa* Penn.

Family : Scrophulariaceae

Name : Kianglo (L)

Distribution : From Kashmir to China and Asia.

Habitat : On sandy & open places, Zanskar valley between 3,420-3,610 m.

Plant Characters : A herbaceous annual herb. Stem branched from the base, slender, glabrous, 10-15 cm long. Leaves 4-5 mm, palmatifid, sessile, toothed. Flowers usually in terminal spikes, bracteate. Capsule loculicidal, oblong. Seeds brown, ribbed, grooved.

Flowering & fruiting : July to August.

Uses (T) : The plant is used in eye complaints in Ladakh.

127. *Euphrasia officinalis* Linn.

Family : Scrophulariaceae

Name : Eyebright (E), Kangchuk (L)

Distribution : Afganistan to Europe, Kashmir, Kumaun and Sikkim.

Habitat : Sandy & damp places, Suru, Nubra and Lahaul valleys around 3,020 m.

Plant Characters : A small branched annual herb. Stem wiry, 40 cm high, branched, often necked below. Leaves ovate, saw-toothed. Flowers white and purple tinged, 2-lipped flower with yellow throats or lilac with purple vein, borne in terminal spike clusters.

Flowering & fruiting : July to September.

Uses (T) : Infusion of dried herb has soothing effect on eyes in conjunctivitis.

(Li) The chief constituents are choline and phytosterols like B-sitosterol and stigmasterol. Very useful for complaints of eyes and nose.

128. ***Euphrasia vulgaris*** Benth.

Family : Scrophulariaceae

Name : Bepamatso (L)

Distribution : Pakistan to Nepal, South Asia and Kashmir.

Habitat : On alpine slopes, Changthang valley around 4,705 m.

Plant Characters : An annual herb. Stem branched from base, pubescent. Leaves ovate-cordate, sessile; serrate upto 10-20 cm. Flowers borne in terminal spikes. Bracts serrate. Capsule exserted, loculicidal, oblong. Seeds oblong, ribbed, hairy.

Flowering & fruiting : July to August.

Uses (T) : The whole plant is used for medicinal purposes.

129. ***Ferula jaeschkeana*** Vatke.

See Sr. No. 29, Chapter VII.

Uses (T) : The roots are used in rheumatism and popularly as poultice.

(Li) : The stem gum resin is used in septic wounds and toothache. The bio constituents are jaeschkeanadiol, jaeschjerin etc.

130. ***Fraxinus xanthoxyloides*** (G. Don.) DC.

Family : Oleaceae

Name : Ash (E), Thuma (L)

Distribution : From Pakistan to Nepal.

Habitat : Common in association with *Juniperus recurva,* Lahaul valley around 3,790 m.

Plant Characters : A large deciduous shrub or small tree. Leaves opposite pinnate, 6-14 cm long, midrib winged. Flowers in dense heads, appearing before or with the young leaves on shoots of the previous year.

Flowering & Fruiting : July to September.

Uses (T): The plant is of medicinal value as tonic.

131. ***Fritillaria roylei*** Hk.

Family : Liliaceae

Name : Chikmo or Kakoli or A-bhi-sa (L)

Distribution : Pakistan to Uttranchal and J&K.

Habitat : Alpine slopes, Zanskar and Lahaul valleys around 4,600 m.

Plant Characters : A perennial bulbous herb. Bulb depressed. Scales membranous. Stem upto 20-60 cm long, pointed, whorled. Flowers yellowish-green, campanulate, pendulous, and usually solitary. Stamens half the length of perianth. Capsule obovoid, obtusely angled.

Flowering & fruiting : July to September.

Uses (T) : Bulbs are used against congestion of lungs, asthma, burns, stomach disorders and also used as health recovering tonic by the Amchis.

(Li) The bulb constitutes the Chinese drug Szechuan-Pei-Mu, used as antipyretic, expectorant and lactagogue. They are used with dried orange skin for asthma, bronchitis, burns, stomach disorder and tuberculosis.

132. ***Galium aparine*** Linn.

Family : Rubiaceae

Name : Goose grass (E), Ranche (L)

Distribution : Pakistan to Himachal Pradesh and temperate Himalayas.

Habitat : On stony slopes, Suru valley around 3,120 m.

Plant Characters : A delicate, trailing or climbing annual herb. Stem weak. Leaves very variable, sessile, 6-8 in whorls, linear or narrowly obovate oblong. Flowers greenish-white tinged with green in axillary stalks. Fruits clothed with spreading hooked bristles. Capsule long, seeds compressed.

Flowering & fruiting : June to September.

Uses (T) : The extract of leaves is useful against urinary complaints.

(Li) The major constituents of the plant are valepotriates, polyphenolic acids, anthraquinones, tannins.

133. ***Galium pauciflorum*** Bunge.

Family : Rubiaceae

Name : Phomongo (L)

Distribution : From Pakistan to Himachal Pradesh, Uttranchal and Kashmir.

Habitat : On stony slopes, Changthang and Spiti valleys around 4,650 m.

Plant Characters : An annual herb. Stem slender, smooth, filiform or capillary. Leaves sessile, opposite, 4-6 cm in a whorl, oblanceolate, cuspidate. Flowers white, minute, solitary, axillary, terminal. Fruits covered with hooked hairs.

Flowering & fruiting : July to August.

Uses (T) : The dried parts of plant are used to cure intestinal parasites and reduce fever.

134. *Galium serpylloides* Royle ex Hk. f.

Family : Rubiaceae

Name : Pimantso (L)

Distribution : From Pakistan to Kashmir, Himachal Pradesh and West temperate Himalayas.

Habitat : On high & dry slopes, Nurba valley around 3,034 m.

Plant Characters : A small decumbent, annual herb. Stem short, much branched from the base upto 15-20 cm long. Leaves usually 3-4 in each whorls, oblong with acute apex. Flowers yellow, minute, solitary in axillary and terminal peduncles, 3-flowered. Fruits usually hairy.

Flowering & fruiting : July to September.

Uses (T) : The leaves are diuretic, purgative and also used to cure kidney complaints.

135. *Gaultheria trichophylla* Royle.

Family : Ericaceae

Name : Winter green

Distribution : Kashmir to Himachal Pradesh.

Habitat : Moist alpine slopes, Lahaul valley between 3,490-3,6580 m.

Plant Characters : A prostrate or procumbent, dwarf, woody herb or undershrub. Leaves subsessile, oblong, margins obscurely crenate with long dark brown hairs, base cuneate. Flowers reddish-white, solitary, axillary. Calyx ovate-oblong. Corolla campanulate. Capsule bluish-black.

Flowering & fruiting : June to September.

Uses (T) : An important medicinal plant widely used in traditional system of medicines.

(Li) The chief constituents are gaultherine and salicylic acid.

136. *Gentiana algida* Pallas. var. ***nubigena*** (Edgew) Kush.

Family : Gentianaceae

Name : Tiktas (L)

Distribution : From Kashmir to South Western China, North Asia.

Habitat : Open moist slopes, Suru, Changthang and Lahaul valleys around 3,930 m.

Plant Characters : A small tufted, perennial herb. Rootstock often covered with old leafy bases. Stem suberect. Basal leaves linear-oblong, spathulate, blunt, thick; cauline leaves lanceolate. Flowers dark blue or purple-tinged, large and blue ribbed outside, shortly pedicellate, borne in terminal and axillary stalked clusters. Calyx tubular with purple tinged. Corolla funnel shaped triangular. Capsule stalked.

Flowering & fruiting : Mid July to September.

Uses (T) : The flowers are used against inflammation of pharynx, bronchitis, cough, hoarseness of throat, excess sputum, toxic and epidemic fevers. Taste bitter and potency cooling.

137. *Gentiana aquatica* Linn.

Family : Gentianaceae

Name : Tiktas (L)

Distribution : Pakistan to Kashmir, Nepal, South West China and Asia.

Habitat : Wet & marshy areas, Suru valley around 2,850 m.

Plant Characters : An annual herb. Radical leaves numerous, weak, little divided, cauline leaves small, ovate, marginate, recurved, ovate-obovte, erect. Flowers blue, terminal, solitary. Corolla infundibuliform, acute. Capsule short, stipulate.

Flowering & fruiting : July to September.

Uses (T) : The roots are used in preparation of health tonic.

138. *Gentiana carinata* (D. Don.) Griseb.; *G. coronata* Royle.

Family : Gentianaceae

Name : Pang-gyan khar-bo (L)

Distribution : From Pakistan to Kashmir and Uttranchal.

Habitat : Higher passes, Nubra valley around 4,535 m.

Plant Characters : A small, tufted, annual herb. Stem branched, 5-8 cm long. Leaves oblong-lanceolate with acute apex; cauline leaves sessile. Flowers light-blue to bluish-white, in terminal clusters. Corolla tube narrow funnel shaped. Capsule compressed.

Flowering & fruiting : July to August.

Uses (T) : The flowers are used against inflammation of pharynx, bronchitis, toxic and epidemic fever, swelling of throat, excess sputum and cough.

139. *Gentiana humilis* auct. non. Steven ex Bunge.

Family : Gentianaceae

Distribution : Kashmir to West China & Central Asia.

Habitat : On higher slopes, Changthang valley around 4,702 m.

Plant Characters : A tufted, weak, pale green annual herb. Stem branched from the base. Radical leaves numerous ovate-elliptic; cauline oblong-elliptic not recurved or marginate. Flowers white to bluish-white. Corolla less than 0.5 inch, capsule short-obovoid, compressed.

Flowering & fruiting : July to mid September.

Uses (T) : The flowers are used against fever.

140. *Gentiana squarrosa* Ledeb.

Family : Gentianaceae

Name : Ziang (L)

Distribution : From Kashmir to Nepal, South West China and Asia.

Habitat : On open alpine slopes, Zanskar valley around 4,120 m.

Plant Characters : A weak annual herb. Stem branched from the base. Cauline leaves small, obovate, recurved. Flowers solitary, terminal. Calyx teeth mucronate, recurved. Corolla 1-2 cm long, tubular. Capsule short, ovoid, compressed. Seeds ellipsoid.

Flowering & fruiting : July to August.

Uses (T) : Roots are used as diuretic.

141. *Gentiana stracheyi* (Clarke) Kitam

Family : Gentianaceae

Name : Ziang (L)

Distribution : Pakistan to Uttranchal, South Asia.

Habitat : Open moist slopes, Zanskar valley around 4,120 m.

Plant Characters : An erect, perennial herb. Stem much branched from the base

reaching upto 50-60 cm. Leaves oblong-linear, scarcely connate at the base. Flowers terminal with long pedicel. Calyx tube 1-1.5 cm, strongly 4-keeled, funnel shaped. Corolla fimbricate at the apex, glands at the base, very obscure. Capsule stalked. Seeds subglobose.

Flowering & fruiting : July to August.

Uses (T) : The root decoction is used as a tonic.

142. ***Gentianella moorcroftiana*** (Wall. ex Griseb.) Airy Shaw.

Family : Gentianaceae

Name : Spang-Gain-karpo (L)

Distribution : From Kashmir to Baltistan, Lahaul and Kumaun.

Habitat : Open slopes & moist places, Suru and Lahaul valleys around 3,650m.

Plant Characters : A slender, erect, annual herb. Stem erect-ascending, branched from the base. Leaves, lanceolate or oblong, elliptic, narrowed below. Flowers blue to white, solitary in terminal or axillary racemes, long pedicillate, tubular-funnel shaped.

Flowering & fruiting : July to September.

Uses (T) : The whole plant is used as antitoxin and febrifuge.

143. ***Gentianella paludosa*** (Hk. f.) Harry. Smith.

Family : Gentianaceae

Name : Pallutso or Cags-tig nag-po (L)

Distribution : From Pakistan to South West China.

Habitat : Moist & alpine slopes, Changthang and Spiti valley around 3,605m.

Plant Characters : An erect, slender, usually branched, annual or biennial herb. Stem usually branched reaching upto 60 cm. Leaves narrowly elliptic, blunt, 2-4 cm; cauline leaves ovate-elliptic. Flowers usually solitary, long stemmed, blue to white, terminal. Calyx green, enlarging to 3 cm. Corolla-tube rounded to broadly elliptic spreading lobes; throat of corolla without hairs or scales and stamens included. Capsule oblong-ovoid, stipitate. Seeds numerous, blackish-brown.

Flowering & fruiting : July to September.

Uses (T) : The whole plant has got high medicinal value, which is used as febrifuge, useful in fever due to bile disease, gall bladder, inflammation of wounds, fever and for preparing of health tonic for old people in Ladakh.

144. ***Geranium collinum*** Steph. ex Willd.

Family : Geraniaceae

Name : Legkatin (L)

Distribution : Western Nepal to South West Tibet and Kashmir.

Habitat : Dry or moist slopes, Suru and Nubra valleys around 3,046 m.

Plant Characters : A prostrate or ascending, hoary, hairy to pubescent, annual herb. Stem branched. Leaves with obtusely cut slender lobes. Stipules small, ovate, acute. Flowers white or purplish, peduncled. Pedicels pubescent or glandular hairy. Sepal shortly awaned, filaments ciliate at the base.

Flowering & fruiting : July to mid September.

Uses (T) : A decoction of leaves and roots is useful against stomach complaints.

(Li) : A hydrolysable tannide – kollinin isolated from leaves.

145. ***Geranium pratense*** Linn.

Family : Geraniaceae

Name : Meadow Cranesbill (E), Gugchuk or Ga-dur (L)

Distribution : Pakistan to Western Nepal and temperate Eurasia.

Habitat : Stony wet slopes, Leh, Suru and Lahaul valleys around 4,560 m.

Plant Characters : An erect, perennial herb. Stem 70 cm high, branched, hairy. Leaves orbicular, 7-9 partite, palmately lobed, densely hairy, segments acute, incised. Flowers in pair, bluish to bluish-violet, 2.5-3.5 cm cross. Bracts small and slender. Sepals long awned. Filaments ciliate. Capsule glandular-hairy.

Flowering & fruiting : July to August.

Uses (T) : The leaf extract is used as febrifuge, analgesic, used against fever from influenza and infectious disease, pneumonia, inflammation of lung and veins swelling of the limbs, as tonic and to check dysentery and diarrhoea in Amchis system of medicine.

(Li) : The plant contains astringent compound gallic acid, ellagic acid etc.

146. ***Geranium sibiricum*** Linn.

Family : Geraniaceae

Name : Eyamlomentok (L)

Distribution : Pakistan to Central Nepal, Kashmir and temperate Eurasia.

Habitat : Sandy slopes & wastelands, Nubra valley between 2,980-3,420 m.

Plant Characters : A prostrate, slender, perennial herb. Stem much branched, hoary. Leaves orbicular-reniform, 5-partite, segments acute. Peduncles slender, usually 2-flowered. Sepals long awned, exceeding the notched petals.

Flowering & fruiting : July to September.

Uses (T) : An extract of leaves is useful in dysentery.

147. ***Geranium tuberaria*** Camb.

Family : Geraniaceae

Name : Yusiang (L)

Distribution : From Pakistan to Himachal Pradesh & Asia.

Habitat : On rocky slopes, Changthang & Suru valleys between 4,390-4,760m.

Plant Characters : An erect, hairy, perennial herb. Stem erect, hairy and glandular. Leaves orbicular-raniform, 5-7 partite with pinnatifid segments, petiole very slender. Peduncles 2-flowered, pinkish, terminal umbelled or trifid, hairy or glandular. Calyx obtuse, hairy. Petals retuse. Seeds smooth.

Flowering & fruiting : Mid June to August.

Uses (T) : A decoction of plant parts is used to treat kidney stone.

148. *Geum elatum* Wall. ex G. Don.

Family : Rosaceae

Distribution : From Kashmir to Himachal Pradesh & South East Asia.

Habitat : Sandy moist slopes, Lahaul valley between 3,760-3,870 m.

Plant Characters : A suberect, rhizomatous, perennial herbs, upto 45 cm tall. Radical leaves 10-25 cm long, pinnatisect; segments crenate or sharply toothed, often lobed; lateral segments sessile, ovate or oblong, alternately large and small terminal segments much larger. Flowering stems with few leaves and 1-6 floweres. Calyx lobes deltoid-ovate, acute, hairy. Petals bright yellow, orbicular. Achenes hairy.

Flowering & fruiting : July to September.

Uses (T) : The whole plant is used as tonic.

149. *Gnaphalium stewartii* Clarke. ex Hk. f. & T.

Family : Asteraceae

Name : Peo (L)

Distribution : Pakistan to Kashmir, Nepal & Kumaun.

Habitat : Sandy areas, Zanskar valley around 3,650 m.

Plant Characters : A woolly, perennial herb. Stem tufted, 5-10 cm leafy. Leaves linear, erect, or recurved acute radical rosulate, sessile. Flower-heads small, lax or dense leafy, yellow, in axillary clusters. Involucral bracts linear-oblong acute, dark brown. Achenes subsilky or scabrid with hairy pappus.

Flowering & fruiting : July to September.

Uses (T) : The whole herb is found effective to cure skin disease.

150. *Goodyera repens* (Linn.) R. Br.

Family : Orchidaceae

Name : Creeping lady's tresses (E)

Distribution : Kashmir to Kumaun & South East Tibet.

Habitat : Frequent on alpine slopes, Lahaul valley around 3,680 m.

Plant Characters : A decumbent or creeping herb, upto 25 cm high. Rhizomes creeping, with stolons and few hairy roots. Stems erect, stiff, glandular-hairy, with leaves crowded near the base and with small bracts like leaves appressed to the stems above. Basal leaves rosulate, ovate or oval, narrowed into a long petiole. Flowers white, tinge with pink globose, pointed.

Flowering & fruiting : July to September.

Uses (T) : The whole herb is having medicinal properties as a tonic.

151. *Gypsophila cerastioides* D. Don.

Family : Caryophyllaceae

Name : Chiktakpa (L)

Distribution : From Kashmr to Western Tibet.

Habitat : Rocky places, Suru valley around 3,220 m.

Plant Characters : A low growing perennial herb. Roots very stout, woody. Stem pale with decumbent brahches. Leaves hairy, ovulate. Flowers white streaked with purple, borne in rounded branched clusters. Capsule small.

Flowering & fruiting : Mid June to August.

Uses (T) : Sweet in taste and potency cooling. The plant is used against cold and cough.

152. *Hackelia uncinata* (Royle ex Benth.) Fischer.

Family : Boraginaceae

Distribution : South West China & Kashmir to Himachal Pradesh.

Habitat : Moist slopes, Lahaul valley around 4,870 m.

Plant Characters : An erect, perennial herb 50-70 cm high. Basal leaves petiolte, ovate; cauline leaves shortly petiolate, ovate to elliptic, acuminate rounded at the base, minutely hairy. Flowers blue with yellow ring in the centre in elongated, lax racemes. Calyx divided to the base ovate, pubescent. Corolla campanulate-rotate; tube equaling the calyx. Nutlets ovoid, broadly winged.

Flowering & fruiting : July to September.

Uses (T) : The whole plant is of medicinal value against cold and cough.

153. *Halenia elliptica* D. Don.

Family : Gentianaceae

Name : Cags-tig ra-mgo-ma (L)

Distribution : From Kashmir to Bhutan.

Habitat : Moist places, Zanskar valley around 3,549 m.

Plant Characters : An erect, annual herb. Stem 10-25 cm tall, solitary, cylindrical. Leaves elliptic, opposite, sub-obtuse. Flowers in axillary cymes. Corolla pale blue, spur equal to petals. Capsule 3-9 mm. Seeds small with concavity at the hilum.

Flowering & fruiting : July to September.

Uses (T) : The entire herb is used against hot disorders of the gall bladder, stomach and epidemic fever.

154. *Halerpestis tricuspis* (Maxim) Hand.

Family : Ranunculaceae

Name : Chitaka or Churuk-balak or Chu-rag-sBl-lg (L)

Distribution : Afganistan to Kashmir and Himachal Pradesh.

Habitat : Marshy land, Leh & Changthang valleys between 3,680-5,059 m.

Plant Characters : A prostrate stoloniferous stemless annual herb. Leaves long petioled, palmately cut, more or less orbicular, crenately incised or shallowly lobed. Flowers yellow in terminal spikes. Achenes in oblong terminal heads.

Flowering & fruiting : July to September.

Uses (T) : The whole plant is used against stomach complaints by local Amchis.

(Li) : The whole plant is used in stomachache, gastro-intestinal problems.

155. *Heracleum lanatum* Michx.

Family : Apiaceae

Name : Rasal (L)

Distribution : From Kashmir to Himachal Pradesh.

Habitat : On grassy meadows, Lahaul valley around 4,690 m.

Plant Characters : An erect, robust, sparsely hairy to pubescent, branched herb, 1.0-1.5 m high. Leaves lobed or pinnatifid; leaflets or lobes oblong, irregularly toothed, glabrous above, grey-pubescent beneath, bracts 1-3, linear or obscure. Bracteoles 5-8, linear. Rays 10-40, 3-8 cm long. Flowers white. Fruits obovate, pubescent.

Flowering & fruiting : June to September.

Uses (T) : The plant is used to check excessive menses and other type of bleeding.

(Li) : Roots are the chief source of xanthotoxin, a photosensitising coumarin and bergapten. The plant is quite useful in the treatment of leucoderma and psoriasis.

156. *Heracleum pinnatum* C. B. Clarke.

Family : Apiaceae

Name : sPru-ma (L)

Distribution : From Pakistan to Himachal Pradesh & Kashmir.

Habitat : Rocky & dry slopes, Leh, Suru and Spiti valleys between 3,450-4,530 m.

Plant Characters : A tall, perennial herb. Stem upto 100 cm, branched, white pubescent. Leaves large, pinnate with 1-3 pairs of widely spaced small, ovate leaflets. Flowers white, terminal, 5-7 cm across. Rays 8-20 in number, 3-5 cm long. Fruit elliptic, broadly winged lateral and 3-dorsal ribs.

Flowering & fruiting : June to August.

Uses (T) : The roots are used to treat inflammation and pain caused by vulnerable fever, checks haemorrhage and abdominal cramps.

(Li) : The plant is reported to contain coumarone, which can be converted into Xanthotoxin-a-chemical used in preparation of sun tan lotions and possessing anti-leucodermal properties.

157. *Herminium monorchis* (Linn.) R. Br.; *Ophrys monorchis* Linn.

Family : Orchidaceae

Name : Musk orchid (E), Peo (L)

Distribution : Pakistan to Central Nepal, Tibet, Asia & Europe.

Habitat : Moist places, Nubra and Suru valleys around 3,205 m.

Plant Characters : A tuberous, slender herb. Stem 15-25 cm long. Leaves 3.6 cm two, elliptic-oblong, upper leaves bract-like. Flowers greenish-yellow in lax spike, honey-scented. Petals and sepals somewhat covering but not forming a hood.

Flowering & fruiting : Mid June to September.

Uses (T) : Bulbs are useful source of drug which is used as tonic as a substitute of Orchid. The bulbs are useful against kidney complaints.

158. *Hippophae rhamnoides* Linn. spp. ***turkestanica*** Rousi.

Family : Elaeagnaceae

Name : Seabuckthorn (E), Tsermang or Sastalulu (L)

Distribution : From Pakistan to Himachal Pradesh, Central Asia & North West Himalayan regions.

Habitat : River belts & wastelands, Leh, Nubra, Zanskar, Suru, Changthang and Lahaul valleys between 2,600-4,000 m.

Plant Characters : A dwarf, very thorny, perennial shrub or small tree. Stem woody, erect, long upto 10-15 ft with silvery-waxy covering on the old shoots. Leaves variable, oblong blunt to green above and silvery scaly on both sides. Flowers very small greenish or yellowish. Male in clusters Female dull yellow, short stalked. Fruits subglobose, succulent, orange or red.

Flowering & fruiting : June to September.

Uses (T) : The whole plant has got enormous medicinal properties. The fruits, leaves and seeds are considered as an anti-aging, anti-cold and energetic. The fruits are antitusassive, blood purifier, heals peptic ulcer, oral mucocitosis, lungs disorders, cuts and wounds. Used against pulmonary disorders, inflammation of pulmonary tract expelling phlegm, healing peptic ulcer, clotting of blood, improve digestion, useful for spleen disorders, lungs, inflammation in blood and pain to blood disorders.

(Li) : Seabuckthorn is considered as a storehouse of various vitamins viz. C, B, E, K, Carotenoids which indicates its strong antioxidant property. The berries are useful in gum bleeding, regeneration of tissues, to help in repair of mucus membrane and other organs. In addition seeds and fruits contain valuable fatty oil which is the rich source of beta-carotene, Vitamin E, rare fatty acids and phytosterols having great demand in pharmaceutical industries. It has been reported that Chinese developed over 200 medicinal products and drugs from roots, leaves, fruits, seeds and bark. Seabuckthorn has got a valuable importance in Ladakh and other parts of trans Himalayan regions after its utilization on commercial scale during year 2001, when Field Research Laboratory has developed, patented and commercialized multi-vitamin herbal beverage, Jam and Herbal tea from Seabuckthorn.

159. *Hippophae salicifolia* D. Don.

Family : Elaeagnaceae

Name : Seabuckthorn (E), Chharma (L)

Distribution : Pakistan to Himachal Pradesh, Uttranchal & Asia.

Habitat : On slopes, Lahaul valley around 4,390 m.

Plant Characters : An erect shrub or small tree. Leaves 5-10 cm long, linear-lanceolate, softly white-pubescent, above when young, densely white-tomentose or stellately-hairy beneath, margins recurved. Fruits ovoid, yellow in axillary clusters.

Flowering & fruiting : June to September.

Uses (T) : This species is also used in traditional system of medicine.

(Li) : All the parts have got medicinal value. Like *H. rhamnoides* this plant is also rich in various vitamins and carotenoids.

160. *Hippophae tibetana* Schlecht.

Family : Elaeagnaceae

Name : Seabuckthorn (E), Tsermang or Sastalulu (L)

Distribution : Pakistan to Himachal Pradesh, Uttranchal & North West China.

Habitat : Along river belts & wastelands, Zanskar and Spiti valleys around 4,230 m.

Plant Characters : A small, dwarf, less thorny, perennial shrub. Leaves variable, oblong blunt to green. Flowers very small greenish or yellowish. Fruits subglobose, succulent, orange or yellow. Seeds light black to brown.

Flowering & fruiting : June to September.

Uses (T) : It is a good source of vitamins and carotenoids.

161. *Hippuris vulgaris* Linn.

Family : Hippuridaceae

Name : Dam-bu ka-ra-machog (L)

Distribution : Western Tibet to whole Northern Hemisphere in frigid and cool temperate water.

Habitat : Water sides of Indus river; Leh & Changthang valley 3506-3765 m.

Plant Characters : An aquatic, glabrous, perennial herb. Leaves in whorls of 4-6-12, linear or oblong. Flowers minute, solitary, sessile; hermaphrodite or unisexual. Drupe ellipsoid, smooth.

Flowering & fruiting : June to September.

Uses (T) : The leaves and stem are used against inflammation of lung, liver, bones, nerves and septic wounds. Febrifuge.

(L) : The plant contains caffeic acid, ferulic acid and traces of flavonoids.

162. *Humulus lupulus* Linn.

See Sr. No. 30, Chapter VII.

Uses (T) : A drug lupin is extracted from flowers used as a tonic, nervine and diuretic. The dried flowers are used to add flavour in beer.

163. *Hyoscyamus niger* Linn.

Family : Solanaceae

Name : Khursani Ajwain (H), Henbane (E), Gay -lantang or Rhang phrom lang-thang-rtse (L)

Distribution : Pakistan to South West China, North Africa, Kashmir and Kumaun.

Habitat : Wastelands, Suru valley between 2,940-3,610 m.

Plant Characters : An erect, robust, biennial herb. Basal leaves stalked, toothed, petioled, often absent at flowering; cauline leaves ovate, sessile. Flowers dull yellow, bell shaped with purple veins and dark purple centres. Calyx enlarged and becomes papery rigid and spine-tipped in fruit. Capsule encircled by the globular enlarged calyx. Seeds many, minute, oval.

Flowering & fruiting : July to September.

Uses (T) : Seeds are used to subside pain due to pathogenic diseases, toothache, pharyngitis and used in treating malignant ulcers. Smoke of leaves is used against asthma and fumigation of seeds considered useful in tooth infection by local Amchis of Ladakh.

(Li) : On destructive distillation the leaves yield a poisonous empyreumatic oil. It is used to conteract the griping action of purgative and to-relieve spasms in the urinary sedative. Seeds are used for extraction of alkaloids and powdered seeds used are applied to relieve toothache. The plant is a source of popular drug 'Hyoscyamine'.

164. *Hyoscyamus pusillus* Linn.

Family : Solanaceae

Name : Khursani Ajwain (H), Henbane (E) & Gaylantang (L)

Distribution : Pakistan to South West China, Kashmir.

Habitat : Wastelands, Suru valley around 3,560 m.

Plant Characters : An erect, low coarse annual or biennial herb. Stem about 30 cm high, pubescent. Leaves very irregularly lobed or sub-entire, sinuate toothed, ovate petioled. Calyx companulate, hispid, corolla yellow. Capsule 5 inch. Seeds many, small.

Flowering & fruiting : July to September.

Uses (T) : Leaves are smoked against asthma and seeds are smoked in tooth infection by locals.

165. *Hypecoum leptocarpum* Hk. f. & T.

Family : Papaveracaceae

Name : Parpata (L)

Distribution : Kashmir to South West China, Bhutan and Uttranchal.

Habitat : Grazing grounds, Zanskar valley around 3,540 m.

Plant Characters : An annual herb. Stem branched from the base, procumbent. Leaves several 3-7 pinnetisect. Flowers pale purple in terminal branched clusters. Capsule 1-2 cm slender.

Flowering & fruiting : July to September.

Uses (T) : A cooling nature, bitter taste and qualities are cool, blunt and light. In Amchi system of medicine the whole plant is used to treat stomach pain and diseases rims-tshad and dug-shad.

(Li) : The plant contains 0.4% alkaloid, protopine the major compound.

166. *Hypericum perforatum* Linn.

Family : Hypericaceae

Name : St John's Wort (E)

Distribution : From Kashmir to Bhutan, Europe & North Africa.

Habitat : Rock crevices, Lahaul valley around 3,740 m.

Plant Characters : An erect, branched, perennial herb, upto 60 cm tall. Leaves oblong or ovate-oblong, pellucid-punctate. Flowers yellow in terminal corymbs. Sepals narrowly lanceolate, acute. Petals black dotted on the the margins. Capsule egg-shaped, rusty brown.

Flowering & fruiting : June to September.

Uses (T) : A paste of leaves is used in treatment of wounds and cuts. Infusion of fresh flowers in olive oil is used externally for sores, wounds, ulcers, swellings and rheumatism.

(Li) : The herb contains hypericin, pseudohypericin, flavanoids - quercitrin, rutin, isoquercitrin etc. The essential oil contains caryophylene, pinene, limonene, hypericins etc. It has been used for excitability neuralgia, febrrositic, sciatica, wounds and specifically for menopausal neurosis.

167. *Hyssopus officinalis* Linn.

See Sr. No. 31, Chapter VII.

Uses (T) : The aerial parts are considered as stimulant, carminative and used in pulmonary, digestive, toothache, sprains etc.

(Li) : Used in formulation of various herbal medicines in Ayuevedic, Unani and Tibetan preparation. It is reported to be useful in throat inflammation and bronchial asthma. The herb is reported to contain tannins, carotein, ursolic acid having great demand in pharmaceutical industries.

168. *Inula falconeri* Hk. f.

Family : Asteraceae

Distribution : Pakistan to Kashmir, Central Nepal and West Asia.

Habitat : Dry, rocky slopes, Suru valley around 3,019 m.

Plant Characters : An annual, glabrous partially hairy herb. Stem slender, glabrous, simple, reaching upto 30-50 cm high. Lower leaves long, linear or oblong. Cauline leaves oblong. Flower-heads small 4-6 mm diam. Involucral bracts linear, acute. Achenes minute sparsely pubescent.

Flowering & fruiting : July to September.

Uses (T) : Roots are useful against sprains.

169. *Inula obtusifolia* Ker.

Family : Asteraceae

Name : Rupmuk (L)

Distribution : Pakistan to Central Asia & Uttranchal.

Habitat : Dry & rocky slopes, Leh and Spiti valleys around 3,879 m.

Plant Characters : A much branched, prostrate, perennial herb. Stem many, arising from woody stocks, 6-12 inch, often rigidly flexuous, leafy to the top. Leaves ovate to oblong with rounded base, hairy beneath, obscurely toothed. Flower-heads yellow, solitary or few, 1.5-3 cm across. Ray florets yellow. Involucral bracts linear, acute. Achenes silky-hairy. Pappus-hairs pale.

Flowering & fruiting : July to September.

Uses (T) : A decoction of leaves is used for internal wounds.

170. *Inula racemosa* Hk. f.

Family : Asteraceae

Name : Puskarmool (H), Manu (L)

Distribution : Afganistan to Central Nepal, Kashmir, Kumaun & temperate regions.

Habitat : Cultivated in Leh, Kargil and Lahaul valleys around 3,870 m.

Plant Characters : An erect, robust, hairy, perennial herb. Leaves leathery, rough above, densely hairy beneath, toothed; radical leaves long-stalked, elliptic-lanceolate, cauline leaves ovate-oblong, semiamplexicaul. Flower-heads yellow, many in racemes, borne on spike like clusters. Achenes glabrous.

Flowering & fruiting : July to September.

Uses (T) : The plant is having great demand in traditional systems of medicines. Roots are considered to be useful in bronchial asthma when mixed half of the amount with Kuth. Used for blanching *rLung* (wind disease) and *thak* (blood), inflammation due to *pad kan* (phlegm) fever and pain in upper body region.

(Li) : The roots are used in gastrointestine and rheumatism. It has been used as anthelmintic for children and also antiseptic, expectorant and diuretic.

Alantolactone, B sitosterol and its glucosides have been reported in alcoholic extract of roots.

171. *Inula rhizocephala* Shrenk. var. ***rhizocephaloides*** (C.B. Clarke) Kitam.

Family : Asteraceae

Name : Riamko (L)

Distribution : Pakistan to Kashmir.

Habitat : Rocky slopes, Nubra and Zanskar valleys around 4,790 m.

Plant Characters : A small prostrate annual or perennial herb, 6-12 cm in diameter. Leaves pressed to the ground, radical, rosulate, obovate, spathulate. Flower-heads many, yellow, densely clustered at the center, ray florets yellow. Fruits, achenes glabrous ribbed. Differs in having no ray-florets and hairless achenes.

Flowering & fruiting : July to September.

Uses (T) : The dried roots are used against colds, cough and chest pain.

172. *Inula royleana* DC.

Family : Asteraceae

Name : Rupmak (L)

Distribution : From Pakistan to Kashmir, Northern Himalayas.

Habitat : On rocky slopes, Zanskar & Suru valleys between 3,050-3,250 m.

Plant Characters : An erect, stout, perennial herb. Stem usually branched, hairy, glandular. Lower leaves elliptic, blunt; upper leaves elliptic with enlarged clasping base. Flower-heads with golden yellow flowers. Ray florets many. Achenes long with pale red pappus, glabrous.

Flowering & fruiting : July to September.

Uses (T) : Shoots are used by Amchis in dermatitis.

173. *Juniperus communis* Linn. var. ***saxatilis*** Pallas.

See Sr. No. 14, Chapter XIII.

Uses (T) : The fruits are used as febrifuge, anti-inflammatory, purgative; useful for constipation, gout, menstrual problem, inflammation of lung, liver, spleen, kidney and gall bladder.

(Li) : The plant contains resin, a bitter substance juniperine and organic acids. Useful against gonorrhoea, and leucorrhoea and in some cutaneous diseases.

174. *Juniperus macropoda* Boiss.

See Sr. No. 15 Chapter XIII.

Uses (T) : Same as in case of *J. communis.*

(Li) : The berries contain glycosides (B sitosterol) and are useful in colic, cough, indigestion.

175. *Juniperus recurva* Buch,-Ham.

See Sr. No. 16, Chapter XIII.

Uses (T) : Same as in case of *J. communis.*

176. *Jurinea ceratocarpa* (Decne.) Benth. ex Clarke.

Family : Asteraceae

Name : Chholmong or Ispangchot (L)

Distribution : From Pakistan to Kashmir.

Habitat : Drier areas & rocky places, Leh and Spiti valleys around 3,250 m.

Plant Characters : A more or less stemless, perennial herb with thick rootstock. Leaves linear, oblong or lanceolate, entire, pinnate lobed, grey green above, white woolly beneath. Flower-heads pinkish purple in a cluster surrounded by involucre of leaves. Bracts lanceolate long pointed.

Flowering & fruiting : July to September.

Uses (T) : The decoction of leaves is used for Asthma and bronchitis.

177. *Jurinella macrocephala* (Royle) Aswal *et* Goel.

Family : Asteraceae

Name : Dhoop or Guggul (L)

Distribution : From Kashmir to Garhwal.

Habitat : Rock crevices, Lahaul valley around 3,710 m.

Plant Characters : A stemless, perennial herb with woody rootstock. Leaves oblong-lanceolate, spreading all radical, deeply pinnatifid, white lanate above, densely white tomentose beneath. Heads 4.0 cm across, numerous, rosulate, short-peduncled; peduncles stout, tomentose, often cottony at the base. Pappus brownish barbellate, subequal and deciduous.

Flowering & fruiting : June to September.

Uses (T) : The roots are considered to be stimulant and given in fever after the child birth. Decoction of the roots is given in the colic.

178. *Lactuca lessertiana* (DC.) Clarke.

See Sr. No. 82, Chapter VIII.

Uses (T) : The whole plant is used against skin diseases.

179. *Lagotis cashmeriana* (Royle ex Benth.) Rupr.

Family : Scrophulariaceae

Name : Hong-Len (L)

Distribution : From Kashmir to kumaun, Himachal Pradesh and Nepal.

Habitat : Moist alpine slopes, Lahaul valley around 4,105 m.

Plant Characters : A small branched, fleshy, perennial herb. Rootstock stout with thick, fleshy root fibres. Radical leaves elliptic or oblong, oblong-ovate, petioled, obtuse, bracts, ovate, toothed, longer than calyx. Flowers dark blue tubular, borne in dense cylindrical spikes.corolla deep blue, lower lip cleft.

Flowering & fruiting : July to September.

Uses (T) : The whole plant is used as febrifuge, anti-inflammatory; dries up dead blood cells. Used against inflammation of liver, lungs and intestines.

180. *Lagotis kunawurensis* (Royle. ex Benth.) Rupr. Mem.; *Gymnandra kunawurensis* Royle ex Benth.

Family : Scrophulariaceae

Name : Tsermo or Hong-len (L)

Distribution : Pakistan to South East Tibet and Kashmir.

Habitat : On alpine pastures and open damp places, Lahaul valley around 4,350 m.

Plant Characters : A perennial fleshy herb. Rootstock stout. Radical leaves ovate or oblong, lanceolate, obtuse, crenate-toothed, thick; basal elliptic; cauline sessile, ovate, sub-acute or toothed. Flowers purple capitate. Bracts broadly ovate, longer than the calyx. Corolla purple, lower lip 3-cleft, large than calyx. Fruit ovoid.

Flowering & fruiting : July to September.

Uses (T) : The entire herb is used against hot disorders of liver, lungs, small intestine, blood, disturbed fever and helps in purification of blood.

181. *Lancea tibetica* Hk. f. & T.

Family : Scrophulariaceae

Name : Raikse or Chagna or sPa-yang rtsa-ba (L)

Distribution : From Pakistan to South West China and Tibet.

Habitat : Moist places, Leh & Suru and Lahaul valleys around 3,496 m.

Plant Characters : A small stemless herb. Leaves oblong-ovate to spathulate, opposite, entire, narrowed to a short half clasping leaf-stalk. Flowers bright mauve, 2-lipped, borne in central clusters. Calyx tube short with elliptic lobes. Fruits enriched by persistent enlarged calyx.

Flowering & fruiting : July to September.

Uses (T) : The extract of whole plant is to be recommended by the Amchis as a tonic. Used against pus in the lung, cough, cut of nerves, wounds and follow of pus, fruit is used for cardio disorders and retention of menses.

(Li) : The plant contains oleanolic acid and two lignans : phillygenol and lantibeside.

182. *Lappula barbata* (M Bieb.) Gurke.; *Myosotis barbata* M Bieb.

Family : Boraginaceae

Distribution : From Pakistan to Gharwal and Kashmir.

Habitat : On sandy wastelands, Changthang valley around 3,670 m.

Plant Characters : An erect, hispid, perennial, herb. Leaves oblong, narrowed at the base, lower petioled. Calyx lobes oblong. Nutlets scabrous on all sides with 2 rows of short glochidia round the margin.

Flowering & fruiting : July to mid September.

Uses (T) : A decoction of roots is sometimes used to reduce fever.

183. *Leontopodium leontopodinum* (DC.) Hand. Mazz.; *Antennaria leontopodina* DC. Prodr. syn. *L. alpinum* Cass. DC.

See Sr. No. 53, Chapter X.

Uses (T) : The ash of whole plant is used against septic wounds.

184. *Leontopodium nanum* (Hk. f. & T.) Hand.; *Antehharia nanna* Hk. f.

See Sr. No. 54, Chapter X.

Uses (T) : Same as in *Leontopodium leontopodinum*

185. *Lespedeza juncea* (Linn.) Pers.

Family : Fabaceae

Distribution : From Himachal Pradesh to Kashmir & Garhwal.

Habitat : Common on open slopes, Lahaul valley 3,550 m.

Plant Characters : A pubescent, branched, undershrub, upto 1.0 m tall. Leaves trifoliolate. Leaflets obovate, obtusc ot truncate, mucronate, glabrous to subglabrous above, densely silky below. Flowers in 2-4 flowered, pedunculate umbels; peduncle and pedicels variable in length, generally 4-6 mm long. Corolla pink to pale yellow. Pods sessile silky.

Flowering & fruiting : July to September.

Uses (T) : The plant is of medicinal value against cold and cough.

186. *Leucopoa albida* (Turcz) V. Kercz & Bobr.

Family : Poaceae

Name : Fukche (L)

Distribution : From Pakistan to Nepal, temperate Eurasia & Asia.

Habitat : Sandy slopes, Zanskar valley around 3,850 m.

Plant Characters : An annual tufted, grass. Stem sheathed to the top. Leaves linear firm, panicle ovoid, lower branches simple, 4-6 flowered. Glumes ovate to ovate-lanceolate, hyaline. Grain linear-oblong.

Flowering & fruiting : July to September.

Uses (T) : The plant is useful in intestinal pains.

187. *Linum perenne* Linn.

Family : Linaceae

Name : Wild linseed

Distribution : Kashmir to Western Tibet.

Habitat : Rare on moist shady slopes, Lahaul valley around 3,650 m.

Plant Characters : A perennial herb upto 75 cm tall. Branches ascending, glabrous. Leaves alternate, linear. Flowers blue, in 3-5 flowered cymes, cymes upto 25 cm long in fruits. Sepals oblong-ovate, acute. Petals free, obovate. Capsule subglobose. Seeds compressed, shining dark brown.

Flowering & fruiting : July to mid September.

Uses (T) : The plant is bitter, acrid, emollient, expectorant and diuretic.

188. *Lloydia serotina* (Linn.) Reichenb.; *Anthericum serotinum* Linn.

Family : Liliaceae

Name : Kangkar or rTsa-awa (L)

Distribution : From Pakistan to South West China & Europe.

Habitat : Alpine rocks, Nubra valley around 2,970 m.

Plant Characters : A small, bulbous, herb. Bulb elongated base rhizomatus. Leaves 5-15 cm, convulate; cauline leaves linear. Flowers few or solitary, white, nectary small. Capsule obcordate.

Flowering & fruiting : July to September.

Uses (T) : The bulbs are used to purify the blood.

189. *Lomatogonium rotatum* (Linn.) Fries ex Nym.

Family : Gentianaceae

Name : Tsemrang (L)

Distribution : From Pakistan to Kashmir.

Habitat : On moist places, Zanskar valley around 3,350 m.

Plant Characters : An erect, annual herb. Stem branched from the base, 5-20 cm tall. Radical leaves obovate, persistent, and cauline narrowed, at the base. Flowers blue in cymes. Pedicel long. Corolla tube very short, blue with green nerves and seeds many.

Flowering & fruiting : July to September.

Uses (T) : Roots are to be used sometimes to reduce fever.

190. *Lonicera spinosa* (Jacq. ex Decne.) Walp.

See Sr. No. 56, Chapter XII.

Uses (T) : The flowers are dried in shade and the decoction of the same is used in asthma and headache.

191. *Lychnis himalayensis* (Rohrb.). Edgew ex Hk. f.

Family : Caryophyllaceae

Name : Gayangarlo (L)

Distribution : From Pakistan to Himachal Pradesh and Kashmir.

Habitat : Sandy places, Nubra valley around 2,950 m.

Plant Characters : A small, slender, annual herb. Stem slender, elongated, branched. Leaves mostly basal, narrow-lanceolate or toothed margin. Flowers purple, 1-3 in terminal cluster. Calyx oblong nerved. Carpophore short.

Flowering & fruiting : July to September.

Uses (T) : Used for medicinal purposes against cold and cough.

192. *Lychnis nutans* Royle ex Benth.

Family : Caryophyllaceae

Name : Lappu (L)

Distribution : From Pakistan to Kashmir.

Habitat : On moist places, Zanskar valley around 3,350 m.

Plant Characters : A small, annual herb. Stem many from the base, decumbent, glabrous. Leaves shortly petioled, ovate or elliptic-lanceolate or orbicular. Flowers purple, few in panicles. Calyx inflated, and petals purple. Seeds reniform.

Flowering & fruiting : July to mid September.

Uses (T) : A decoction of roots is effective in dysentery.

193. *Lycium ruthenicum* Murray ex Dunal.

Family : Solanaceae

Name : Umila (L)

Distribution : From Himachal Pradesh to Nepal, China and Kashmir.

Habitat : On dry places, Zanskar & Nubra valleys 3,180-3,970 m.

Plant Characters : A spinous, glabrous perennial herb. Stem many from the base. Leaves linear, often fascicled at the nodes. Flowers small, few. Calyx campanulate. Corolla funnel shaped. Berries small, globose.

Flowering & fruiting : July to September.

Uses (T): The leaves have got diuretic properties.

194. *Lycopsis arvensis* Linn.; *Anchusa ovata* Lehm.

Family : Boraginaceae

Name : Foling (L)

Distribution : From Afganistan to temperate Eurasia & Himachal Pradesh and Kashmir.

Habitat : On dry places, Zanskar valley around 3,450 m.

Plant Characters : An annual, herb. Stem upto 20 cm tall. Leaves linear, oblong, sessile, lanceolate, hispid, upper acute. Flowers in axillary and terminal cymes, blue. Calyx hispid, oblong-linear. Corolla tube cylindric. Stamens attached near the base of the corolla tube. Fruits nutlets.

Flowering & fruiting : July to September.

Uses (T) : The plant is of medicinal value as tonic.

195. *Malaxis muscifera* (Lindl.) Ktze.

Family : Orchidaceae

Name : Jeewak (L)

Distribution : Western Himalayas from Himachal Pradesh to Kashmir & Afganistan.

Habitat : Frequent in meadows, Chotadhara-Batal 3,520 m. Lahaul valley.

Plant Characters : A glabrous, small, terrestrial herb with tuberous roots. Stems tuberous at the base. Leaves two, oblong or ovate, obtuse, base sheathing. Flowers minute, yellowish-green, in lax-flowered, 10-20 cm long racemes. Bracts equaling or shorter than the pedicels. Lip ovate, acute, with thick margin.

Flowering & fruiting : July to August.

Uses (T) : The roots are used as a general tonic.

196. *Malcolmia africana* (Linn.) R. Br.

Family : Brassicaceae

Name : Chungpo (L)

Distribution : From Kashmir to Punjab and Western Tibet.

Habitat : Moist places, Suru valley around 3,360 m.

Plant Characters : An erect, leafy, annual herb. Stem branched, 25-40 cm high. Leaves oblong, lanceolate, petioled, 3-5 cm long, toothed. Flowers rose coloured in lax racemes. Petals obovate-oblong, twice as long as the persistent sepals. Pod pubescent, 4-8 cm long.

Flowering & fruiting : July to September.

Uses (T) : Taste sweet and the aerial parts are used against fever.

197. *Malva verticillata* Linn.

See Sr. No. 92, Chapter VIII.

Uses (T) : The fruits are used to treat dropsy, thirsts, diarrhoea, infected sores and dilates all vessels in the body.

(Li) Malvalic acid has been identified in the oils from leaves and seeds. The ash of dried leaves is given in scabies, vomiting and whooping cough.

198. *Mattiastrum thomsonii* (Clarke.) I. M. J.; *Paracaryum thomsonii* Clarke ex Hk. & f.

Family : Boraginaceae

Distribution : From Pakistan to Bhutan, Tibet.

Habitat : Open alpine slopes, Nubra valley around 4,890 m.

Plant Characters : A hairy perennial herb. Leaves petioled, ovate-lanceolate, densely hairy. Flowers small. Segments projecting out of corolla tube. Nutlets small, white glochidate.

Flowering & fruiting : Mid June to August.

Uses (T) : Leaf juice is used as a diuretic and in the treatment of intestinal worms.

199. *Meconopsis aculeata* Royle

See Sr. No. 58, Chapter X.

Uses (T) : The entire plant is used against ulcers, disorders of lungs, liver and inflammation, pharyngitis, backache and disorders of spinal cord.

(Li) : The roots are considered poisonous and narcotic. The plant is used in medicine for backache, colic, renal pain and tonic. The main alkaloid is protopine

200. *Meconopsis bikramii* Aswal.

See Sr. No. 59, Chapter X.

Uses (T) : Same as in *Meconopsis aculeata*

201. *Medicago lupulina* Linn.

See Sr. No. 118, Chapter IX.

Uses (T) : The herb is used to treat hot disorders of lungs and coughs.

202. *Mentha longifolia* Linn. Hudson.

See Sr. No. 37, Chapter VII.

Uses (T) : The whole plant is considered as anti-dysenteric in Ladakh. Horsemint is considered carminative, antiseptic and stimulant. A decoction of the plant is used in fever and heat apoplexy.

203. *Microula tibetica* Benth.

Family : Boraginaceae

Distribution : From Western Nepal to China, Tibet.

Habitat : Open slopes, Zanskar valley around 3,290 m.

Plant Characters : A tufted pubescent, stemless perennial herb. Radical leaves elliptic-spathulate, rosulate. Flowers several bracteate in dense cymes. Calyx campanulate. Corolla tube short. Fruits nutlets-4, diamond shaped scabrous with glochidia.

Flowering & fruiting : July – September.

Uses (T) : The whole herb is considered to be useful in pulmonary diseases.

204. *Minuartia kashmirica* (Edgew.) Mattf.

Family : Caryophyllaceae

Distribution : From Kashmir to Tibet.

Habitat : Common on stony slopes & crevices of rocks, often forming large patches or cushions, Lahaul valley around 3,690 m.

Plant Characters : An erect or loosely tufted perennial herbs, 10-30 cm high. Leaves entire, linear-subulate, spreading, 1-3 nerved. Flowers white in long-peduncled, umbellate or subumbellate, many nerved. Sepals ovate-lanceolate, acuminate. 3-nerved, shorter than the petals. Capsule ovoid. Seeds honeycombed, distinctly tuberculate.

Flowering & fruiting : July – September.

Uses (T) : The whole plant is used as liver tonic.

205. *Morina longifolia* Wall.

Family : Dipsacaceae

Name : Kim (L)

Distribution : Alpine Himalayas from Kashmir to Bhutan.

Habitat : Open & drier slopes, Suru valley around 2,990 m.

Plant Characters : An erect, tall, spinous perennial herb. Stem usually unbranched, slender, 30-40 cm high. Leaves star-shaped with shallow spines, lobes and a long pointed spiny apex. Stem leaves fused in a sheath at the base. Flowers white to rose pink, in a long interrupted spike. Corolla tube long slender somewhat 2-lipped. Calyx lobes absent.

Flowering & fruiting : July to September.

Uses (T) : The root extract is said to be useful against worm infestation and boils.

206. *Morina persica* Linn.

Family : Dipsacaceae

Distribution : Alpine Himalayas from Kashmir to Himachal Pradesh.

Habitat : Common on dry slopes, Lahaul valley around 3,670 m.

Plant Characters : An erect, tall, spinous perennial herb. Leaves whorled, oblong, sinuately pinnatifid, doubly spinous-toothed. Flowers white or tinged with purple, crowded in the axils of the upper leaves forming an interrupted spikes. Calyx-lobes obovate, entire. Corolla-tube 2.5-3.5 cm long; lobes entire.

Flowering & fruiting : June to September.

Uses (T) : The root extract is said to be useful in worm infestation.

207. *Myosotis silvatica* Ehrh. ex Hoffm.

Family : Boraginaceae

Name : Wood Forget-me-not (E)

Distribution : Himalayas from Kashmir to Himachal Pradesh.

Habitat : Common on alpine meadows and on slopes, Lahaul valley around 3,950 m.

Plant Characters : An erect herb upto 45 cm high. Basal leaves petiolate, oblong-elliptic to oblong-lnaceolate, rounded or obtuse at apex; middle cauline leaves sessile; upper ones narrower and shorter. Flowers bluish-purple to bluish-white in short terminal racemes. Nutlets blackish-brown, ovoid, glabrous, smooth, shining.

Flowering & fruiting : June to August.

Uses (T) : The root extract is used as diuretic by the amchis.

208. *Myricaria squamosa* Desv.; *M. hoffmeisteri* Klotz.

See Sr. No. 20, Chapter XIII.

Uses (T) : The whole plant is used as febrifuge, localize poison, anti-tussive, dries pus, ripens pimples.

209. *Nepeta clarkei* Hk. f.

See Sr. No. 38, Chapter VII.

Uses (T) : The whole plant is useful against septic wounds.

210. *Nepeta coerulescens* Maxim.; *N. thomsoni* Benth. ex Hk. f.

See Sr. No. 39, Chapter VII.

Uses (T) : The decoction of whole plant is used against dysentery.

211. *Nepeta discolor* Royle ex Beth.

See Sr. No. 40, Chapter VII.

Uses (T) : The leaf decoction is used in cough and cold.

212. *Nepeta erecta* Benth.

See Sr. No. 41, Chapter VII.

Uses (T) : The decoction of whole plant is useful in dysentery.

213. *Nepeta eriostachya* Benth.

See Sr. No. 42, Chapter VII.

Uses (T) : The whole plant is used for eye complaint.

214. *Nepeta floccosa* Benth.

See Sr. No. 43, Chapter VII.

Uses (T) : Decoction of leaves is used in cough and cold.

215. *Nepeta glutinosa* Benth.

See Sr. No. 44, Chapter VII.

Uses (T) : The decoction of dried leaves is used against diarrhoea.

216. *Nepeta laevigata* (D. Don.) Hand-Mazz.

See Sr. No. 45, Chapter VII.

Uses (T) : The decoction of whole plant is useful in pneumonia.

217. *Nepeta leucolaena* Benth. ex Hk. f.

See Sr. No. 46, Chapter VII.

Uses (T) : Leaf decoction is effective against skin diseases.

218. *Nepeta longibracteata* Benth.

See Sr. No. 47, Chapter VII.

Uses (T) : A decoction of whole plant parts is useful in stomach pains.

219. *Onosma hispidum* Wall.ex Don.

Family : Boraginaceae

Name : Ratanjot, Gajawan (H); Kaume (L)

Distribution : Himalayas from Kashmir to Kumaun & Afganistan.

Habitat : Rare on dry rocky slopes, Suru, Lahaul and Spiti valleys between 4,650-4,790 m.

Plant Characters : An erect bristly haired perennial with stout rootstock. Stem many, arising from the rootstock, 25-40 cm. Basal leaves oblanceolate, roundish, obtuse; cauline leaves shorter and narrower. Flowers many in rounded cluster, creamish-white to pale yellow. Nutlets smooth.

Flowering & fruiting : June to September.

Uses (T) : In Amchi system of medicine the root is used as blood purifier, blood disease, cold, cough, lung and pulmonary problems.

(Li) : The roots contain fat soluble edible dye, mucilage and salts of sodium, potassium, calcium. It has been reported to be used as cardiac tonic and stimulant.

220. *Origanum vulgare* Linn.

See Sr. No. 49, Chapter VII.

Uses (Li) : The plant contains an essential oil, tannins, resin, sterols and flavonoids. Both carvacrol and thymol are antibacterial and antifungal. The plant is quite useful as carminative and stimulates the flow of bile.

221. ***Orobanche hansii*** Ker.

Family : Orobanchaceae

Name : Gro-shang-rtse (L)

Distribution : From Himachal Pradesh to Kashmir & Tibet.

Habitat : Rare parasite on cultivated fields & moist areas, higher passes, Nubra and Lahaul valleys, 3,180 m.

Plant Characters : A scapigerous stout biennial or perennial parasitic herb. Scales upto 12 mm long, transparent. Flowering spike dense. Flowers blue. Bracts ovate, shorter than the corolla, segments bifid to the middle. Filaments glabrous.

Flowering & fruiting : July to mid September.

Uses (T) : The leaves, stems, flowers and fruits are used to restore kidney heat, promots the functioning of kidney and checks haemorrhage.

222. ***Oxyria digyna*** (Linn.) Hill.

See Sr. No. 99, Chapter VIII.

Uses (T) : A decoction of leaves is used to improve digestion; gastritis, inflammation of stomach and purgative.

223. ***Oxytropis lapponica*** (Wahl.) Gay.

See Sr. No. 127, Chapter IX.

Uses (T) : Leaf extract has got antiseptic properties.

224. ***Oxytropis microphylla*** (Pallas) DC.

See Sr. No. 50, Chapter VII.

Uses (T) : The leaves are used in treating infectious fever, inflammation, subsides swelling, disorders, haemorrhage, fractured bones,. It is also used as antiseptic, febrifuge etc.

225. ***Papaver nudicaule*** Linn.

See Sr. No. 61, Chapter X.

Uses (T) : It is used for medicinal value against pains.

226. ***Parnassia laxmanni*** Pallas ex Schultes.

Family : Saxifragaceae

Distribution : From Himachal Pradesh to Kashmir & Pakistan.

Habitat : Common along streams & on moist slopes, Lahaul valley around 3,520 m.

Plant Characters : An erect, glabrous, annual herb, 10-20 cm tall. Leaves ovate or ovate-oblong; basal leaves petioled; cauline sessile, usually solitary in lower half of the stem. Flowers white, solitary, terminal. Petals spathulate-obovate, with slightly fimbricate margins. Staminoides broadly dilated upwards, 3-lobed.

Flowering & fruiting : June to September.

Uses (T): The whole plant is useful as diuretic.

227. *Pedicularis brevifolia* D. Don.

Family : Scrophulariaceae

Name : Luramenthag (L)

Distribution : From Pakistan to Kashmir.

Habitat : Moist, alpine meadows, Suru valley around 2,970 m.

Plant Characters : A perennial herb. Root-stock stout. Stem simple, erect, cauline leaves opposite or whorled, oblong, pinnatifid. Flowers in spikes. Corolla tube twice as long as calyx, purple. Capsule oblong, nearly straight, acute.

Flowering & fruiting : July to mid September.

Uses (T) : The dried inflorescence are used as sedative.

228. *Pedicularis cheilanthifolia* Schrenk.

Family : Scrophulariaceae

Name : Kikimo (Lug-ru-ser-po) (L)

Distribution : Distributed from Pakistan to South East China and Burma.

Habitat : Found on moist places and alpine slopes; Changthang and Nubra valleys 4,150 m.

Plant Characters : A perennial herb. Stem usually many, hairy, spreading, 10-20 cm. Leaves petioled, 3-6 in whorls, pinnatifid. Flowers pinkish white borne in lax or dense terminal clusters. Corolla-tube longer than calyx, densely hairy, toothed, calyx with ribs. Capsule acuminate. Seeds punctuate, pale.

Flowering & fruiting : July to mid September.

Uses (T) : The whole herb is useful in stomach complaint.

229. *Pedicularis longiflora* Rudolph. var. ***tubiformis*** (Klotz) Pennell.

Family : Scrophulariaceae

Name : Phakchang or Lugru-ser-po (L)

Distribution : From Pakistan to South West China.

Habitat : Moist alpine slopes & wastelands, Leh valley between 3,460-3,970m.

Plant Characters : A perennial herb with many stems arising from the root. Leaves oblong, pinnatifid, 3-5 cm, numerous, petioled. Flowers golden-yellow, in terminal cluster with long corolla tube. Capsule long, obliquely oblong-lanceolate, acute, exerted. Seeds minute.

Flowering & fruiting : July to August.

Uses (T) : A decoction of plant parts is used as diuretic. The flowers are useful in treating hot disorders and inflammation of the liver and gall bladder, excessive seminal discharge and oedema.

230 *Pedicularis mollis* Wall ex Benth.

Family : Scrophulariaceae

Name : Lugru-mar-po (L)

Distribution : From Afganistan to Kashmir, Nepal & Burma.

Habitat : Damp alpine slopes, Changthang valley around 4,220 m.

Plant Characters : A low densely woolly plant with many stems, reaching upto 30-70 cm. Leaves 3-5 cm, ovate-oblong, pinnately lobed. Flowers darkned or purple borne in long spikes with the upper whorls of flowers densely crowded. Bracts pinnately lobed. Capsule ovoid, tip exerted, finally acuminate. Seeds broad, flattened, punctuate, pale.

Flowering & fruiting : July to mid September.

Uses (T) : Flowers are used as antitoxin, febrifuge, anti-diarrhoeal and antiseptic, hot disorders and inflammation of stomach.

231. *Peganum harmala* Linn.

Family : Zygophyllaceae

Name : Wild Rue (E), Sepan (L)

Distribution : From Afganistan to Kashmir, Asia, Europe & North Africa.

Habitat : Wastelands dry places, Leh valley around 3,430 m.

Plant Characters : A perennial much branched bushy herb. Stem branched from the base, branches flattened and spreading. Leaves pinnately cut into long linear lobes. Flowers greenish white, solitary, terminal. Corolla 4-5 ovate-oblong. Stamens many (15) with many antherless filaments. Fruits globular, seeds dark brown.

Flowering & fruiting : July to September.

Uses (T) : The seeds are narcotic and used as fever, stomach complaint, anthelmintic, antiseptic, disinfectious, eye disorders, narcotic, measles, incense, asthma, rheumatism, diuretic, tonic, appetizer and joint pains, aphrodisiac, urinogenital complaints, lactation, menstrual complaints.

(Li) The plant is aphrodisiac, abortifacient, amoebicidal, anodyne, diuretic. The seeds are narcotic and contain harmaline, harmidine and harmalol.

232. *Perovskia abrotanoides* Kiril.

See Sr. No. 51, Chapter VII.

Uses (T) : A decoction of leaf is used for acidity of stomach, painful stool, urination.

233. *Phlomis bracteosa* Royle ex Benth.

Family : Lamiaceae

Name : Jemtikli (L)

Distribution : From Kashmir to Himachal Pradesh & Afganistan.

Habitat : Moist slopes, Lahaul valley, 3,420 m.

Plant Characters : An erect, simple or branched, pubescent herbs, 30-90 cm high. Leaves petioled, ovate, obtuse, crenate, broadly cordare or truncate, both surfaces finely tomentose with straight or mixed stellate hairs. Flowers in axillary and terminal dense clusters. Calyx hispid, teeth ciliate. Corolla purplish-blue or white. Nutlets obovoid.

Flowering & fruiting : July to September.

Uses (T) : It has got medicinal properties against cold and cough.

234. *Physalis alkekengi* Linn.

Family : Solanaceae

Name : Strawberry Tomato, Winter Cherry (E), Shoklo (L)

Distribution : From Japan to South East Europe.

Habitat : Irrigated land & moist places, Suru valley, 2,920 m.

Plant Characters : A diffuse perennial herb. Stem branched from the base, 40-70 cm high. Leaves petioled, ovate, entire margin. Flowers solitary or paired, pale-yellow or white borne in axillary cymes. Calyx 5, persistent, enlarged in fruiting stage and covered the entire fruits. Fruit a berry red-orange. Seeds small, yellow.

Flowering & fruiting : July to September.

Uses (T) : The fruits are used in the treatment of urinary disorders, gout and rheumatism The medicine is prepared through ripe red or orange coloured fruits. They are crushed in powder, mixed with mustard oil, heated for a few minutes and than applied and rubbed gently on affected part of body. The fruits are applied in high fever.

(Li) : The fruits as well as leaves contain an amorphous bitter principle. The fruits contain also Vit. 'C', a carotenoid pigment (Physalien) and probably an alkaloid. They are reported to be diuretic, febrifuge, hydrogogue and vermifuge.

235. *Physochlaina praealta* (Decne.) Miers.

Family : Solanaceae

Name : Langthang (L)

Distribution : From Kashmir to West Tibet, Pakistan & Central Nepal.

Habitat : Alpine stony slopes along roadside, Leh & Nubra valleys between 3,210 -3650 m.

Plant Characters : An erect, leafy, perennial, robust herb. Stem grooved, 40-100 cm. Leaves petioled, ovate-oblong, entire or wavy margin. Flowers dull-yellow, funnel shaped in terminal clusters. Calyx bell shaped, 5-lobed, enlarging in fruits. Capsules rhomboid, pointed. Seeds many, scrobiculate , reticulate.

Flowering & fruiting : July to September.

Uses (T) : Seeds are used as vermifuge and leaves are narcotic and used in ulcer and eye diseases. Used against pain in stomach and intestine due to worm, toothache and inflammation of pulmonary track.

(Li) : The plant is excellent source of atropine, which, at present is being imported from abroad. The leaves are narcotic and possess mydriatic properties causing dilatation of the pupil of the eye, as belladonna.

236. *Picrorhiza kurrooa* Royle ex Benth.

Family : Scrophulariaceae

Name : Kutki or Kour (H) & Hanglang (L)

Distribution : Alpine Himalayas from Kashmir to Sikkim & South West China.

Habitat : Rare on moist slopes & alpine grasslands, Lahaul valley around 4,150 m.

Plant Characters : A low, more or less hairy perennial herb. Rootstock thick as the little finger, 6-10 inches long, clothed. Leaves radical, spathulate sharply serrate. Flowers white or pale blue-purple in raceme. Corolla a short stamened. Capsule long, ovoid.

Flowering & fruiting : July to mid September.

Uses (T) : The roots are best source of bitter tonic, which is useful in stomachache, fever, laxative, dysentery, liver diseases cold, cough and heart problems by the local Amchis.

(Li) : The plant is cooling, cardio tonic, antipyretic, anthelmintic, anaemia, promote appetite, secretion of bile, useful in 'kapha', fever, urinary discharge, asthma, blood trouble, influenza, jaundice, menstrual disorders, joints pain, piles, nose problem, liver protactant, good for gums and teeth and snake bites. A brown resinous glucoside Picrorhizin, its aglycone picrorhizetin, Kutkin, B-sitosterol have been reported from the rhizomes.

237. *Pimpinella tenera* (Wall. ex DC.) Benth ex. Clarke.

Family : Apiaceae

Distribution : From Kashmir to Himachal Pradesh.

Habitat : Rare in rock-crevices, Lahaul valley, 4,820 m.

Plant Characters : A slender, branched, glabrous, annual herb, 15-20 cm high. Basal leaves ovate in general outline, 3.5-8.0 cm long and 2-6 cm broad excluding the petioles 2-10 cm long; lamina ternate-pinnate, ultimate divisions linear-oblong, obtuse or mucronate; cauline leaves similar. Umbels terminal or axillary. Rays 4-7, unequal, 0.3-1.6 cm long. Flowers yellow. Fruits ovate, narrowed upwards, glabrous, laterally compressed.

Flowering & fruiting : July to August.

Uses (T) : The whole plant is used as carminative and diuretic.

238. *Plantago depressa* Willd.

Family : Plantaginaceae

Name : Tha-ram (L)

Distribution : From Kashmir to Bhutan & Tibet.

Habitat : On moist places, Leh and Spiti valley between 3,500-3,650 m.

Plant Characters : A stemless, annual or perennial herb. Rootstock short, stout. Leaves in rosettes, erect, elliptic. Scapes glabrous or puberulous. Spikes narrow, with lax distant greenish-white flowers. Capsule twice as long as the calyx, very pale.

Flowering & fruiting : July to September.

Uses (T) : An infusion of whole plant is quite effective in liver complaint, skin diseases, stomachache, eye diseases, cough and gastro-intestinal problems. The seeds are used fro heating effect.

(Li) : It is antispasmodic, antiseptic, expectorant, carminative, anthelmintic and stimulant, catarrh and convulsive and whooping-coughs. The plant contains acteoside and aucubin.

239. *Plantago major* Linn.

Family : Plantaginaceae

Name : Ripple Grass (E)

Distribution : Temperate Himalayas from Kashmir to Bhutan.

Habitat : Common as weed in cultivated fields or vegetable garden; Leh valley around 3,205 m.

Plant Characters : A glabrous, perennial herb. Rootstock small, stout. Leaves radical, ovate to oblong-ovate, stalk slightly winged margin entire, apex obtuse. Scapes many upto 40 cm long, carrying long cylindrical spike. Flowers lax at the base and dense towards the apex. Capsule pyramidal with a trapping apex. Bracts equaling the calyx. Seeds ovoid-black.

Flowering & fruiting : July to September.

Uses (T) : The plant is used against stomach pain, stimulant, diuretic, tonic and are used as remedy for dysentery and diarrhoea. by Amchis.

(Li) : The chief constituents are choline, salicylic acid, aucubin. It is used in disorder of epidermis and in headache, fever, inflammation, weakness, wounds, cuts, earache and toothache. The leaves are considered cooling, alternative, febrifuge, diuretic, astringent and vulnerary. The seeds are considered demulcent, stimulant, diuretic, tonic, remedy for dysentery and diarrhoea.

240. *Podophyllum hexandrum* Royle

Family : Berberidaceae

Name : Himalayan Mayapple (E), Vankakri or Vanvaigan (H) Demokusu, Papra, Ol-mose (L)

Distribution : From Kashmir to Sikkim & Afganistan to South West China.

Habitat : Rare on forest areas & cultivated fields, Zanskar valley around 3, 755 m.

Plant Characters : A perennial scapigerous herb. Stem modified into underground rhizomes. Leaves 2, blade rounded in outline, 10-25 cm, deeply cut into 3 ovate toothed lobes, sometimes further lobed. Flowers solitary, terminal white or pink, cup-shaped. Sepals 3, soon falling; stamens 6. Fruit large, scarlet or reddish berry, pulpy, orange or red seeds.

Flowering & fruiting : May to August.

Uses (T) : The entire plant is used for gynaecological diseases like menstrual irregularity, diseases of the uterus and improves rLung and blood circulation, helps delivery of the baby and placenta. The roots are used against skin problems, hardening of skin and skin diseases by local Amchis. The young and ripe fruits are edible and is useful against high altitude mountain sickness.

(Li) : The rhizomes and roots of the plant, which constitute the drug "podophyllin" as podophyllotoxin, is considered a cholagogue, purgative, emetic and a bitter tonic. Podophyllin is also used in veterinary medicine as a cathartic for dogs and cats. Roots have potential chemical, which is effective in skin cancer and radiation.

241. ***Polygonatum multiflorum*** (Linn.) All.

Family : Liliaceae

Name : Mitha dudia (H), Soloman's seal (E), Ra-mnye (L)

Distribution : From Kashmir to Kumaun and temperate Eurasia.

Habitat : Forest areas, Lahaul valley 3,490 m.

Plant characters : An erect to suberect, unbranched herb, upto 70 cm tall. Stem grooved. Leaves oblong, sessile, narrowed at the base, glaucous beneath. Peduncles 1-3 flowered. Tapels 1.25 cm long, constricted in the middle, white with green lobes.

Flowering & fruiting : July to August.

Uses (T) : The rhizomes are used in treatment of kidney problems and as health tonic.

(Li) : The plant contains saponins, flavonoids and vitamin 'A'. The rhizome has astringent and demulcent properties.

242. ***Polygonatum verticillatum*** (Linn.) All.

Family : Liliaceae

Name : Mitha dudia (H), Soloman's seal (E), Ra-mnye (L)

Distribution : Pakistan to South East Tibet, Kashmir, Sikkim & Manipur.

Habitat : Open forest areas, Lahaul valley between 3,020-3,250 m.

Plant Characters : An erect, robust, perennial herb. Rootstock thick creeping. Stem glabrous, grooved reaching upto 70 cm high. Leaves whorled, sessile, linear, 5-7 cm long, ciliate on margins. Flowers white with green tips in the axils of branched cluster of 2-3. Berry 6-10 seeded.

Flowering & fruiting : July to September.

Uses (T) : The roots restore bodily strength, and are useful in the treatment of pain in the kidneys, nervine tonic, appetizer, weak digestive heat, first stage dropsy and in *rLung* wind diseases.

(Li) : The rhizomes are valued as **salep**, a strength-giving food. The plant possesses diuretic properties.

243. ***Potentilla ambigua*** Camb.

Family : Rosaceae

Distribution : From Afganistan to Himachal Pradesh.

Habitat : On wastelands places & open slopes, Changthang valley around 4,310 m.

Plant Characters : A perennial herb. Stem tufted, silky, leafy. Stipules leafy. Leaflets 3 obovate or orbicular-cuneate. Flowers solitary, peduncled. Petals larger than calyx, yellow. Achenes many smooth.

Flowering & fruiting : July to August.

Uses (T) : The whole plant has got medicinal value as health tonic.

244. *Potentilla anserina* Linn.

Family : Rosaceae

Name : Silver weed (E), Toma (L)

Distribution : From Pakistan to Bhutan & North temperate Zone.

Habitat : Common on irrigated lands and open slopes, Leh valley around 3,260 m.

Plant Characters : A slender, creeping, perennial herb. Leaves pinnate, arising in a tuft, from the rootstock. Leaflets ovate, deeply toothed, silvery beneath. Flowers large, solitary on a long pedicel, yellow. Achene hairy.

Flowering & fruiting : July to September.

Uses (T) : The plant is spasmolytic, tonic, and vulnerary, used in the form of tea (or in tonic) against diarrhoea, leucorrhoea and cramps.

(Li) : The herb is considered to possess astringent, spasmolytic, tonic and vulnerary properties. It contains anthocyanins like cyaniding and delphinidin and tannins - choline, betaine, histidine.

245. *Potentilla atrosanguinea* Lodd.; *P. argyrophylla* (Wall. ex Lehm.) Griers & Long.

Family : Rosaceae

Name : Chisheng (L)

Distribution : From Afganistan to Kashmir & Sikkim.

Habitat : Alpine pastures & irrigated lands, Leh valley around 4,760 m.

Plant Characters : A veriable, perennial herb. Stem thick, erect or spreading, 15-45 cm long. Leaves trifoliate, elliptic-ovate, toothed, basal leaves long stalked. Inflorescence lax branched. Flowers large, orange or dark red; corolla obcordate. Achenes smooth.

Flowering & fruiting : July to August.

Uses (T) : The whole herb is used in medicine for fever. The leaves are analgesic and used against wounds.

246. *Potentilla fruticosa* Linn.

Family : Rosaceae

Name : Shrubby Cinquefoil (E), Khiamgar (L)

Distribution : From Pakistan to Bhutan, Kashmir & Sikkim.

Habitat : On open slopes, Changthang valley around 4,360 m.

Plant Characters : An low-spreading, silky, perennial shrub. Stem branched. Leaves pinnate with 3-7 small leaflets, crowded, lanceolate. Leaflets ovate-lanceolate, acute. Stipule conspicuous. Flowers bright yellow. Petals obovate, much longer than calyx; achenes minute, haired.

Flowering & fruiting : July to August.

Uses (T) : Leaf decoction is useful to reduce fever.

247. *Potentilla multifida* Linn.

Family : Rosaceae

Name : Thatko (L)

Distribution : From Pakistan to Kashmir.

Habitat : On rocky slopes & near water sources, Zanskar and Spiti valleys around 4,240 m.

Plant Characters : A perennial herb. Stem branched from the base. Leaves pinnate. Leaflets linear, lanceolate, tomentose. Flowers yellow pedicelled, in terminal cymes. Corolla smaller than calyx. Achenes few.

Flowering & fruiting : June to September.

Uses (T) : The seeds are used in sleeplessness by local Amchis.

248. *Prangos pabularia* Lindl.

See Sr. No. 152, Chapter – IX.

Uses (T) : All the plants parts have got medicinal properties. The roots are used as diuretic. Seeds are considered carminative. Roots and seeds are also used in energetic drugs. Fruits are used to cure stomach complaint also considered energetic by local Amchis. As ethno-veterinary use it is believed to increase milk yield in early lactation in cattle.

(Li) : It is held in considerable repute in indigenous medicine for its roots and fruits. The roots are considered diuretic and are used as external application in the treatment of itches, carminative, laxative, diuretic, aphrodisiac, stimulant, stomachache and liver tonic, it promotes expulsion of foetus. An infusion of roots is given for indigestion and in irregularity of menses. Osthol, merancin, its hydrate, psoralen, xanthotoxol, xanthotoxin, imperatorin and prangenin have been isolated from roots, leaves and fruits.

249. *Primula macrophylla* D. Don. (excl. syn. *P. moorcroftiana*). *P. stuartii* Wall. var. *perpurea* (Royle) Hk. f.; *P. nivalis* Pall. var. *macrophylla* (D. Don.) Pax.

See Sr. No. 69, Chapter X.

Uses (T) : A decoction of leaves and rootstocks is used to treat cough in Ladakh.

250. *Primula rosea* Royle

See Sr. No. 71, Chapter X.

Uses (T) : The dried rootstock is used to cure muscular pain.

251. *Ranunculus brotherusii* Freyn.

Family : Ranunculaceae

Name : Maokiang or Ice-tsva rigs-dang-po mo-rigs sam-mchog (L)

Distribution : From Pakistan to South West China.

Habitat : Open slopes & drier places, Leh and Lahaul valleys around 3,850m.

Plant Characters : A small erect herbaceous plant. Stem 5-15 cm. Upper leaves cut into acute segments. Flowers small yellow, terminal with silky sepals. Fruits achenes round, smooth.

Flowering & fruiting : July to August.

Uses (T) : The flowers restore digestive heat, regulate body heat and is beneficial in treating ulcers, diphtheria, last stage of dropsy and tumors of cold nature. An extract of whole plant is used for blood purification and in body pain by local Amchis.

252. *Ranunculus laetus* Wall. ex D. Don.

Family : Ranunculaceae

Distribution : Afganistan to South West China & C. Asia.

Habitat : On damp places and shrubberies, Zanskar valley between 2,950-3,250 m.

Plant Characters : An erect perennial herb. Stem branched, 30-60 cm tall. Leaves rounded in outline, deeply 3-5 lobed, the lobes oblanceolate, pointed and further 3-lobed, toothed, hairy. Flowers yellow, solitary in branched leafless inflorescence. Achenes smooth, somewhat flattened.

Flowering & fruiting : July to September.

Uses (T) : The leaves are used as tonic and stimulant by locals in Ladakh.

(Li) : It is said to be closely allied to and perhaps a variety of the European species, *R. acris* Linn. well known for its acrid and toxic properties.

253. *Ranunculus lobatus* Jacq. ex Camb.

Family : Ranunculaceae

Distribution : Afganistan to Kashmir & Central Asia.

Habitat : Moist places, Zanskar valley around 3,550 m.

Plant Characters : A small decumbent, glabrous herb. Stem leafy, 5-10 cm long. Radical leaves at the base; cauline leaves short petioled, with toothed margins. Flowers small. Petals broadly obovate. Achenes oval, slightly compressed, beaked.

Flowering & fruiting : July to mid September.

Uses (T) : The decoction of the plant is used against toothache and gum inflammation by local tribes of Ladakh.

254. *Ranunculus trichophyllus* Chaix.; *R. aquatilis* Linn. var. *trichophyllus* Hk. f. & T.

Family : Ranunculaceae

Name : Rengo (L)

Distribution : From Kashmir to Sikkim & South East Tibet.

Habitat : Moist places & fast moving water, Suru valley around 2,860 m.

Plant Characters : A submerged, perennial herb. Leaves dark green, much dissected, stipules auricled. Flowers minute, white, carried above water. Achenes wrinkled.

Flowering & fruiting : June to August.

Uses (T) : The decoction of whole plant is useful against diarrhoea. It is also used in fever by Amchis.

255. *Ranunculus tricuspis* Maxim.

Family : Ranunculaceae

Name : Sharchang or Chu-rug sbal-lag (L)

Distribution : Afganistan to Kashmir & Central Asia.

Habitat : Along river sides & water streams, Leh valley around 3,560 m.

Plant Characters : A small hairless, herb. Stem creeping, 20-35 cm, rooting from the node. Leaves dissected, hairy, variable. Flowers yellow, solitary. Petal 5-8 and sepals spreading or refluxed.

Flowering & fruiting : July to August.

Uses (T) : The leaves, stem, flowers and fruits are used in treating hot disorders of the tendons and ligaments.

256. *Rheum moorcroftianum* Royle

Family : Polygonaceae

Name : Rhubarb (E), Revanchini (H)

Distribution : From Pakistan to Kashmir & Kumaun.

Habitat : Alpine slopes, Lahaul valley between 4,235-4,960 m.

Plant Characters : A stemless, stout, pubescent, perennial herb with woody rootstock. Leaves all radical, thick, coriaceous, orbicular, margins undulate usually cordate, 25-40 cm across, often pubescent, enlarged in fruit. Tapels greenish. Anthers purple. Fruits ovoid, narrowly winged.

Flowering & fruiting : July to September.

Uses (T) : Roots are used in the indigenous system of medicine.

(Li) : The roots contain anthaquinone derivaties, glucoside, emodin, rhein. The drug is purgative in nature.

257. *Rheum spiciforme* Royle

Family : Polygonaceae

Name : Rhubarb (E), Revanchini (H), Lachhu (L)

Distribution : From Himachal Pradesh to Kashmir & Western Nepal.

Habitat : Alpine stony slopes, Lahaul and Changthang valleys between 4,155-4,880 m.

Plant Characters : A stemless, stout, glabrous, perennial herb with thick woody rootstock. Leaves all radical, thickly coriaceous, orbicular, broadly ovate or cordate, glabrous, nerves very prominent. Racemes radical, dense-flowered or oblong, broader than the disk.

Flowering & fruiting : June to August.

Uses (T) : The purple coloured roots are used in the indigenous system of medicine.

258. *Rheum webbianum* Royle.

Family : Polygonaceae

Name : Himalayan Rhubarb (E), Revanchini (H), Lachhu or Chu-rtsa (L)

Distribution : From Pakistan to Western Nepal.

Habitat : On open slopes & shrubberies, Leh & Zanskar valley between 3,105-3,920 m.

Plant Characters : A stout, erect, perennial herb. Roots thick. Stem veriable, 30-150 cm. Lower leaves long petioled, orbicular-cordate 15-60 cm across.

Panicles axillary and terminal. Flowers pale yellow small in cluster. Fruits broadly oblong notched at both ends.

Flowering & fruiting : July to September.

Uses (T) : The roots, stems, leaves and leaf-stalks are purgative and are beneficial in treating indigestion, abdominal diseases, astringent, boils, purgative, wounds and flatulence. The roots are diuretic, purgative, febrifuge; used against indigestion, wounds and gastritis etc.

(Li) : The roots are an excellent source of anthaquinones, useful as laxative.

259. *Rhodiola crenulata* (Hk. f. & T.) Ohba.

Family : Crassulaceae

Name : Stone crop (E)

Distribution : India to Afganistan, Central Nepal.

Habitat : Frequent in rock-crevices. Lahaul valley around 4,935 m.

Plant Characters : A perennial herb upto 20 cm tall. Rhizome thick, subcylindrical. Flowering stems simple, usually fastigitate, glabrous, terete. Leaves alternate, imbricate, more or less indistinctly petiolate. Inflorescence terminal, 20-40 flowered, dense or compact, forming corymbose fascicles, surrounding by the upper cauline leaves and bracts.

Flowering & fruiting : July to August.

Uses (T) : The roots have got medicinal properties and are used against lung complaints.

260. *Rhodiola heterodonta* (Hk. f. & T.) A. Boriss.

Family : Crassulaceae

Name : Rose root or Stone crop (E), Shrolo (L)

Distribution : Afganistan to Central Nepal.

Habitat : Moist places & higher passes, Nubra and Changthang valleys around 3,960 m.

Plant Characters : A perennial succulent plant with thick and fleshy rootstock. Leaves flat, lanceolate-triangular to pentagonal-ovate, serrated margins. Flowers yellow borne on dense domed clusters. Seeds often reddish.

Flowering & fruiting : July to September.

Uses (T) : The roots are useful in treatment of hot disorder of the lungs, infectious common cold with fever, improves physical strength, heals ruptured pulmonary capillaries, dispels bad breath and fetid body odour. It is also used for inflammation of lungs, other lung problems and contagious diseases.

261. *Rhodiola imbricata* Edgew.; *Sedum roseum* (Linn.) Scop.

Family : Crassulaceae

Name : Rose root or Stone crop (E), Shrolo (L)

Distribution : Pakistan to Central Nepal.

Habitat : Rocky slopes, wet places & higher passes, Leh,Changthang and Spiti valleys around 5,420 m.

Plant Characters : A succulent, perennial herb 10-35 cm with rose scented massive rootstock. Leaves 1.3-3 cm long, oblong to narrow, elliptic, nearly entire. Flowers pale yellow in congested clusters, surrounded by an involucre of leaves. Fruits of 4-5, many seeded.

Flowering & fruiting : July to September.

Uses (T) : The roots are used in medicine against lung problems, cold, cough, fever, loss of energy and pulmonary complaints. It is also used in preparation of health tonic and it restores memory. It is also used for inflammation of lungs, due to contagious diseases.

262. *Rhodiola tibetica* (Hk. f. & T.) Fu.

Family : Crassulaceae

Name : Stone crop (E), Shrolo (L)

Distribution : Afganistan to Lahaul, Kashmir & Ladakh.

Habitat : Common on moist alpine passes, Leh and Changthang valleys around 5,860 m.

Plant Characters : An erect, perennial herb. Leaves oblong or ovate-oblong, almost entire. Flowers rose-purple, in 5-20-flowered, terminal corymbs. Follicles not narrowed gradually at the base. Seeds oblong-ellipsoid, compressed.

Flowering & fruiting : July to September.

Uses (T) : The roots are used in medicines as a health tonic.

263. *Rhododendron anthopogon* D. Don. spp. ***hypenanthum*** (Balf. F.) Cullen.

See Sr. No. 29, Chapter XIII.

Uses (T) : The plants are used in preparation of incense. The plant is used in treatment of gnorrhoea and stomach ailments. The leaves are also used by local people for preparing tea and are sold under the name "Talisapatra".

264. *Rhododendron campanulatum* D. Don.

See Sr. No. 30, Chapter XIII.

Uses (T) : The root is a popular remedy for boils. Leaves are said to be used in chronic rheumatism and sciatica.

(Li) : The plant is reported to contain ericolin, ursolic acid, quercetin having demand in pharmaceutical industries.

265. *Rosa macrophylla* Lindl.

See Sr. No. 77, Chapter X.

Uses (T) : The flowers are used in medicine for stomach pain.

266. *Rosa webbiana* Wall. ex Royle.

See Sr. No. 78, Chapter X.

Uses (T) : The flowers and fruits are used against fever due to poison, food poisoning, stomachic, inflammation of liver, hepatitis, jaundice.

(Li) : The fruits are edible and are rich in vitamin 'C'.

267. *Rubia cordifolia* Linn.

Family : Rubiaceae

Name : Manjith (H), Indian Madder (E) & bTsod (L)

Distribution : Africa to Asia, Australia & India.

Habitat : Climbing on shrubs near moist areas, Suru and Spiti valleys around 3,980 m.

Plant Characters : A herbaceous climbing perennial. Roots perennial. Leaves smooth or scabrid, whorled, petioled, ovate, acute. Flowers small, white or greenish. Anther globose. Stem 4 angled. Fruits black, fleshy with red juice.

Flowering & fruiting : July to August.

Uses (T) : The stems and roots are effective against blood diseases, chest complaints, discoloration of skin and leucoderma, inflammation, menstrual disorders, parapysis, stomachache, tonic, ulcers, urinary complaints.

(Li) : The roots are used for heating, laxative, antidysentric, antipyretic, analgesic, anthelmintic, rheumatism, joints pain. These roots contain purpurin, munjistinn, pupuroxanthin and anthraquinones.

268. *Rubia tibetica* Hook. f.

Family : Rubiaceae

Name : Manjith (H), Indian Madder (E)

Distribution : Africa to Asia & India.

Habitat : Frequient on slopes, Lahaul valley around 4,210 m.

Plant Characters : An erect, scabrid herb with quadrangular stems and short branches. Leaves sessile, 2-6 at each nodes, ovate, suborbicular, elliptic-ovate or lanceolate, coriaceous, scabrid at the margins. Flowers solitary and in the axillary or terminal cymes. Calyx globose, enclosing the ovary and adherent to it. Corolla gamopetalous, pubescent, tube very short, lobes 5, ovate-lanceolate, acuminate. Seeds suberect, albuminous.

Flowering & fruiting : July to September.

Uses (T) : Same as in *Rubia cordifolia.*

269. *Rumex acetosa* Linn.

Family : Polygonaceae

Distribution : Kashmir to Kumaun.

Habitat : Moist slopes, Lahaul valley around 3,710 m.

Plant Characters : An erect, unbranched or rarely branched, glaucous, dioecious, perennial herb upto 1.0 m. tall. Radical leaves long-petioled, oblong, base cordate, hastate or sagitate; cauline leaves with hastate or sagitate base. Flowers greenish-purple in racemes.

Flowering & fruiting : June to August.

Uses (T) : The leaves are used as fresh relish. The plant is used as a laxative.

(Li) : The The whole plant is having oxalates, free oxalic acid, anthaquinone.

270. *Rumex patientia* Linn. spp. ***tibeticus*** Rech. f.

See S. No. 128, Chapter VIII.

Uses (T) : The whole herb is used as febrifuge, heals wounds, purgative, vermifuge. Used against inflammation of lungs and liver, constipation, influenza, wounds, pimples skin disease and worms.

271. *Sagina saginoides* (Linn.) Karsten

Family : Caryophyllaceac

Distribution : Kashmir to Tibet.

Habitat : On rocky & sandy slopes, Lahaul valley around 3,510 m.

Plant Characters : A caespitose, small, glabrous, perennial herb, 5-10 cm tall. Leaves linear, glabrous, united at the base in a scarious sheath. Flowers white, solitary on axillary and terminal peduncles. Sepals free. Petals shorter than the sepals. Capsule pale straw coloured, shining as long as the sepals.

Flowering & fruiting : June to September.

Uses (T) : The whole plant is useful as tonic.

272. *Saussurea bracteata* Decne.

Family : Asteraceae

Name : Pseudobrahma Kamal (E), Span-rtsa-Do-bo or Jar-bag (L)

Distribution : From Pakistan to Kashmir & Central Asia.

Habitat : Dry-rocky slopes, Changthang and Leh valleys around 5,330 m.

Plant Characters : A dwarf, perennial herb. Stem scaberulous reaching upto 8-20 cm. Leaves linear-lanceolate, coarsely toothed. Flower-heads solitary, silky, large. Involucral bracts black, acuminate. Pappus white, bristled.

Flowering & fruiting : Mid June to August.

Uses (T) : The flower buds are made into paste and is applied to cure boils.

273. *Saussurea glaciales*

Family : Asteraceae

Name : Pansipa (L)

Distribution : From Pakistan to Kashmir & Ladakh.

Habitat : Alpine pastures, Suru valley around 3,430 m.

Plant Characters : An almost stemless densely woolly tufted herb giving an appearance of a snow ball. Leaves variable, obovate to oblong, entire or obscurely toothed. Flower-heads reddish-purple surrounded by spreading leaves. Involucral bracts lanceolate, papery.

Flowering & fruiting : July to September.

Uses (T) : A decoction of flower buds is used against intestinal complaints.

274. *Saussurea gnaphalodes* (Royle ex DC.) Sch. Bip. *S. sorocephala* (Shrenk) Sch. Bip.

Family : Asteraceae

Name : Yuliang (L)

Distribution : Pakistan to South West China and Lahaul to Kashmir & Tibet.

Habitat : On high altitude screes, Leh valley around 5,530 m.

Plant Characters : A stemless tufted perennial herb. Leaves variable, ovate to oblong, entire or toothed, covered with woolly hair. Flower-heads reddish purple in dense domed cluster, embedded in dense woolly hairs. The whole plant often looks like a cotton ball.

Flowering & fruiting : Mid July to September.

Uses (T) : The whole plant is used in kidney problems.

275. *Saussurea jacea* (Klotz.) Clarke.

Family : Asteraceae

Name : Pashuk (L)

Distribution : From Afganistan to Kashmir.

Habitat : On sandy places & irrigated lands, Suru valley around 3,130 m.

Plant Characters : An erect, perennial herb. Stem branched from the base, 30-60 cm high, very leafy. Leaves ovate with entire margin. Flower-heads conspicuous, dark margined. Bracts rigid, adpressed, the outer ovate, blunt or acute and the inner longer lanceolate.

Flowering & fruiting : July to September.

Uses (T) : The extract of the aerial parts is used as a diuretic.

276. *Saussurea lappa* (Decne.) Sch.-Bip.; *S. costus* (Falc.) Lipsch.

Family : Asteraceae

Name : Costus (E), Kuth (H), Rustha (L)

Distribution : Endemic to Kashmir and cultivated in Kashmir, Himachal Pradesh & other Himalayan region.

Habitat : Being cultivated in Lahaul valley and introduced in Leh upto 3,850m.

Plant Characters : An erect, robust perennial herb. Root possessing a characteristic penetrating odour; radical leaves with long lobately winged; flowers dark blue-purple in axillary and terminal clusters; achene curved, compressed.

Flowering & fruiting : July to September.

Uses (T) : The roots are used in bronchitis alongwith Manu in equal quantity by the amchis.

(Li) : The roots are being used in asthma, dysntry, rheumatism, ulcer, incense and this use of drug has received ample confirmation by several scientific investigations. The roots are hot, bitter, pungent, fattening, aphrodisiac, carminative, analgesic, anthelmintic, emmenagogue, alternative, cures leucoderma, erysipelas, itching, ringworm, 'tridosa' bronchitis, vomiting, scabies, vata, headache, hysteria, and epilepsy, stimulant to brain, cures liver, kidney diseases, chest pain and joint pains paralysis, asthma, cough, inflammations, ophthalmia, old fever. The root contains an alkaloid Saussurine and tannins.

277. *Saussurea obvallata* (DC.) Edgew.

Family : Asteraceae

Name : Brahma Kamal (E), Span-rtsa-Do-bo (L)

Distribution : From Pakistan to South West China; rare in Nepal, but prominent in "The valley of Flowers" Garhwal.

Habitat : On rocks & stream, Zanskar valley around 5,330 m.

Plant Characters : An erect, robust, perennial herb (30-50 cm). Leaves oblong-lanceolate, blunt, toothed; lower petioled and upper sessile. Flower-heads purple in dense umbel like cluster surrounded by large pale-yellow boat shaped pappery bracts.

Flowering & fruiting : July to September.

Uses (T) : The paste of flower buds is applied to boils. The whole plant is used against cold, cough, influenza, fever, intestinal pain and arthritis by local Amchis.

278. *Saxifraga cernua* Linn.

Family : Saxifragaceae

Name : Bri-rta sa-'dzin dman-pa (L)

Distribution : From Pakistan to Kashmir, Central Nepal, Central Asia & North America.

Habitat : Open slopes & moist places, Suru valley around 4,450 m.

Plant Characters : A perennial herb. Stem 10-20 cm high. Radical leaves reniform, lobed, upper leaves trifid. Flowers yellow, solitary-terminal. Calyx-tube campanulate with ovate-linear lobes.

Flowering & fruiting : July to September.

Uses (T) : The plant has got medicinal value as tonic.

279. *Saxifraga jacquemontiana* Decne.

Family : Saxifragaceae

Name : Sasomantso (L)

Distribution : From Pakistan to Kashmir & South West China.

Habitat : Higher passes & dry slopes, Lahaul valley around 5,250 m.

Plant Characters : A small perennial herb forming a dense mat. Stem branched, 3-7 cm long. Leaves sessile, broadly oblong, glandular-pubescent, forming rosettes at the tips. Flowers yellow, subsessile. Petals broadly elliptic. Capsule ovoid.

Flowering & fruiting : July to August.

Uses (T) : The entire plant is useful in the treatment of inflammation of the nerves.

280. ***Saxifraga pulvinaria*** Harry Smith.

Family : Saxifragaceae

Name : Yenhua or Sum-cu-tig (L)

Distribution : Pakistan to Central Nepal & temperate Eurasia.

Habitat : Rocky slopes & pastures, Zanskar valley around 4,980 m.

Plant Characters : A densely tufted perennial herb forming firm cushions. Leaves densely imbricate, obovate-elliptic or broadly oblong, blunt in tiny rosettes. Flowers white, sessile, solitary, terminal. Petals narrow, margins undulate. Capsule broadly ovoid.

Flowering & fruiting : July to August.

Uses (T) : The entire herb is used to treat hot disorders of the liver and gall bladder.

281. ***Saxifraga stenophylla*** Royle.; *S. flagellaris* Willd.

Family : Saxifragaceae

Name : Yenhau (L)

Distribution : Pakistan to Central Nepal.

Habitat : Snow melted areas, Leh valley around 5,550 m.

Plant Characters : A small perennial herb. Runners often red, long upto 8 cm. Stem usually red with few leaves. Leaves linear-lanceolate in rosette and without marginal bristles. Flowers small, yellow, in terminal cluster. Petals longer than sepals.

Flowering & fruiting : July to September.

Uses (T) : The whole plant is used against spreading of putrid blood and pus in lungs.

282. ***Scorzonera virgata*** DC. (*S. divaricata* Hook. f.)

Family : Asteraceae

Name : Rtsa-mkhris (L)

Distribution : Pakistan to Himachal Pradesh & Kashmir.

Habitat : Wastelands & sandy places, Leh valley around 3,450 m.

Plant Characters : A slender, much-branched, perennial herb forming a large clump. Stem usually upto 50 cm, many. Leaves grey green, entire 2-4 mm broad with inrolled margins. Flower-heads yellow, solitary on long peduncles.

Flowering & fruiting : July to September.

Uses (T) : The plant extract is useful to cure intestinal problems.

283. ***Scrophularia dentata*** Royle ex Benth.

Family : Scrophulariaceae

Name : Hamchi (L)

Distribution : Himalayas from Kumaun to Lahaul-Spiti.

Habitat : Moist slopes, Changthang valley around 4,560 m.

Plant Characters : An erect, annual or perennial herb. Stem glabrous below, glandular above nearly black when dry. Leaves small, petioled, obovate-oblong, incised towards the base. Flowers in cymes with short pedicel. Staminodes small narrow.

Flowering & fruiting : July to September.

Uses (T) : A decoction of the leaves is used to stimulate appetite.

284. ***Scrophularia koelzii*** Pennell.

Family : Scrophulariaceae

Name : Figwart (E), Yarma (L)

Distribution : Himalayas from Kumaun to Lahaul-Spiti & Ladakh.

Habitat : Stony slopes, Suru and Spiti valleys around 3,060 m.

Plant Characters : An erect, perennial herb, glabrous below and glandular above. Stem many, arising from the root-stock, 20-40 cm high. Leaves shortly petioled, oblong to spathulate. Flowers dark purple borne in slender spike like clusters. Calyx with with rounded lobes. Styles exserted.

Flowering & fruiting : July to September.

Uses (T) : Seeds are considered to be useful in joint pain.

285. ***Scutellaria prostrata*** Jacq. ex Benth.

Family : Lamiaceae

Name : Haunching (L)

Distribution : From Kashmir to Kumaun & Tibet.

Habitat : Open rocky slopes, Leh valley around 4,875 m.

Plant Characters : A low spreading, much branched perennial herb. Rootstock stout. Stems many 15-35 cm, dwarf, prostrate. Leaves ovate coarsely toothed, soft hairs on both sides. Flowers yellow tipped, with violet in short dense leafy terminal cluster. Corolla tube much longer than calyx.

Flowering & fruiting : June to September.

Uses (T) : The whole plant is used to reduce fever.

286. ***Sedum ewersii*** Ledeb.

See Sr. No. 129, Chapter VIII.

Uses (T) : The plant is considered to be useful against dysentry.

287. ***Selinum elatum*** (Edgew.) Hiroe

Family : Apiaceae

Distribution :Himachal Pradesh to Kashmir & Kumaun.

Habitat : On alpine meadows, Lahaul valley around 3,495 m.

Plant Characters : An erect, glabrous to sparsely hairy herb, 50-100 cm high. Lower leaves long-petioled, 2-3 pinate, sheathing, upto 20 cm long; upper leaves gradually shorter; pinne of the upper leaves sessile, deeply incised; segments lanceolate to oval. Rays 12-30. fruits dorsal and intermediate, ridges narrowly winged.

Flowering & fruiting : July to September.

Uses (T) : The roots are used as tonic.

288. ***Senecio chrysanthemoides*** DC.

Family : Asteraceae

Name : Heching rGu-drus (L)

Distribution : Pakistan to South West China, Kashmir & Sikkim.

Habitat : Open slopes, Zanskar and Spiti valleys around 3,850 m.

Plant Characters : An erect, perennial herb. Stem erect usually brached from the base 40-100 cm. Lower leaves with a large toothed terminal lobe and few irregularly toothed lateral lobes, the upper leaves with toothed lobes. Flower-heads yellow in terminal branched clusters with 10-12 strap-shaped ray-florets. Involucral bracts lanceolate to oblong, acute.

Flowering & fruiting : July to September.

Uses (T) : The entire plant is used against joint ruptured capillaries, colic pain due to inflammation of intestinal tract, dysentery.

(Li) : The alkaloid , seneciphylline, has been isolated in a yield of 0.01 percent from dry powdered roots.

289. ***Senecio tibeticus*** Hk. f.

Family : Asteraceae

Name : Niyamgar (L)

Distribution : Pakistan to Kashmir, Ladakh & China.

Habitat : Sandy slopes & higher passes, Changthang valley around 4,350 m.

Plant Characters : A dwarf, annual herb. Stem branched reaching upto 25 cm high, pubescent. Leaves hoary, radical leaves petioled, oblanceolate, lyrate-pinnatifid. Flower-heads several. Involucral bracts linear-lanceolate, cottony. Achenes glabrous with white pappus.

Flowering & fruiting : June to August.

Uses (T) : The leaves are used as diuretic.

290. *Silene vulgaris* (Moench.) Garcke; *Silene cucubalus* Wibel.

Family : Caryophyllaceae

Name : Bladder Campion (E)

Distribution : Pakistan to Central Nepal, Western Asia, Western Europe & North Africa.

Habitat : Moist slopes & meadows, Suru valley around 2,720 m.

Plant Characters : A hairless, perennial herb. Stem branched, 30-80 cm long. Leaves pale green to glaucous, ovate to lanceolate, acute. Flowers greenish white in lax cluster. Calyx globular encircled by the inflated calyx.

Flowering & fruiting : July to September.

Uses (T) : The whole plant is useful against skin diseases.

291. *Solanum nigrum* Linn.

See Sr. No. 132, Chapter VIII.

Uses (T) : The leaves and fruits are used against skin diseases locally by tribes. The herb has antiseptic and antidysentric properties. The plant is also diuretic and laxative.

(Li) : The plant is rich source of riboflavin, nicotinic acid and vitamin 'C'. In China the leaves are applied to wounds and sores. The fruits or berries are laxative, aphrodisiac, tonic, diuretic, useful in appetizing, heart ailments, nostril complaints, swelling, cough, asthma, vomiting, bronchitis, fever, urinary discharge, favour conception, throat trouble, facilitate delivery, dysentery and kidney problems vide.

292. *Solidago virgaurea* Linn.

Family : Asteraceae

Name : Golden rod (E)

Distribution : Europe, temperate Asia, America, Kashmir to Himachal Pradesh & Tibet.

Habitat : On moist forest lands, Lahaul valley around 3,460 m.

Plant Characters : An erect, glabrous or pubescent herbs upto 80 cm tall. Stems simple, rarely branched, often woody at the base. Leaves glabrous or pubescent on midrib and nerves; lower ones petioled, ovate or oblong, toothed; upper ones sessile, narrower, entire. Heads yellow in rusty brown racemes, combined into leafy panicles. Achenes ribbed, glabrous to hairy.

Flowering & fruiting : July to September.

Uses (T) : The leaves are used as antiseptic and diuretic.

293. *Sophora moorcroftiana* (Benth.) Benth. ex Baker.; *Caragana moorcroftiana* Benth.

See Sr. No. 45, Chapter XIII.

Uses (T) : Seeds are said to be used in medicine for blood purification.

294. *Statice macrorrhabdos* Boiss.; var. *thomsoni* Clarke.

Family : Plumbaginaceae

Distribution : Western Tibet to Baltistan & Ladakh.

Habitat : On drier rocks, Leh valley around 3,620 m.

Plant Characters : An erect, annual shrub. Leaves all radical, fleshy, rosulate, spathulate-orbicular obtuse or emarginated rarely minutely mucronate, sessile. Bracts ovate, mucronate, coraceous with scarious margin. Calyx tube funnel shaped, truncate scarious, 5-ribbed. Petals divided to the base of the lamina into two lanceolate-linear divisions.

Flowering & fruiting : July to August.

Uses (T) : The whole plant is used in stomach problems.

295. *Swertia cordata* (G. Don.) Clarke.

Family : Gentianaceae

Name : Heising or Tikta (L)

Distribution : From Kashmir to Tibet.

Habitat : Rare on grassy meadows, Lahaul valley around 3,650 m.

Plant Characters : A branched, glabrous herb upto 75 cm tall. Leaves veriable in size, broadly ovate, acute. Flowers in axillary and terminal leafy panicles. Calyx and corolla 5-lobed. Corolla yellowish-white; margins marked with pale purple streaks; one circular gland on each lobe.

Flowering & fruiting : July to August.

Uses (T) : The whole plant is considered to be effective in fever and headache. The plant is also used as tonic.

296. ***Swertia petiolata*** Royle ex D. Don.

Family : Gentianaceae

Name : Heising or Tikta (L)

Distribution : Afganistan to Western Nepal & South East Tibet.

Habitat : Open moist slopes, Suru, Zanskar and Lahaul valleys around 3,350m.

Plant Characters : An erect, perennial herb. Stem branched, reaching upto 50 cm. Basal leaves lanceolate, long stalked and clasping the stem at the base, uppermost leaves narrow-lanceolate, acute. Flowers pale yellow or nearly white in terminal clusters. Corolla tube narrow elliptic. Seeds not winged.

Flowering & fruiting : July to September.

Uses (T) : The whole plant is considered to be effective in fever and headache. The plant is also used as tonic.

(Li) : The plant contains xanthone derivatives, ursolic acid, B-sitosterol.

297. ***Swertia thomsonii*** Clarke. ex Hk. f. & T.

Family : Gentianaceae

Name : Tikta (L)

Distribution : Afganistan to Western Nepal & Tibet.

Habitat : On moist places, Suru valley around 3,150 m.

Plant Characters : An erect, rhizomatus, robust, annual herb. Leaves lower ones petioled, elliptic-oblong, upper ones sessile, linear. Flowers small, numerous, pale yellow to white. Capsule oblong.

Flowering & fruiting : July to September.

Uses (T) : The whole plant is considered to be effective in fever and headache. The whole plant is used as febrifuge and anti-inflammatory, used against fever, inflammation of liver and stomach, fever from contagious diseases and wounds.

298. ***Tanacetum dolichophyllum*** (Kitam.) Kitam.; *T. longifolium* Wall. ex DC.

See Sr. No. 61, Chapter VII.

Uses (T) : The dried leaves and flowers are quite effective against intestinal worms.

299. ***Tanacetum fruticulosum*** Ledeb.

See Sr. No. 62, Chapter VII.

Uses (T) : The whole plant is used against stomachache.

300. *Tanacetum gracile* Hk. f. & T.

See Sr. No. 63, Chapter VII.

Uses (T) : The dried leaves and flowers are used against intestinal worms.

301. *Tanacetum nanum* Clarke.

See Sr. No. 64, Chapter VII.

Uses (T) : The leaves are used medicinally against intestinal worms.

302. *Tanacetum tibeticum* Hk. f. & T. ex. Clarke.

See Sr. No. 65, Chapter VII.

Uses (T) : The plant is considered as febrifuge. Dried leaves and flowers are used against intestinal worms.

303. *Taraxacum officinale* Wigg.

Family : Asteraceae

Name : Dandelion (E), Han or Khur-mang (L)

Distribution : Distributed and cultivated in the temperate Himalayas, Meghalaya, S. India, Pakistan to Central Nepal & China.

Habitat : Wastelands & along water streams, Leh and Lahaul valleys around 3,480 m.

Plant Characters : A scapigerous, small herb. Roots with milky latex. Leaves stalkless, long with backward directed triangular tooth-edges lobes. Flowers yellow in a head on long stalk. Seed flat and ribbed with hair.

Flowering & fruiting : Mid June to August.

Uses (T) : Roots are used as diuretic, tonic and as blood purifier in human beings. The whole plant is used as febrifuge, analgesic, hot disorders of rLung, mKhris-pa, Bad-kan and blood, chronic fever, dyspepsia due to improper diet, gall bladder.

(Li) : Both fresh and dried rhizomes and roots constitute the drug, commonly known as "Taraxacin" which is administered as liquid extract. It is also used as a blood purifier, kidney, liver complaints, diuretic, stomachache, kidney and liver disorders, hepatic stimulant and tonic. The bioactive constituents are taraxacin, taraxacerin phytosterols, taraxasterol and homotaraxasterol.

304. *Taraxacum sikkimensis* Wigg.

Family : Asteraceae

Name : Khur-mang karpo (L)

Distribution : From Pakistan to Central Nepal, China & Western Tibet.

Habitat : On water stream & moist places, Changthang valley around 3,750m.

Plant Characters : A scapigerous, small, perennial herb. Stem erect, small, milky latex Leaves sessile, oblanceolate, entire pinnatified, acute, toothed. Heads white, solitary, on long stalks.

Flowering & fruiting : June to August.

Uses (T) : The whole plant is used in medicine as febrifuge, analgesic, digestive.

305. *Thalictrum alpinum* Linn.

Family : Ranunculaceae

Name : Alpine Rue (E)

Distribution : Kashmir to Western Tibet, Pakistan, Bhutan, North Alpine & Arctic regions.

Habitat : Moist slopes, Changthang and Spiti valleys around 3,780 m.

Plant Characters : A perennial delicate plant. Stem many from the base upto 10-15 cm long. Leaves ternate, further divided into 3-5 lobes. Flowers greenish-purple with long spreading stamens with purple thread like filaments, borne in spike like clusters. Achenes ribbed.

Flowering & fruiting : July to August.

Uses (T) : The roots and leaves are used against fever and body pain.

306. *Thalictrum foetidum* Linn.

Family : Ranunculaceae

Name : Lesser Meadow Rue (E), Haichinsah or sNgo-sprin (L)

Distribution : Pakistan to Central Nepal & temperate Eurasia.

Habitat : Forest areas & cultivated fields, Leh valley around 3,750 m.

Plant Characters : A much branched leafy perennial herb. Stem 30-100 cm long. Leaves glandular-hairy, 3-5 times divided into toothed leaflets. Flowers numerous, greenish-yellow, with numerous hanging yellow stamens. Corolla greenish. Achenes flattened, ribbed with an enlarged lanceolate beak.

Flowering & fruiting : July to September.

Uses (T) : The decoction of roots and leaves is used as diuretic and for reducing fever.

307. *Thermopsis barbata* Benth.

Family : Fabaceae

Distribution : Kashmir to Kumaun.

Habitat : On moist slopes, Lahaul valley around 3,580 m.

Plant Characters : A branched densely pilose, perennial herbs with woody rootstock. Leaves sessile, 3-foliate, leafltes lanceolate to elliptic, entire, subglabrous to densely hirsute; stipule foliaceous. Inflorescence racemose, 1-3 flowers at each node. Corolla deep purple. Pods elliptic, pilose when young, glabrescent, 1-6 seeded.

Flowering & fruiting : July to September.

Uses (T) : The plant is bitter and sweet and is quite useful against any kind of pain.

(Li) : The plant contains lupinine, cytosine etc.

308. *Thlaspi alpestre* Linn. ex. Hk. f. & T. Anders.

Family : Brassicaceae

Name : Bumbuk or Bre-ga (L)

Distribution : Temperate Europe, temperate Himalayas from Sikkim to Kashmir.

Habitat : Alpine slopes, Zanskar and Lahaul valleys around 4,340 m.

Plant Characters : A small perennial tufted herb. Stem 10-20 cm long, erect, glabrous, simple or branched. Radical leadves orbicular; cauline leaves obovate, auricled. Flowers large white borne in in racemes, racemes much elongated in fruits. Pods triangular, obcordate, winged. Seeds 4-8 in each lobe.

Flowering & fruiting : July to September.

Uses (T) : The fruits are useful in treating cough, kidney problems of hot nature and vaginal discharge. A decoction of leaves and flowers is used in fever and headache. The whole plant is used as febrifuge, anti-inflammatory, digestive.

(Li) : Used against inflammation of lungs and kidney, seminal and vaginal discharges, kidney problem, pus in lungs and appendicitis.

309. *Thlaspi arvense* Linn.

Family : Brassicaceae

Name : Breg-pa (L)

Distribution : Cultivated in sub-alpine Himalayas from Europe and Asia.

Habitat : On alpine slopes, Lahaul Valley, 3,670 m.

Plant Characters : An erect, branched, annual herb. Basal leaves oblanceolate, petiolate, toothed; upper leaves lanceolate-oblong, sessile, auricled at base, dentate to subentire. Flowers white, in 25-70 florets racemes. Silique deeply notched, seeds ellipsoid, dark brown.

Flowering & fruiting : June to September.

Uses (T) : The plant is used as substitute of *T. alpestre*.

(Li) : The seeds contain glucoside sinigrin, linoleic acid, oleic acid, palmitic acid.

310. *Thymus serpyllum* sensu Hk.f. (non Linn.) var. *T. linearis* Benth.

See Sr. No. 66, Chapter VII.

Uses (T) : An infusion of whole plant is quite effective in liver complaint, skin diseases, stomachache, eye diseases, cough and gastro-intestinal problems.

(Li) : It is antispasmodic, antiseptic, expectorant, carminative, anthelmintic and stimulant, catarrh and convulsive and whooping-coughs; they are also used as sedative in radiculalgia and epilepsy. The chief constituents are thymol a great demand in pharmaceutical industries.

311. *Tragopogon dubius* Scop.

Family : Asteraceae

Name : Goatsbread (E), Tharnoo (L)

Distribution : Temperate Himalayas from Pakistan to Nepal, China & Tibet.

Habitat : Along cultivated fields, Suru valley around 2,850 m.

Plant Characters : A hairless, perennial herb. stem branched, 20-40 cm high. Leaves broad in the middle, sheathing bases longer. Involucral bracts long. Flower-heads 2-4 cm broad, campanulate, yellow. Fruit with a long beak as long as the nuts.

Flowering & fruiting : June to August.

Uses (T) : The plant latex is considered to be useful in healing wounds.

312. *Tragopogon gracilis* D. Don.

Family : Asteraceae

Name : Goatsbread (E), Kuzgringli (L)

Distribution : Temperate Himalayas from Nepal to China & Tibet.

Habitat : Common on dry slopes, Lahaul valley around 3,840 m.

Plant Characters : An erect, glabrous, branched, biennial-perennial herbs. Stems with milky latex. Leaves alternate, narrowly linear with a broad sheathing base. Radical leaves sometimes as long as the stems. Heads yellow, solitary, terminal or occasionally scape with two heads.

Flowering & fruiting : June to September.

Uses (T) : The plant latex is used in medicines.

313. *Tragopogon pratense* Linn.

Family : Asteraceae

Name : Goatsbread (E), Tharno (L)

Distribution : Himalayas from Nepal to Tibet.

Habitat : Frequent on moist slopes, Lahaul valley around 3,750 m.

Plant Characters : An erect or decumbent, glabrous, branched, biennial-perennial herbs, upto 50 cm tall. Stems woth milky latex. Leaves alternate, gradually narrowed from a broad sheathing base. Heads yellow, solitary, terminal. Achenes muricate, long beaked.

Flowering & fruiting : June to September.

Uses (T) : The tender heads are often eaten raw having medicinal properties.

314. *Tribulus terrestris* Linn.

Family : Zygophyllaceae

Name : Gokhroo (H), Land caltrop or Puncture vine (E), gZe-ma, Rasha, Kokulla (L)

Distribution : Afganistan to South West China & Burma.

Habitat : Cultivated fields & wastelands, Leh, Nubra and Lahaul valleys 3,490 m.

Plant Characters : An annual, prostrate herb. Stem branched, slender, decumbent, hairy. Leaves pinnate compound, leaflets 4-12 ovate to obovate. Flowers solitary, yellow in long raceme. Calyx ovate to linear. Stamens 10. Fruits of 5 stiff achene fused together; each achene triangular.

Flowering & fruiting : July to August.

Uses (T) : Fruits are considered useful in liver complaints, leucoderma, impotency, cold, cough, urinary disorders and weakness, infusion useful in gout. Promotes urination. Cooling, aphrodisiac, and diuretic. The seeds are used to activates kidney function, diuretic, dries pus.

(Li) : The fruits are credited with diuretic and tonic properties. It forms a constituent of the well known Ayurvedic medicine Dashamoolarishtha and Amritha Prasa Ghritha, prescribed for several diseases. Fruits are useful in urinary disorders and sexual weakness, infusion useful in gout. The plant contains carboline alkaloids, steroidal sapogenins, saponins, flavonoids.

315. *Ulmus wallichiana* Palchon.

See Sr. No. 48, Chapter XIII.

Uses (T) : The bark is used as herbal shampoo for washing hair.

316. *Urtica hyperborea* Jacq. ex Wedd.

See Sr. No. 144, Chapter VIII.

Uses (T) : It is said to be used against rheumatism and joint pain.

317. *Valeriana jatamansi* Jones.

Family : Valerianaceae

Name : Indian valerian (E), Jatamansi / Mushkbala (E), Tagar (L)

Distribution : From Kashmir to Kumaun & Nepal.

Habitat : On moist, shaddy places, Lahaul valley around 3,780 m.

Plant Characters : A pubescent, perennial, herbs 15-50 cm tall with thick, aromatic rootstock. Radical leaves ovate or cordate, sinuate or crenulate, long-petiolate; cauline leaves few, much smaller, sessile, uppermost often 3-fid or sect. Flowers in lax corymbose cymes or dense corymbs. Achenes tomentose.

Flowering & fruiting : July to September.

Uses (T) : The roots are used against epilepsy, liver diseases, diuretic, antiseptic, etc.

(Li) : Rhizomes contain monoterpine derivative called "Valepotriates", useful as tranquillizers and sedatives.

318. *Verbascum thapsus* Linn.

Family : Scrophulariaceae

Name : Cow's lungwort, Common Mullein, Flannel Mullein (E), Dandashal (L)

Distribution : From Kashmir to Bhutan & Aka hills in Meghalaya, Afganistan to South West China and temperate Eurasia.

Habitat : On wastelands, Suru, Lahaul and Spiti valleys around 2,750 m.

Plant Characters : A very distinctive, tomentose perennial plant. Stem simple reaching upto 70- 150 cm high. Upper leaves woolly with base of blades continuing in a wing down stem. Flowers golden-yellow, sessile, in spikes. Corolla-tube short with spreading lobes.

Flowering & fruiting : July to September.

Uses (T) : The leaves and seeds are used against asthma and chest pain. Leaves and seeds are smoked to cure breathing problems by local tribes. The dried leaves are smoked to relieve irritations of the upper respiratory tract, asthma and spasmodic coughs.

(Li) : Mullein contains mucilage, flavonoids, triterpenoid saponins and tannins. In many parts of Europe, it is a popular remedy for frost-bites, bruises and piles. Inhaling of smoke of leaves is said to be helpful for asthmatic patients.

319. *Veronica koelzii* Pennell.

Family : Scrophulariaceae

Name : Sopia Bod-skyes ba-sha-ka (L)

Distribution : From Kashmir to South West China, Tibet & Central Asia.

Habitat : Drier rocky slopes, Suru valley around 2,790 m.

Plant Characters : An annual herb. Stem simple or branched, 5-25 cm. Leaves 2-4 cm long, pubescent. Flowers sessile on terminal heads, sub-capitate. Corolla small, purple or violet-purple. Capsule ovoid.

Flowering & fruiting : July to September.

Uses (T) : A decoction of leaves is locally used to cure cold and cough.

320. *Veronica macrostemon* Bunge ex Ladeb.

Family : Scrophulariaceae

Name : Shimiogar (L)

Distribution : From Pakistan to Bhutan, Nepal & temperate Eurasia.

Habitat : On alpine slopes, Changthang valley around 4,980 m.

Plant Characters : A hoary pubescent, perennials with slender rootstock. Stem branched. Leaves pubescent on both sides, sessile, ovate, obtuse. Flowers in dense terminal racemes. Calyx villous. Stamens exerted.

Flowering & fruiting : June to August.

Uses (T) : A decoction of leaves is used to cure skin diseases.

321. V*incetoxicum hirundinaria* Medik.

Family : Asclepiadaceae

Name : sNgo dug mo nyung (L)

Distribution : Pakistan to South East Tibet, Kashmir to Sikkim in Himalayas.

Habitat : Dry slopes & roadside, Leh and Suru valleys around 4,250 m.

Plant Characters : An erect, pubescent, perennial shrub. Stem 6-10 inches high. Leaves broadly cordate or ovate, acute, cymes sessile; flowers corymbose, corolla yellow. Capsule, long, seeds black, winged, woolly.

Flowering & fruiting : July to September.

Uses (T) : The fruits are useful to cure dysentery and development of hot disorder of the gall bladder.

322. *Viola biflora* Linn.

Family : Violaceae

Name : Banaksha (Banafasha) (L)

Distribution : From Kashmir to Himachal Pradesh & Nepal.

Habitat : Ocasional on rock crevices, Lahaul valley around 3,710 m.

Plant Characters : An erect or decumbent, glabrous or pubescent, annual herb upto 30 cm high. Leaves reniform to broadly ovate, cordate, crenate, glabrous to hirsute. Stipules ovate, denticulate, sometimes leafy. Flowers yellow, solitary-terminal. Sepals linear-obtuse. Lower petioles streaked with blakish-purple. Capsule oblong.

Flowering & fruiting : July to September.

Uses (T) : Used in indigenous systems of medicine as a health tonic.

323. *Viola kunawarensis* Royle.

Family : Violaceae

Name : Leomentok (L)

Distribution : From Kashmir to Central Nepal & Afganistan.

Habitat : Moist slopes, Changthang and Suru valleys around 5,330 m.

Plant Characters : A small, perennial herb. Leaves elliptic to lanceolate, long stalked. Flowers 5-10, violet with a short blunt spur borne on short leafless stalks. Petals darker veined. Capsule elliptic, acute.

Flowering & fruiting : July to August.

Uses (T) : Used as medicinal as health tonic.

324. *Viscum album* Linn.

Family : Loranthaceae

Name : Mistletoe (E), Kahbung (L)

Distribution : From India to China & temperate Eurasia.

Habitat : Rare parasite on *Juglans regia*, Lahaul valley around 3,640 m.

Plant Characters : A dioecious plant. Stems dichotomously or umbellately branched from the base. Leaves lanceolate or elliptic to obovate-lanceolate, obtuse or rounded, entire, abruptly narrowed towards the base, sessile or subsessile. Inflorescence terminal on the top of each internodes between the two normal leaves and on the bifurcations of the stems. Male inflorescence 3-flowered; female usually 3-5 flowered. Fruit sessile, globose.

Flowering & fruiting : July to September.

Uses (T) : The whole plant is reputed as a cardiotonic, antiseptic, diuretic and purgative.

(Li) : The plant is reported to contain fatty acid, nicotine, flavonoids, phytosterols.

325. *Waldheimia glabra* (Decne.) Regel.; *Allardia glabra* Decne.

See Sr. No. 68, Chapter VII.

Uses : The whole plant is used against septic wounds.

326. *Waldheimia stoliczkai* (Clarke.) Osten f.

See Sr. No. 69, Chapter VII.

Uses (T) : Flowers are used in medicine for incense.

327. *Waldheimia tomentosa* (Decne.). Regel.; *Allardia tomentosa* Decne.

See Sr. No. 70 Chapter VII.

Uses (T) : A paste is prepared from leaves and flowers which useful against septic wounds in Ladakh.

328. *Youngia glauca* Edgew.

Family : Asteraceae

Distribution : Himalayas from Himachal Pradesh to Kashmir.

Habitat : Common on sandy slopes, Lahaul valley around 3,620 m.

Plant Characters : An erect, corymbosely branched, glabrous herbs upto 15 cm tall. Stems many, dichotomously branched. Leaves few; radical pinnatifid; cauline leaves sessile, linear, entire. Heads .5-1.0 cm across, terminal. Flowers yellow. Achenes ribbed, pale, longer than the white pappus.

Flowering & fruiting : June to September.

Uses (T) : The whole herb is used as health tonic.

329. *Youngia tenuifolia* (Willd.) Babcock & Stebbins.

Family : Asteraceae

Distribution : Himalayas from Lahaul to Kashmir.

Habitat : Rare on rock crevices, Lahaul valley around 3,725 m.

Plant Characters : An erect, corymbosely branched, perennial herbs upto 20-50 cm tall. Radical leaves pinnately parted into lanceolate, linear or filiform, acute or acuminate segments, segments distant, spreading, unequal, entire or toothed. Flowers pale or bright-yellow. Achenes black or brown unequally ribbed. Pappus white or yellowish-white.

Flowering & fruiting : July to September.

Uses (T) : The whole herb is used as health tonic.

❏❏❏

7

Aromatic Plants

The aromatic plants and aroma chemicals play a vital role directly as well as indirectly in the day to day life of man since its appearance on the earth. The sense of smell is much more developed in animals in comparison to human in respect of detection of odours emitting from bodies of enemies, for location of their own kind and for distinguishing between desirable and undesirable food. The sense of smell plays a significant role in man in the identification of right type of food. Most of the fruits when ripe emit a pleasant aroma. Pleasant smell of flowers attracts insects and thus helps in cross-pollinations. Infact aroma and aromatic flora have strong linkage with faunal diversity. The aromatic plants have been used as flavour for food and drinks, for imparting fragrance to human, for beautification of surroundings of gardens and homes, around religious sites, as incense and mosquito repellant, in cosmetics and in medicines.

Similarly, aromatherapy based on aroma chemicals and essential oils, has been known to Mankind since ages. The French chemist Dr. Rene-Maurice Gattefosse is considered to be the father of modern aromatherapy. In true sense, essential oils are not really oils (fatty oils), they are in fact volatile chemicals. These oils play a vital role in the development of plant. Many essential oils attract beneficial insects and scare off harmful ones. In addition, when the oils evaporate, the vapour creates a barrier around the plant, protecting it from the heat and cold. Essential oils may be found in different parts of plant such as roots (Angelica), fruits (Caraway), woody portion (Sandal wood), bark (Cinnmom), resin (Sandal wood), skin of fruits (Citrus fruits). Unlike chemotherapy aromatherapy acts in accordance with holistic principles; it awakens and strengthen vital energies and self-healing capabilities of the patient. Now-a-days the essential oils have been extensively used as perfumes & perfume products, in pharmaceuticals and flavouring foods. Although there is sharp competition between naturally derived aromatic materials and synthetics, the two are supplementary and must go together.

India is one of few countries in the world where varieties of aromatic plants are growing naturally and may be cultivated due to varied climatic conditions ranging from cold desert Himalayan zones, temperate Himalayas, vast tract of tropical and temperate plains and hot desert areas. The roots of Indian perfumery lie in the hoary past of several thousand years. The uses of different types of natural perfumes, fragrant parts of plants such as herbs, twigs, barks, and flowers to Gods were well documented in our religious literatures. The uses of turmeric and sandalwood paste for imparting fragrance in humun body was also well documented. A favourite and fragrant tree, Kadamba (*Anthocephalus cadamba*) the flowers of which appear in gold balls, was closely associated with the life of Lord Sri Krishna. Perfumery has been one of the Indian arts and our country was famous centre for perfumes and allied products in the past. Method of distillation of essential oils was well developed about 5000 years back in our country. A terracotta distillation apparatus which is supposed to have been developed during Indus valley civilization of Harappa (3000 BC) is kept in Taxila Museum in Pakistan still. In earlier days Kannauj district of Utter Pradesh was famous centre for quality and varieties of perfumes, so as now also. The estimated world production of perfumes is approximately 250,000 tonnes and the worldwide sale of the same is around worth $ 5,000 million. The turnover of all perfumes in our country is around Rs. 300 crores annually while the market size of Indian cosmetics is estimated at Rs 4,500 crores. The day by day the demand of essential oils is increasing keeping its western trends, rapid globalization of fragrances and constant upgradation of technology.

Classification of aromatic plants based on their scented plant parts

There is great variation in fragrance depending upon plants and their parts and so that these plants may be classified as under :

i) **Plants with scented leaves :** *Artemisia dracunculus, Artemisia pallens, Eucalyptus species, Hyssopus officinalis, Juniperus macropoda, Lawsonia inermis, Lavendula officinalis* etc.

ii) **Plants with scented young twigs** : *Abies alba, Cupressus sp, Picea sp, Thuja occidentalis, Tsuga sp* tec.

iii) **Plants with scented flowers :** *Achillea millefolia, Jasminium sp, Primula sp, Rosa sp, Salvia sclarea, Tagets sp, Tanacetum sp* etc.

iv) **Plants with aromatic wood :** *Cupressus sp, Juniperus sp, Thujopsis sp, Veronica sp* etc.

v) **Plants with aromatic bark :** *Boswellia sp, Calophyllum sp, Cinnamomum sp, Juglans regia, Magnolia sp* etc.

vi) **Plants with aromatic gums :** *Abies grandis, Calophyllum sp, Ferula sp, Gardenia sp* etc.

vii) **Plants with underground parts :** *Acorus sp, Allium sp, Curcuma sp, Inula sp, Saussurea sp, Valeriana sp* etc.

viii) **Plants with scented fruits and berries :** *Aegle marmelos, Chaerophyllum sp, Humulus sp, Juniperus sp, Myristica sp, Piper sp* etc.

ix) **Plants with aromatic seeds :** *Apium sp, Carum sp, Bunium sp, Coriandrum sp, Cuminum sp, Myristica sp* etc.

x) **Fruits with scented peel** : Citrus, lemon etc.

xi) **All plant parts scented :** *Anthemis sp, Biebersteinia sp, Chenopodium sp, Mentha sp, Nepeta sp, Origanum sp,* etc.

Classification of Odour : According to French Perfumer of the late nineteenth century, Charles Pisse, the perfumes should correspond with a scale, like notes of music and that one false scent or note would ruin the harmony. The perfumes stimulate the olfactory nerves and influence the sense of smell. There is an octavate as well as octavates of notes in music. Like to keys of instrument, certain odours coincide or blend. As for example the almond, heliotrope and vanilla blend together. There are several chemicals and natural substances that are used in the formulation and production of perfumes. On the above basis the perfumes may be categorized in the followings :-

i) ***Green Notes :*** It is provided by the essential oils of galbanum oil together with aroma chemicals such as phenyal acetaldehyde etc. These are refreshing and evocative of nature.

ii) ***Citrus Notes :*** Oils of lemon, lavender, rosemary, sweet orange, clary sage etc belong to this group. This is as a refreshing note.

iii) ***Floral Notes :*** Provided by natural odour of fragrant flowers.

iv) ***Aldehydic Notes :*** The perfumes constitute fatty aldehydes such as aldehyde C-8, ALDEHYDEC-9 etc.

v) ***Lavender Notes :*** The main ingredients are oils of lavender, bergamot, clary sage, neroli, lemon etc.

vi) ***Chypre Notes :*** This group of perfumes is formulated by mixing sweet, warm spicy, woody and floral fragrances.

vii) ***Spicy Notes :*** This is provided by oils of nutmeg, ginger, pepper and thyme etc.

viii) Formulated by utilizing balsams and resins of kinds of Peru balsam along with musk and civet both natural and artificial. These are sedative in nature.

ix) ***Fougere Notes :*** It is classified as a deep noted sophisticated lavenders. Absolutes of jasmine and rose flowers provide the floral shading and the base note is provided by coumarin, vetiver etc.

x) ***Woody Notes :*** Obtained from the oils of cedar wood, sandol wood, vetiver, patchouli etc.

xi) ***Tobacco Notes :*** This group is formulated with the principal that no single note predominated.

xii) ***Leather Notes :*** This note of tanned leather is provided by the Birch Tar Oil which psychologically extend the masculity of men.

Uses of fragrances and essential oils

Now a days fragrances and flavours are the necessity of life. Elegant cosmetics with novelty and variety, useful, safe and convenient range of perfumery for personal, household, industrial and agricultural uses are many. There are superconcentrated perfumes, perfumes for daily hygiene, body and bath perfumes, cosmetics and deodorants, room freshners etc. Now flavours are also in great demand for flavouring food and food products.

Some major uses of fragrance are as under :

i) ***Attars and fragrant waters :*** Attars are used for perfuming tobacco, soaps and hair oils. Fragrant waters are employed for perfuming sacred places, public, social gathering and in various functions.

ii) ***Agricultural products :*** such as insecticides, insect and animal repellents, veterinary products and cattle products.

iii) ***Household disinfectants & air freshing products :*** Products associated with the killing of household germs, household insecticides, air refreshners and air deodorizers.

iv) ***Industrial products :*** Fragrance is also used in leather and rubber articles, textiles, plastic materials, kraft papers etc.

v) ***Polishing Products :*** In shoe creams, pastes, various types of furniture polishes etc.

vi) ***Cleansing products :*** Use of fragrance for cleansing products is very common such as shampoos, cleansing for various purposes, dry cleaners for garments etc.

viii) ***Washing products :*** Such as various types of soaps, detergent powders, dyes, dry bleaches etc.

viii) ***Scents, Toilet water and Colognes :*** Strong infusion of any fancid fragrance are very pleasant in these items such as various types of perfumes, varieties of bath scents etc.

ix) ***Hygiene products :*** Used in toilet soaps, shaving soaps, creams, medicinal and aromatic soaps, deodorants, etc.

x) ***Beauty products :*** Various types of different types of face powders, moiturising, medicated astringent, lotions etc.

Some major uses of flavour are as under :

i) ***As appetizers :*** Herbs play a major role in appetizers - mint, thymes, marjoram, basil etc.

ii) ***Flavours in butter and cheese :*** Such as tarragon, chives, rose petals, parsley, caraway, etc.

iii) ***Flavours in biscuits, breads, cakes :*** Rosemary, tansy, sage, fennel, mint etc. are the common.

iv) ***Flavouring conserves, desserts and drinks :*** Angelica, coriander, dandelion, rose petals, hyssop etc are the common.

v) ***Flavours in jam, jellies and puddings :*** Aromatic plants such as rose petals, mint, angelica, lavender etc play a major role.

vi) ***Flavouring pickles, sauces and vinegars :*** Fennel, sweet almond, parsley, tarragon, chive etc.

vii) ***Flavouring food items :*** Suitable flavoured are thyme, marjorum, basil, coriander, oregano, caraway etc.

Field studies

The extensive field exploration reveals that Ladakh and Lahaul-Spiti of cold desert Himalaya abound in aromatic plants and approximately 16 percent of total flora are aromatic in nature. Many of them have been used traditionally for various uses viz. *Artemisia sp, Nepeta sp* : in traditional medicine, *Juniperus sp, Waldheimia sp* : as incence, *Aquilegia sp, Primula sp, Rosa sp*: fragrant flowers, *Bunium sp, Carum sp, Mentha sp* : flavouring agents etc. by the local people of this region.

The aromatic plant species have been described in alphabetical order alongwith families, english (E), hindi (H) / local names (L), distribution, habitat, brief botanical notes and their uses in the enumeration. (Table 4).

Enumeration of Aromatic Flora

1. ***Achillea millefolium*** Linn. Boiss.

See Sr. No. 02, Chapter VI.

Uses : The flavouring agent is used in local dishes and in tea. The aerial parts contain 0.010% essential oil. The chief constituent are azulenes.

2. ***Aquilegia fragrans*** Benth.

See Sr. No. 4, Chapter X.

Uses : The fragrant flowers are used for incense by the locals.

3. *Artemisia annua* Linn.

See Sr. No. 30, Chapter VI.

Uses : It is cultivated for an essential aromatic oil, which contains a drug '*Artemisinin*' a remedy of malaria. All the plant parts are aromatic in nature and it contains about 0.5 % aromatic oil.

4. *Artemisia biennis* Willd.

See Sr. No. 31, Chapter VI.

Uses : The whole plant contains an essential oil of high aromatic flavour.

5. *Artemisia brevifolia* Wall. ex DC. *Artimisia maritima* Linn.

Family : Asteraceae

Name : Kirmala (H), Worm seed (E), Burtse /Phur-nag (L)

Distribution : Western Tibet, Kashmir to Kumaun and Garhwal.

Habitat : Dry places, stony slopes & along roadsides, Leh, Suru, Zanskar, Changhtang and Spiti valleys between 4,230-4,805 m.

Plant Characters : A much branched highly aromatic perennial plant (20-45 cm). Leaves ovate, stalked, much dissected into linear blunt segments. Flower-heads small, yellowish to reddish in axillary cluster of spike.

Flowering & fruiting : Mid July to August.

Uses : A strong aromatic oil can be extracted from this plant which is a source of drug '*Santonin*' used as antiworm. About 0.3 % aromatic oil can be extracted from it.

6. *Artemisia desertorum* Spreng.

See Sr. No. 33, Chapter VI.

Uses : The whole plant yields a low flavoured essential oil.

7. *Artemisia dracunculus* Linn. *A. glauca* Pallas ex Willd.

Family : Asteraceae

Name : Little Dragon Mugwort (E), Shersing or Burtse (L)

Distribution : West Asia, Kashmir to Kumaun & Himachal Pradesh.

Habitat : On wastelands, Nubra, Zanskar and Lahaul valleys around 3,340 m.

Plant Characters : An erect, strong aromatic, perennial herb. Stem grooved, smooth, upto 40-80 cm. Leaves entire, linear-oblong, acute. Flower-heads globular, short-stalked, forming long axillary raceme. Involucral bracts glaucous with papery margin.

Flowering & fruiting : Mid June to August.

Uses : It has got flavoured aromatic oil called '*Tarragon* oil' which is used for flavouring dishes. 0.3 % essential oil has been extracted from the semi dried leaves.

8. *Artemisia gmelinii* Web. ex Stechm.

See Sr. No. 35, Chapter VI.

Uses : Highly aromatic oil, about 0.01 % can be extracted from semi dried plant.

9. *Artemisia laciniata* Willd.

See Sr. No. 36, Chapter VI.

Uses : The plant yields a low quality aromatic oil.

10. *Artemisia nilagirica* (Clarke) Pamp.

See Sr. No. 37, Chapter VI.

Uses : The quality aromatic oil is used as antiseptic.

11. *Artemisia parviflora* Roxb.

See Sr. No. 38, Chapter VI.

Uses : It contains aromatic oil.

12. *Artemisia persica* Boiss.

See Sr. No. 39, Chapter VI.

Uses : The aromatic oil may be used in medicines.

13. *Artemisia salsoloides* Willd.

See Sr. No. 40, Chapter VI.

Uses : The highly aromatic oil is used as insecticide.

14. *Artemisia scoparia* Waldst. & Kit.

See Sr. No. 41, Chapter VI.

Uses : The aromatic oil can be used as antiseptic.

15. *Artemisia sieversiana* Ehrh ex Willd.

See Sr. No. 42, Chapter VI.

Uses : The aromatic oil is having strong smell. Approximately 0.01% can be extracted from fresh plants which is said to be used in medicines.

16. *Artemisia stracheyi* Hk. f. & T. ex Clarke.

See Sr. No. 43, Chapter VI.

Uses : The essential oil may be extracted.

17. *Artemisia tournefortiana* Reichb.

Family : Asteraceae

Name : Khamchu (L)

Distribution : Afganistan, Westerm Himalaya from Tibet to Kashmir.

Habitat : As weed in cultivated fields as well as alpine slopes, Leh, Nubra and Lahaul valleys around 3,670 m.

Plant Characters : A tall, aromatic, annual plant. Stem stout, thyrsoidly branched, 1-2 m high. Leaves 8-13 cm long, pinnatisect, linear, oblong to ovate. Flower-heads ovoid, erect clustered in dense axillary panicles.

Flowering & fruiting : July to September.

Uses : The strongly aromatic oil is used as medicine against worms. About 0.01% oil can be easily extracted from the fresh leaves.

18. *Biebersteinia odora* Stephan ex Fischer.

Family : Geraniaceae

Name : Khardung (L)

Distribution : From Pakistan to Kashmir, Himachal Pradesh and Uttranchal.

Habitat : Rocky places, Leh and Spiti valleys upto 4,890 m.

Plant Characters : A strong aromatic, perennial herb with densely tufted rootstock. Leaves usually fern like, linear; leaflets broadly ovate, deeply lobed. Flowers deep yellow in terminal cluster. Sepals elliptic, pubescent. Petals orbicular.

Flowering & fruiting : July to August.

Uses : The leaves contain very low quantity & about 0.1% of aromatic oil.

19. *Bunium persicum* (Boiss.) Fedtsch. *Carum bulbocastanum* Clarke.

See Sr. No. 36, Chapter VIII.

Uses : The seeds are highly aromatic and yield a fragrant oil.

20. *Carum carvi* Linn. var. ***gracile***

See Sr. No. 42, Chapter VIII.

Uses : Seeds are used as condiment for flavouring local dishes and has got medicinal properties. The major chemical constituent of the essential oil is carvone having great demand in pharmaceutical industries.

21 *Chenopodium botrys* Linn.

See Sr. No. 46, Chapter VIII.

Uses : The whole plant is aromatic in nature which contains 0.1% aromatic oil.

22. *Chrysanthemum pyrethroides* (Kar. & Kir.) B. Fedtsch. *Chrysanthemum richteria* Benth.

Family : Asteraceae

Name : Serpan (L)

Distribution : From Pakistan to Kashmir and Ladakh.

Habitat : On stony slopes, Leh and Spiti valleys between 4,560-5,230 m.

Plant Characters : An erect, woolly-hairy, grey, aromatic plant. Stem branched, many arising from the rootstock. Leaves very much dissected, woolly, pinnatifid, radical petioled, toothed lobes; cauline sessile. Flower-heads solitary, terminal. Ray florets white, disc florets yellow. Achenes ribbed. Pappus coraceous.

Flowering & fruiting : July to August.

Uses : The flowers are constituents of incense and contain 0.01% essential oil.

23. *Dracocephalum heterophyllum* Benth.

Family : Lamiaceae

Name : Zinkzer (L)

Distribution : Pakistan to South West Tibet.

Habitat : Stony slopes in large patches, Leh, Changthang and Spiti valleys between 4,650-4,930 m.

Plant Characters : A semi-prostrate, highly aromatic perennial herb. Stem ascending. Leaves leathery, stalked, oblong ovate with heart shaped base. Flowers white hoary in large dense leafy cylindrical spikes. Corolla tubes inflated above. Calyx glabrous or pubescent. Nutlets obovate, truncate, rusty brown.

Flowering & fruiting : July to September.

Uses : It contains good quality of essential oil which can be used in fragrance and cosmetics. It contains about 0.2% - 0.3% essential oil.

24. *Dracocephalum moldavicum* Linn.

Family : Lamiaceae

Name : Kirmir (L)

Distribution : Kashmir to Pakistan to South West Asia and Kashmir

Habitat : Higher passes & meadows, Changthang & Suru valleys between 4,620-4804 m.

Plant Characters : An erect, annual, herb. Stem branched. Leaves sessile or petioled 1-2 inches, narrowed into short slender, lanceolate, obtusely serrate or subpinnatifid. Spike long and leafy in distant whorls. Corolla tubular, blue. Calyx teeth broadly ovate mucronate.

Flowering & fruiting : July to September.

Uses : It contains little quantity of essential oil.

25. *Dracocephalum stamineum* Kar. & Kir.

Family : Lamiaceae

Name : Zinkzer (L)

Distribution : From Pakistan to South West Tibet and China.

Habitat : On stony slopes, Leh valley around 4,320 m.

Plant Characters : A low, semi prostrate, aromatic perennial herb. Stem branched from the base. Leaves small, petioled, orbicular-cordate or reniform deeply crenate, toothed margin. Flowers mauve in dense terminal heads. Calyx sessile often coloured.

Flowering & fruiting : June to August.

Uses : The small quantity fragrant oil is present.

26. *Elaeagnus angustifolia* Linn. *E. hortensis,* M Bieb.

See Sr. No. 7, Chapter XIII.

Uses : The flowers are highly fragrant.

27. *Elsholtzia densa* Benth.

Family : Lamiaceae

Name : Philongtso or Bye-rug nag-po (L)

Distribution : From Pakistan to China, Nepal and South Asia.

Habitat : Cultivated fields as a weed, Nubra, Suru and Lahaul valleys between 3,049-3,585 m.

Plant Characters : An erect, aromatic, annual herb. Stem quadranglar, branched from the base, slender to very stout. Leaves shortly petioled, oblong-lanceolate or elliptic, acute or obtuse, serrate. Flowering spikes oblong, villous, interrupted. Flower many, small. Corolla lilac. Nutlets long, ellipsoidal.

Flowering & fruiting : July to September.

Uses : The tender leaves are used to flavour the local dishes. The plant contains 0.3% essential oil, the major components are ketone, borneol, camphene etc.

28. *Elsholtzia eriostachya* (Benth.) Benth.

Family : Lamiaceae

Name : Betso or Bye-rug ser-po (L)

Distribution : Pakistan to South West China and India.

Habitat : Damp & drier places, Zanskar and Lahaul valleys around 3,603 m.

Plant Characters : An erect, mint-like strong scented, slender sparsely hairy, annual herb. Stem usually simple. Leaves shortly petioled, oblong or ovate-oblong, obtuse, denticulate. Spikes solitary, villous; flowers minute, yellow. Bracts short ovate. Fruits nutlets, ellipsoid.

Flowering & fruiting : July to September.

Uses : Leaves and flowering parts are used as condiment and flavouring agent.

29. *Ferula jaeschkeana* Vatke.

Family : Apiaceae

Name : Wild hing or Chuklam (L)

Distribution : Afganistan to Pakistan, Turkistan, Himachal Pradesh, Central Asia.

Habitat : Open slopes, Suru Lahaul valleys around 3,120 m.

Plant Characters : An erect, glabrous, tall, perennial herb. Leaves 2-3 pinnate, closely crenate or doubly crenate or almost serrate, broad oblong, toothed. Flowers yellow. Fruits broadly oblong, reddish, winged.

Flowering & fruiting : June to September.

Uses : The root produces mild fragrant essential oil.

30. *Humulus lupulus* Linn.

Family : Urticaceae

Name : Hops (E)

Distribution : Cultivated in USA, S. America, Europe, Australia & India. It is successfully growing in Himachal Pradesh and J&K.

Habitat : Cultivated as cash crop, Leh and Lahaul valleys around 3,650 m.

Plant Characters : A climbing shrub reaching a height of 2-3 m. Leaves stalked, hairy, deeply lobed, looks hand like with toothed margin, opposite, stipulate; serrate. Male flowers small, greenish yellow, arranged in panicle cymes; female flowers in cone-like inflorescence; fruit achene.

Flowering & fruiting : July to September.

Uses : The dried flowers yield a strong aromatic oil which is used to add flavour in beer. It contains approximately 0.1% thick essential oil.

31. *Hyssopus officinalis* Linn.

Family : Lamiaceae

Name : Hyssop (E), Juffa or Tengu (L)

Distribution : Kashmir to Uttranchal.

Habitat : Dry slopes and along road sides, Lahaul and Spiti valleys around 3,980 m.

Plant Characters : An erect or decumbent, aromatic woody herb, 30-60 cm high. Stems much branched from the base. Leaves sessile, oblong linear or lanceolate, obtuse, entire. Spikes axillary or terminal, whorls many-flowered, calyx funnel-shaped. Corolla blue to bluish-purple. Nutlets trigonous, dark-brown.

Flowering & fruiting : July to September.

Uses : The aromatic plant is used against headache in aromatherapy. The essential oil contains B-pinene, L-pinene, camphene etc having tremendous application in pharmaceutical industry.

32. *Inula racemosa* Hk. f.

See S. No. 170, Chapter VI.

Uses : The roots are used to extract aromatic oil.

33. *Juniperus communis* Linn. var. ***saxatilis*** Pallas; ***communis*** non. Linn. Hk. f.

See S. No. 14, Chapter XIII.

Uses : The fragrant leaves and bark are used to make dhoop in houses and religious places by budhists. The plant yields 0.25 % of essential oil of pleasent smell. The chemical constituent is juniperin.

34. *Juniperus indica* Boiss.

Family : Cupressaceae

Name : Black Juniper (E), Shukpa (L)

Distribution : Pakistan to temperate Zones, Kashmir, Himachal Pradesh, temp & sub-arctic Europe, Asia.

Habitat : Drier areas & cultivated lands Lahaul valley around 3,450 m.

Plant Characters : A meadium sized shrub. Leaves crowded, needle like, acute apex. Male cones ovoid, recalling leaf-buds. Fruits globose, blue black, 3-seeded.

Flowering & fruiting : April to September.

Uses : The fragrant leaves and bark are used as incense by budhists.

35. *Juniperus macropoda* Boiss.

See Sr. No. 15, Chapter XIII.

Uses : The fragrant leaves and bark are used to make dhoop in houses and religious places by budhists.

36. *Juniperus recurva* Buch,-Ham

See Sr. No. 16, Chapter XIII.

Uses : Same as in case of *J. macropoda.*

37. *Mentha longifolia* Linn. Hudson.

Family : Lamiaceae

Name : Jangali Pudina (H), Horse Mint (E), Pholoing (L)

Distribution : Temperate Himalayas & West Tibet. from Pakistan to Central Nepal, North & West Asia, Europe, Kashmir, Garhwal, Kumaun and Punjab.

Habitat : Along water channel & moist places, Leh, Nubra and Lahaul valleys between 3,160-3,520 m.

Plant Characters : An erect, very aromatic, perennial herb. Stem hairy, delicate, much branched 30-90 cm long. Leaves mostly basal, sessile, ovate toothed, round to kidney shaped, oblong-lanceolate, 2-7 cm, serrate. Flowers tiny, lilac, borne on terminal spikes. Nutlets reticulate. Fruits of 4, one seeded parts.

Flowering & fruiting : July to September.

Uses : The aromatic parts are commonly used as chatani and as flavouring agent to local dishes. It is having 0.3-0.5 % essential oil which contains piperitone oxide, diosphenol, diosphenolene.

38. *Nepeta clarkei* Hk. .f.

Family : Lamiaceae

Name : Khamyu (L)

Distribution : From Pakistan to Kashmir.

Habitat : Open drier clumps, Suru valley around 3,230 m.

Plant Characters : A low spreading, erect, aromatic, perennial herb. Stem branched reaching upto 60 cm high. Leaves ovate-lanceolate to lanceolate, serrate, shortly stalked. Flowers blue with paler lower lip and a slender corolla-tube borne in dense but widely spaced whorls and forming terminal spikes of 8-15 cm long. Nutlets smooth.

Flowering & fruiting : July to September.

Uses : The whole plant is aromatic in nature and a source of fragrant essential oil.

39. *Nepeta coerulescens* Maxim.; *N. thomsoni* Benth. ex Hk. f.

Family : Lamiaceae

Name : Neimlo or Khora (L)

Distribution : From Kashmir to Sikkim & Tibet.

Habitat : Open slopes, Zanskar valley around 3,830 m.

Plant Characters : A stout, aromatic, erect, perennial herb. Stem with ascending branches, finely pubescent reaching upto 20-40 cn high. Leaves oblong with heart shaped base, sessile, blunt or acute. Flowers purple or blue in dense ovoid or conical leafy spikes. Nutlets large, smooth, orbicular.

Flowering & fruiting : Mid June to August.

Uses : The aromatic plant yields an essential oil.

40. *Nepeta discolor* Royle ex Beth.

Family : Lamiaceae

Name : Shamalolo (L)

Distribution : From Afganistan to Kashmir & Uttranchal.

Habitat : Rocky slopes, Nubra and Suru valleys around 2,980 m.

Plant Characters : A low spreading, glabrous to hairy perennial herb. Stem branched, hairy. Leaves small, broadly ovate-cordate, densely white, woolly beneath. Flowers white-purple in cylindrical spikes. Calyx mauve with awl-shaped teeth. Corolla white or purplish-blue.

Flowering & fruiting : July to September.

Uses : The aromatic plant yields a highly fragrant essential oil.

41. *Nepeta erecta* Benth.

Family : Lamiaceae

Name : Eripantso (L)

Distribution : From Pakistan to Uttranchal.

Habitat : Sandy & open slopes, Changthang valley around 3,280 m.

Plant Characters : A slender, densely white-woolly, aromatic, perennial herb. Stem erect, softy haired. Leaves triangular to oblong, heart-shaped, apex blunt or acute. Flowers blue to blue purple in distant whorls. Calyx curved with triangular lobes.

Flowering & fruiting : June to August.

Uses : An essential oil can be extracted from it.

42. *Nepeta eriostachya* Benth

Family : Lamiaceae

Name : Zim-thik-le (L)

Distribution : From West Tibet to Kashmir & Garhwal.

Habitat : Dry slopes, Changthang valley around 4,680 m.

Plant Characters : A suberect or ascending, hairy, perennial herb. Leaves sessile or shortly petioled, ovate, obtuse at the apex, margins crenate or serrate. Spikes cylindrical, villous, dense flowered; flowers deep blue to purplish-blue. Bracts ovate to obovate-lanceolate, ciliate.

Flowering & fruiting : July to September.

Uses : A strongly aromatic plant yields an essential oil.

43. *Nepeta floccosa* Benth.

Family : Lamiaceae

Name : Shamalolo or Shangukaram (L)

Distribution : Pakistan to Uttranchal, Kashmir & Eastern Himalayas.

Habitat : Stony & dry slopes, Leh valley around 2,920 m.

Plant Characters : An erect, floccose with white wool, lemon scented perennial herb. Stem many, spreading from the base upto 60 cm. Leaves thick, long-petioled, with matted woolly white hairs, ovate-cordate. Flowers in slender panicles, pinkish mauve to blue, widely spaced whorls.

Flowering & fruiting : June - September.

Uses : The dried leaves and shoots are aromatic and used to add flavour in local dishes. It contains about 0.01% mild fragrant essential oil.

44. *Nepeta glutinosa* Benth.

Family : Lamiaceae

Name : Jatukpa (L)

Distribution : Pakistan to Kashmir.

Habitat : Stony slopes, Nubra, Suru & Spiti valleys between 2,980-3,250 m.

Plant Characters : An erect, aromatic, perennial herb. Stem many from the base forming large clumps upto 60 cm. Leaves ovate-cordate, serrated with shallow, triangular lobes, sessile. Flowers violet borne in long leafy spikes. Calyx glutinous. Corolla blue or white slender with exserted style.

Flowering & fruiting : July to September.

Uses : A mild fragrant plant having little essential oil.

45. *Nepeta laevigata* (D. Don.) Hand-Mazz.

Family : Lamiaceae

Name : Jatukpa (L)

Distribution : From Afganistan to South West China.

Habitat : Stony & open dry slopes, Suru valley around 2,970 m.

Plant Characters : A spreading aromatic perennial, glabrous, herb. Stem simple or branched 30-90 cm high. Leaves usually petioled, ovate, acute to triangular lanceolate, saw-toothed. Flowers in dense, terminal spikes, serrate, sessile, blue or blue-purple.

Flowering & fruiting : June to August.

Uses : The aromatic parts can be used for essential oil.

46. *Nepeta leucolaena* Benth. ex Hk. f.

Family : Lamiaceae

Distribution : From Himachal Pradesh to Central Nepal & China.

Habitat : Dry slopes, Zanskar valley around 3,250 m.

Plant Characters : An erect, hoary, slender, annual, aromatic herb. Stem branched from the base, 20-40 cm tall. Leaves small, shortly petioled ovate, cordate, obtuse, whorls distant. Flowers several in long interrupted terminal spikes. Calyx villous, corolla vary small. Nutlets smooth, linear-oblong.

Flowering & fruiting : July to September.

Uses : The plant bears strongly aromatic aerial parts.

47. *Nepeta longibracteata* Benth.

Family : Lamiaceae

Name : Prianku (L)

Distribution : From Pakistan to Himachal Pradesh.

Habitat : On rocky slopes, Changthang and Spiti valleys around 4,450 m.

Plant Characters : A dwarf spreading, prostrate, silky-hairy, aromatic annual herb. Stem simple, diffuse, woolly branched from the base. Leaves petioled, fan-shaped, rounded-toothed. Flowers violet-blue, capitate spikes with cymes sessile. Bracts silky. Corolla blue, funnel-shaped. Nutlets strongly aromatic.

Flowering & fruiting : July to August.

Uses : An aromatic oil of strong fragrance can be extacted from this species. Its semi dried whole herb yields about 0.1% essential oil.

48. *Nepeta podostachys* Benth.

Family : Lamiaceae

Name : Shangukaram (L)

Distribution : From Afganistan to Himachal Pradesh & Kashmir.

Habitat : Rocky slopes, Leh and Zanskar valleys around 4,650 m.

Plant Characters : A slender, perennial herb. Stem many spreading, 40 cm. Leaves small, few, oblong-elliptic, finely toothed, petioled. Flowers white in terminal cylindrical spikes.

Flowering & fruiting : July to September.

Uses : The aerial parts are having strong fragrance.

49. *Origanum vulgare* Linn.

Family : Lamiaceae

Name : Wild oregano / Marjoram (E)

Distribution : From Afganistan to Central Nepal, Tibet & North China.

Habitat : Pastures, Lahaul & Spiti valleys between 2,900-3,580 m.

Plant Characters : A strong aromatic, spreading, perennial herb 30-75 cm high. Stem branched from the base. Leaves broadly ovate. Flowers pinkish or whitish crowded in short dense terminal clusters. Nutlets smooth, brown.

Flowering & fruiting : July to September.

Uses : The aromatic plant parts are used as flavouring agent especially for making pizzas. The aerial parts contain essential oil ranging between 0.25-0.40 from dried parts. The major components are thymol, carvacrol etc.

50. *Oxytropis microphylla* (Pallas) DC.

Family : Fabaceae

Name : sTag-sha nagpo (L)

Distribution : From Kashmir to Pakistan, Nepal & Sikkim.

Habitat : On open slopes and higher areas, Leh and Spiti valleys around 4,425 m.

Plant Characters : An erect, aromatic, stemless, perennial herb. Leaves small, linear-lanceolate, short-petioled, leaflets crowed, linear-oblong, gray-green, gland-dotted. Flowers pinkish-purple to pale-yellow on long stalk. Pod half-bilocular, sessile linear, pubescent, recurved.

Flowering & fruiting : June to September.

Uses : The whole plant bears strong fragrance. The quantity of aromatic oil is about 0.01%.

51. *Perovskia abrotanoides* Kiril.

Family : Lamiaceae

Name : Iskiling /Tarubu (L) Wild lavender (E)

Distribution : From Afganistan to Kashmir & Central Asia.

Habitat : Stony slopes in large clumps, Leh and Nubra valleys, 3,350 m.

Plant Characters : A whitish or greyish hairy, strong aromatic shrub. Stem branched 40-100 cm tall. Leaves 2-5 cm, oblong-ovate, lobed, curved woth brown scales. Upper leaves nearly entire. Inflorescence branched. Flowers small blue-violet in distant whorls. Corolla 2-lipped with the tube longer than calyx.

Flowering & fruiting : Mid June to September.

Uses : It contains a lavender type oil. The whole plant parts yield 0.2- 0.5 % essential oil.

52. *Prangos pabularia* Lindl.

See Sr. No. 152, Chapter IX.

Uses : The roots and other plant parts are highly fragrant. The chemical constituents of the volatile oils are myrcene, camphene traces etc.

53. *Primula macrophylla* D. Don. (excl. syn. *P. moorcroftiana*). *P. stuartii* Wall. var. *perpurea* (Royle) Hk. f.; *P. nivalis* Pall. var. *macrophylla* (D. Don.) Pax.

See Sr. No. 69, Chapter X.

Uses : The flowers are aromatic in nature.

54. *Primula rosea* Royle.

See Sr. No. 71, Chapter X.

Uses : The flowers are fragrant in nature.

55. *Rhododendron anthopogon* D. Don. Spp. ***hypenanthum*** (Balf. F.) Cullen.

See Sr. No. 29, Chapter XIII.

Uses : The leaves are used in preparation of incense.

56. *Rosa macrophylla* Lindl.

See Sr. No. 77, Chapter X.

Uses : The flowers are aromatic and a source of rose water and essential oil.

57. *Rosa webbiana* Wall. ex Royle.

See Sr. No. 78, Chapter X.

Uses : The flowers are aromatic in nature and source of essential oil.

58. *Salvia sclarea* Linn.

Family : Lamiaceae

Name : French clary sage (E)

Distribution : From Europe to Asia.

Habitat : Introduced and cultivated at FRL, 3,550 m.

Plant Characters : A strong aromatic, long, branched annual or biennial herb. Stem branched from the base. Leaves very rough, ovate, elliptic-oblong. Flowers pinkish or whitish crowded in long terminal spikes. Nutlets brown, smooth.

Flowering & fruiting : July to September.

Uses : The aromatic plant has got heavy demand in perfume industry. The flowers contain essential oil known as oil of clary ranging between 0.1-0.2 %. The major components are linalyl acetate, linalool etc.

59 *Saussurea lappa* (Decne.) Sch.-Bip.; *S. costus* (Falc.) Lipsch.

See Sr. No. 276, Chapter VI.

Uses : The roots are highly aromatic and an essential oil of medicinal properties can be extracted from this plant. The oil is in great demand as a fixative in perfumery industry.

60. *Stachys tibetica* Vatke.

See Sr. No. 165, Chapter IX.

Uses : The whole plant has got mild aromatic fragrant and a little amount of essential oil can be extracted from it.

61. *Tanacetum dolichophyllum* (Kitam.) Kitam. (*T. longifolium*) Wall. ex DC.

Family : Asteraceae

Name : Khampa-serpo (L)

Distribution : From Pakistan to West Nepal & South West China.

Habitat : Dry rocky slopes, Changthang & Zanskar valleys around 4,680 m.

Plant Characters : An erect, aromatic perennial herb. Leaves 12-15 cm, grey-green, much-dissected, oblong. Flower-heads yellow borne in dense rounded terminal cluster. Achenes narrowly obovoid.

Flowering & fruiting : July to September.

Uses : The dried leaves and flowers are strongly aromatic in nature and a source of essential oil.

62. *Tanacetum fruticulosum* Ledeb.

Family : Asteraceae

Name : Khamchu (L)

Distribution : Pakistan to Kashmir, West Nepal, South West China & Ladakh.

Habitat : Rocky slopes, Leh and Zanskar valleys around 3,080 m.

Plant Characters : A densely woolly, semi-prostrate, perennial herb. Stems branched, 20-40 cm long. Leaves pinnatisect, segments linear-obtuse. Flower-heads yellow in terminal clusters. Involucral bracts oblong densely woolly.

Flowering & fruiting : July to September.

Uses : The whole plant is aromatic in nature.

63. *Tanacetum gracile* Hk. f. & T.

Family : Asteraceae

Name : Khamchu (L)

Distribution : From Pakistan to Kashmir.

Habitat : Stony slopes & wastelands, Leh and Nubra valleys between 3,280-4,380 m.

Plant Characters : A grey-hairy perennial with many stems arising from the base. Stem 40-60 cm, branched above, densely grey-haired. Leaves pinnately cut, segments linear-obtuse. Flower-heads small, yellow in branched terminal cluster. Fruits obovate with a cup like crown.

Flowering & fruiting : July to September.

Uses : The dried leaves and flowers are fragrant and a source of essential oil. The fresh aerial parts yield about 0.1% - 0.2% aromatic oil.

64. *Tanacetum nanum* Clarke.

Family : Asteraceae

Name : Khamchu (L)

Distribution : From Pakistan to Kashmir & Sikkim.

Habitat : Dry slopes & wastelands, Changthang valley around 4,350 m.

Plant Characters : A hoary, perennial plant. Rootstock branched. Stem many from the base. Radical leaves 1-2 cm, ovate, pinnatisect, segments minute, incurved. Flower-heads capitate. Involucral bracts woolly with scarious margins

Flowering & fruiting : July to August.

Uses : The leaves and aerial parts are aromatic in nature.

65. *Tanacetum tibeticum* Hk. f. & T. ex Clarke.

Family : Asteraceae

Name : Khamchu (L)

Distribution : From Pakistan to Himachal Pradesh & Kashmir.

Habitat : On sandy areas, Leh and Spiti valleys around 3,660 m.

Plant Characters : A white woolly semi-prostrate perennial herb. Stems many arising from a woody rootstock. Leaves 1 cm long, pinnatisect, segments, linear, obtuse. Flower-heads yellow in dense terminal corymbs. Achenes 5-ribbed.

Flowering & fruiting : July to mid September.

Uses : The dried leaves and flowers are aromatic in nature.

66. *Thymus serpyllum* Linn. Sp. Pl. 1753; *T. linearis* Benth.

Family : Lamiaceae

Name : Jangli Ajwain (H), Tumbrak (L), Wild thyme (E)

Distribution : From Afganistan to Central Nepal, Tibet & North China.

Habitat : Pastures, Lahaul & Suru valleys between 2,950-3,780 m.

Plant Characters : A strong aromatic, spreading, tufted perennial herb. Stem branched from the base. Leaves sessile, elliptic-oblong. Flowers pinkish or reddish crowded in short dense terminal clusters giving a look of the field. Nutlets smooth.

Flowering & fruiting : July to September.

Uses : The aromatic plant parts are used as flavouring agent. The aerial parts contain an essential oil known as oil of wild thyme ranging between 0.25-0.40 from dried parts. The major components are thymol, carvacrol, etc.

67. *Valeriana jatamansi* Jones.

See Sr. No. 317, Chapter VI.

Uses : The roots are highly fragrant and used in making incense.

68. *Waldheimia glabra* (Decne.) Regel.; *Allardia glabra* Decne.

Family : Asteraceae

Name : Sasomantso or Palu (L)

Distribution : From Pakistan to Tibet.

Habitat : Common on alpine passes & shady areas, Nubra and Lahaul valleys around 3,950 m.

Plant Characters : A loosely tufted, mat forming, aromatic perennial herb. Stem usually prostrate, branched. Leaves deeply lobed, hairless. Flower-heads sessile with pink to purple ray florets. Involucral bracts blunt with purple papery margins.

Flowering & fruiting : July to August.

Uses : The whole plant is fragrant and considered as holy which is used for making dhoop in houses and religious places.

69. *Waldheimia stoliczkai* Clarke.

Family : Asteraceae

name : Palu or Solo nakpo (L)

Distribution : Pakistan to Tibet.

Habitat : Alpine passes, Changthang and Spiti valleys around 2,950 m.

Plant Characters : An ascending, aromatic, perennial herb. Leaves woolly, oblong, 2-pinnatisect, segments narrow. Heads solitary, terminal, with many pink flowers. Achenes ribbed, yellowish or white pappus-hairs, grooved.

Flowering & fruiting : July to September.

Uses : Flowers are used for incense.

70. *Waldheimia tomentosa* (Decne.). Regel.; *Allardia tomentosa* Decne.

Family : Asteraceae

Name : Makungla or Solo-karpo (L)

Distribution : Pakistan to Western Nepal & Central Asia.

Habitat : Stony drier places, Leh and Lahaul valleys around 4,850 m.

Plant Characters : A loosely tufted, aromatic, perennial herb. Stems ascending, 10-25 cm. Leaves 3-5 long, usually covered with dense soft woolly hairs. Flower-heads whitish pink, solitary, terminal. Involucral bracts broadly lanceolate, densely woolly haired. Achenes grooved.

Flowering & fruiting : July - September.

Uses : The aromatic plant parts are used for making dhoop in Gompas.

□□□

8

Edible Wild Plants

Human interaction with plants can not be particularized to a certain period of time in the history. Peking man, the extinct species, closest to the modern human has been found to use walnuts, hazelnuts, pine nuts, rose hips, roasted Chinese hackberry seeds which is estimated to be more than 3,00,000 years old by the archeologists. The study of fossil teeth of hominids also shows that ancient human beings were omnivorous which establishes man's association with plants since time immemorials.

The process of trial and error method with plant species led to selection of such plants which are appealing and rich in nutritive value by the ancient people. These selected plants were gradually added to the repertoire of the edible plants. Different parts of edible wild plants such as roots, tubers, rhizomes, stems, fruits, inflorescence, flowers, seeds, thalamus, fruiting body, and fronds consumed either raw, roasted, fried, cooked, boiled or in the form of oil, spice, seasoning materials, pickles, jams etc., were selected over the years. Since certain plant species are localized to particular areas, the inventory of edible plants made at different parts of world were dissimilar in nature. Nevertheless, it has been observed that about 3,000 plant species have been used as food throughout the history and about 200 species have been domesticated as food crops so far.

Food has different physical components and vegetables and fruits are considered store house of all protective factors of health. The vitamins, minerals and essential amino acids etc, considered as building and protective substances, are supplied by regular consumption of vegetables and fruits. Vegetables and fruits are important sources of both digestible and indigestible carbohydrates. The digestible carbohydrates are present largely in the form of sugars and starches while indigestible cellulose provides roughage which is important to normal digestion. Similarly these are also an excellent sources of minerals and certain vitamins especially Vit 'A' and 'C'. The precursors of vitamin 'A', including beta-carotene and certain other carotenoids, are to be found particularly in yellow-ornage fruits and vegetables and in the green leafy vegetables.

Ladakh and Lahaul-Spiti of trans Himalayan cold desert are not an exception to it. The prolonged and freezing winter prevailing in cold arid zones forced local inhabitants to search for edible wild plants. At different elevations wild plants start growing as soon as snow melts and temperature is favourable for plant growth. Quite a good number of plants start growing during March - May. Over the years of experience perhaps by hit and trial methods inhabitants of the area identified certain wild flora which are good to be used as vegetables. Generally tender shoots, leaves, stems and underground parts of these plants are being used as fresh or in dehydrated form. Besides some of wild edible plants are subjected to solar dehydration as emergency vegetables for prolonged winter. Because of intense sunlight the leafy/ tender parts are dehydrated under shade having adequate ventilation. Since humidity is low during August- September solar dehydration is quick and satisfactory. Products thus obtained are consumed during winter without any spoilage because of prevailing very low temperature. Presently uses of such plants are restricted to certain communities / areas in spite of their potential in local and national markets as well as nutritive values. Days to days uses of such plants are decreasing and are on the verge on loosing. Infact, there is a great scope of enhancing the acceptability of wild edibles as income generating resource for the trans Himalayan communities and initiating the potential plant resources for human consumption. These attributes have necessiated the identification and documentation of wild edibles to develop effective strategies for their wider consumption.

Keeping in view the rich ethno-edible wealth of cold arid zones of the Himalayas this laboratory has carried out extensive survey of most of the parts covering far-flung and higher passes since 1993 and collected valuable information on edible wild plants with the help of tribals, shepherds, senior citizens and available sources. Plant species of edible wild flora have been described in alphabetical order alongwith families, English (E), hindi (H) / local names (L), distribution, habitat, brief botanical notes and edible parts & uses in the enumeration. (Table- 5). In addition, the nutritional profile of 20 edible wild plants have been appended in table- 5a covering crude protein, fiber, fat, starch and minerals which reveal their nutraceutical potential.

Enumeration of Edible Wild Flora

1. ***Adonis chrysocyathus*** Hk. f. & T.

Family : Ranunculaceae

Name : Semshi (L)

Distribution : Western Himalayas from Kashmir to Himachal Pradesh & Tibet.

Habitat : On higher passes, wet areas; Changthang valley between 4,870-5,125 m.

Plant Characters : A perennial herb. Rootstock scaly, horizontal. Stem branched, erect, leafy, 10-20 cm tall. Radical leaves petioled, pinnate with narrow segments. Flowers golden yellow, solitary and terminal. Corolla 16-24 obovate. Achenes, glabrous, flattened, keeled.

Flowering & fruiting : July to August.

Uses : The leaves are edible and cooked as leafy vegetable.

2. *Allium carolinianum* DC.

Family : Amaryllidaceae

Name : Koshuk or Skotche (L)

Distribution : From Kashmir to Kumaun & Sikkim & Afganistan.

Habitat : Alpine slopes, Suru and Spiti valleys around 3,825 m.

Plant Characters : An erect, perennial herb. Bulb hardly developed. Stem thickened. Leaves basal, flat, narrow to a acute point. Flowers pinkish, pediceled. Capsule turbinate.

Flowering & fruiting : June to August.

Uses : Fresh and tender leaves are edible; dried leaves used as flavouring agent.

3. *Allium humile* Kunth.

Family : Amaryllidaceae

Name : Jangali Piyaz (H), Skotche (L)

Distribution : From Pakistan to Western Nepal & in Temp zones.

Habitat : Open alpine slopes, Leh, Suru and Lahaul valleys around 4,005 m.

Plant Characters : A small bulbous herb. Bulb sub cylindrical. Leaves many basal, flat, obtuse. Scapes acutely angled with rounded tip. Heads many flowered, white.

Flowering & fruiting : July to mid September.

Uses : Used as spice when dried, also used as flavouring agent and dehydrated for winter use.

4. *Allium jacquemontii* Kunth.

Family : Amaryllidaceae

Name : Jangali Piyaz (H)

Distribution : From Pakistan to Himachal Pradesh & Kashmir.

Habitat : Rare on boulders, Lahaul valley around 3,760 m.

Plant Characters : A bulbous herb. Bulb ovoid; outer coats coriaceous, striate, black-brown; inner membranous. Leaves 2-3, linear fistular. Flowering scapes slender, 15-30 cm tall. Flowers rose-coloured in dense flowered umbels. Filaments about 2/3 the length of the tepals; anthers dark voilet coloured.

Flowering & fruiting : June to August.

Uses : Used as flavouring agent.

5. ***Allium loratum*** Baker.

Family : Amaryllidaceae

Name : Jangli Piyaz (H), Skotche (L)

Distribution : From Pakistan to Western Nepal & India.

Habitat : Rocky slopes, Nubra and Zanskar valleys around 4,205 m.

Plant Characters : A small bulbous perennial herb. Bulbs ovoid, outer scales membranous, gray. Leaves 3-6, linear or spear shaped upto 15 cm long. Flowers white, in lax rounded umbel.

Flowering & fruiting : July to September.

Uses : Leaves have got garlic smell, which are used as flavoring agent.

6. ***Allium oreoprasum*** Schrenk.

Family : Amaryllidaceae

Name : Skotche (L)

Distribution : Pakistan to Western Nepal & Asia.

Habitat : On moist slopes, Changthang valley around 4,645 m.

Plant Characters : A bulbous herb. Bulbs tufted, cylindric. Outer scale rigid. Leaves upto 20 cm long, minutely serrulate. Scape terete. Heads 3-5 diam, hemispheric, 6-12 flowered, white rosy flowers.

Flowering & fruiting : July to September.

Uses : Same as in case of *A. humile*.

7. ***Allium przewalskianum*** Regel.

Family : Amaryllidaceae

Name : Wild garlic (E), Skotche (L)

Distribution : From Nepal to Tibet & India.

Habitat : Alpine pastures & wastelands, Nubra valley around 3,400 m.

Plant Characters : A small perennial bulbous herb. Bulbs elongated, clustered, scales finely reticulate, red or brown. Leaves several, hollow, succulent

fragrant, reaching upto 60-100 cm long. Flowers several in heads, pink. Seed small, black, globose.

Flowering & fruiting : July to September.

Uses : Leaves are edible; used as condiment and vegetable. The leaves are made into paste, which is consumed either fresh or dried and made into round balls for winter use. These balls are usually available in local market.

8. *Allium ramosum* Linn.

Family : Amaryllidaceae

Name : Skotche (L)

Distribution : From Pakistan to Nepal, China, Tibet & India.

Habitat : Alpine slopes & pastures, Suru valley around 3,200 m.

Plant Characters : A small bulbous herb. Bulbs elongated, clustered, scales finely reticulate, white or brown or blackish. Leaves several, reaching upto 40-50 cm long. Scape slender. Flowers several in heads, white. Capsule subglobose.

Flowering & fruiting : July to September.

Uses : The leaves are used as condiment or flavouring agent.

9. *Allium schoenoprasum* Linn.

Family : Amaryllidaceae

Name : Chive (E), Skotche (L)

Distribution : Western Himalayas to North America, Kashmir & Kumaun.

Habitat : Alpine slopes, Lahaul valley around 3,450 m.

Plant Characters : A bulbous, perennial herb. Bulbs clustered, narrow, scales membranous. Leaves 4-10 inches long, terete, sheaths elongate, grooved above smooth or scaberulous. Flowers purple. Capsule small, globose, 2-seeded.

Flowering & fruiting : July - September.

Uses : Leaves are edible and are used for condiment.

10. *Allium semonovii* Regel.

Family : Amaryllidaceae

Name : Wild garlic (E), Skotche (L)

Distribution : Pakistan to Himachal Pradesh & many other parts of Northern Himalayas.

Habitat : Moist places, Suru valley around 3,590 m.

Plant Characters : A bulbous, perennial herb. Bulbs cylindrical, covered with old fibrous sheaths. Leaves usually 2, stout, hollow, stout, acute with a long sheath. Flowers pale yellow, in a dense globular umbel with a broad persistent papery spathe.

Flowering & fruiting : July to September.

Uses : The young leaves are used for flavouring agent and pasted, dried leaves used in winter for the same purpose.

11. *Allium stracheyi* Baker.

Family : Amaryllidaceae

Name : Wild garlic (E)

Distribution : From Pakistan to Nepal, Kashmir & Kumaun.

Habitat : Rare on slopes, Lahaul valley around 3,875 m.

Plant Characters : A bulbous small herb. Bulbs clustered, ovoid, striate. Leaves 3-5 linear, obtuse, flattened. Flowering scapes slender, compressed upto 30 cm tall. Flowers yellow or pink in dense flowered terminal umbels; pedicels about as long as the tepals. Filaments much exerted.

Flowering & fruiting : July to August.

Uses : The leaves are employed to the local dishes as flavouring agent.

12. *Allium thomsonii* Baker.

Family : Amaryllidaceae

Name : Skotche (L)

Distribution : From Pakistan to Nepal, China, Tibet, Kashmir & Kumaun.

Habitat : Stony slopes, Nubra valley around 3,075 m.

Plant Characters : A bulbous small herb. Bulbs tufted more or less ovoid. Scales generally brown. Leaves few, 3-6, upto 15 mm long, linear, fleshy. Flowers purple in dense umbel, campanulate.

Flowering & fruiting : July to September.

Uses : The leaves are used as condiment for flavouring agent.

13. *Allium victorialis* Linn.

Family : Amaryllidaceae

Name : Gokpa (L)

Distribution : From Pakistan to China, Kashmir & Kumaun.

Habitat :Rare on moist slopes, Lahaul valley around 3,675 m.

Plant Characters : A bulbous herb. Bulbs cylindrical, 4-6 cm long, coats reticulately fibrous. Leaves 3-6 petiolate, broadly lanceolate to oval, 10-20 cm long, 4-6 cm broad, shorter than the scapes. Flowers white to yellowish white. Filaments longer than the tepals.

Flowering & fruiting : June - August.

Uses : The leaves and scapes are used as vegetable.

14. *Amaranthus spinosus* Linn.

Family : Amaranthaceae

Name : Chaulai (H), Amaranthus (E), Neu (L)

Distribution : Throughout India.

Habitat : Common weed of cultivated field, Leh, Nubra and Lahaul valleys between 2,980-3,120 m.

Plant Character : A thorny herb upto 30-60 cm high. Stem erect or suberect, branched or unbranched, ribbed. Leaves usually egg ot lance-shaped, stalked. Flowers unisexual, greenish white often with purple tinge. Anthers yellow. Grains minute with a single shiny seed.

Flowering & fruiting : June to September.

Uses : The tender shoots and leaves are used as a delicious vegetable and are eaten as potherb. Young leaves may be dried, stored for winter use.

15. *Anaphalis contorta* (D. Don.) Hk. f.

Family : Asteraceae

Name : Telgang (L)

Distribution : South West China to North Western Himalayas.

Habitat : On alpine slopes and grazing grounds, Leh, Suru and Spiti valleys around 4,190-4,650 m.

Plant Characters : A small tufted perennial plant. Stem usually branched from the base. Leaves crowded, obovate, sessile, hairy. Flower-heads white, compact in terminal clusters with shining involucral bracts.

Flowering & fruiting : Mid July to September.

Uses : The tender shoots are harvested from natural habitat and properly washed, boiled and used as vegetable.

16. *Anaphalis triplinervis* (Sims) Clarke var. ***intermedia*** (DC) Airy Shaw *A. nepalensis* (Spreng) Hand. Mazz.

See Sr. No. 13, Chapter VI.

Uses : The young buds are slightly acidic in taste and eaten as raw by children.

17. ***Arabidopsis himalaica*** (Edgew.) Schulz *Sisymbrium himalaicum* (Edgew.) Hk.f.

Family : Brassicaceae

Name : Byi'u laphug (L)

Distribution : Throughout Himalayas.

Habitat : Dry places, Leh and Changthang valleys between 3,890-4,450 m.

Plant Characters : An erect hairy annual herb. Stem 20-50 cm, hairy, branched from the base. Basal leaves often in rosette, oblong spathulate, toothed. Upper leaves smaller, elliptic. Flowers lilac in terminal clusters, small. Fruits linear, hairless.

Flowering & fruiting : July to September.

Uses : The locals used the young leaves for green leafy vegetable.

18. ***Arabidopsis mollissima*** (C.A. Mey.) N. Busch. *Sisymbrium mollissimum* C.A. Mey.

Family : Brassicaceae

Name : Sikiang (L)

Distribution : From Kashmir to Himachal Pradesh.

Habitat : On rocky slopes, Zanskar valley around 3,680 m.

Plant Characters : An erect, perennial herb. Stem solitary or 2-3 from the base, reaching upto 20-30 cm long. Leaves shortly petioled, obovate-lanceolate, cauline leaves oblong and stem clasping. Flowers purple or white borne in racemes. Pod straight.

Flowering & fruiting : Mid June to mid September.

Uses : Tender shoots and leaves are harvested, properly washed and boiled then cooked as vegetable.

19. ***Arabidopsis thaliana*** (Linn.) Heynh. *Sisymbrium thalianum* Linn.

Family : Brassicaceae

Name : Byi'u la phug (L)

Distribution : From Kashmir to Sikkim.

Habitat : Along cultivated field, Zanskar valley between 3,190-3,710 m.

Plant Characters : An erect annual herb. Stem branched, leafy, glabrous. Radical leaves obovate, petioled, 3-4 cm long; cauline leaves sessile, lanceolate. Flowers white in terminal and axillary corymbs, peduncled. Pod compressed, glabrous, 1-nerved.

Flowering & fruiting : July to mid September.

Uses : The young herbs are commonly collected, washed and boiled properly and used as vegetable in early summers.

20. *Arabidopsis wallichii* (Hk. f. & T.) N. Busch. *Sisymbrium wallichii* Hk. f. & T.

Family : Brassicaceae

Name : Byi'u la phug (L)

Distribution : Himalayas from Kashmir to Kumaun.

Habitat : Moist places, Zanskar valley around 3,670 m.

Plant Characters : An erect perennial herb. Stem simple, slender reaching upto 40 cm long. Radical leaves crowded, spreading; cauline leaves linear-lanceolate. Flowers white, changing to pale pink. Pod spreading, curved, glabrous.

Flowering & fruiting : Mid June to mid September.

Uses : The tender shoots and leaves are edible and cooked as pot-herb.

21. *Arabis alpina* Hk.f. & T. Anders.

Family : Brassicaceae

Distribution : Himalayas from Kashmir to Nepal, Alpine & Arctic Europe and North America.

Habitat : Stony slopes, Leh and Changthang valleys between 3,460-3,595 m.

Plant Characters : An annual erect herb. Stem simple, hairy. Leaves oblong, lanceolate, sparsely hairy, entire margin. Flowers 1-2 cm, white in terminal cluster. Sepals glabrous. Petals spathulate. Fruits narrow, linear, hairy stalked.

Flowering & fruiting : Mid June to August.

Uses : The tender shoots and leaves are edible and cooked as leafy vegetable.

22. *Arabis tenuirostris* Schultz.

Family : Brassicaceae

Name : Chiki (L)

Distribution : Himalayas from Kumaun to Sikkim.

Habitat : Moist places, Suru and Lahaul valleys between 2,980-3,120 m.

Plant Characters : A herbaceous annual plant. Stem slender, branched, radical leaves spathulate. Cauline leaves sub-amplexicaule, toothed. Flowers white. Pod compressed, pedicelled.

Flowering & fruiting : July to mid September.

Uses : Leaves have got mustard smell and boiled properly then consumed as vegetable.

23. *Arabis tibetica* Hk. f. & Th.

Family : Brassicaceae

Name : Tiktso (L)

Distribution : Western Tibet, Himalayas from Kashmir to Sikkim.

Habitat : Alpine pastures, Zanskar valley around 4,230 m.

Plant Characters : A small tufted annual herb. Stem branched from the base. Radical leaves rosulate, petioled, obovate or spathulate; cauline leaves sessile, lanceolate, entire or toothed. Flowers numerous, white or pink, shortly stalked. Petals twice long as the sepals. Pod narrow-linear, glabrous.

Flowering & fruiting : July - mid September.

Uses : It is commonly used as pot herb.

24. *Arenaria holosteoides* (C.A. Mey.) Edgew ex Hk. f. & T.

Family : Caryophyllaceae

Name : Chiki (L)

*Distribution :*Afganistan, Himalayas from Himachal Pradesh to Kashmir.

Habitat : Cultivated fileds, Leh, Nubra & Spiti valleys between 3,105-3,590m.

Plant Characters : A slender herb (50-80 cm) with forked branches. Leaves long, stalkless, ribbon like with a broad base. Flowers small, white. Capsule roundish splitting open into 2-6 parts.

Flowering & fruiting : July to September.

Uses : The tender leaves are cooked as vegetable.

25. *Arenaria serpyllifolia* Linn.

Family : Caryophyllaceae

Name : Rtswa-a-krong (L)

Distribution : Throughout the Himalayas, temperate Asia & Europe.

Habitat : Along moist places, Changthang and Spiti valley around 4,230 m.

Plant Characters : An annual herb. Stem branched, pubscent, decumbent. Leaves subsessile, ovate, acuminate, 2-3 nerved. Flowers several in cymes. Bracts leafy. Calyx 3-5 ribbed, pubescent. Capsule 6-valved, ovoid.

Flowering & fruiting : Mid June to July.

Uses : The tender leaves are commonly cooked as vegetables.

26. *Arnebia benthamii* (Wall. ex G. Don.) Johnston

See Sr. No. 27, Chapter VI.

Uses : The roots are source of edible dye which is used for colouring food items such as butter, cheese and ghee.

27. ***Arnebia euchroma*** (Royle) John. I. M Johnston.

See Sr. No. 28, Chapter VI.

Uses : Roots yield edible, fat soluble red dye, which is used by the local people for colouring dishes and sweets etc.

28. ***Arnebia guttata*** Bunge. *A. tibetana* Kurtz.

See Sr. No. 29, Chapter VI.

Uses : Same as for *Arnebia euchroma.*

29. ***Artemisia dracunculus*** Linn. Sp. Pl. 849. 1753; *A. glauca* Pallas ex Willd.

See Sr. No. 7, Chapter VII.

Uses : The leaves are used for flavouring local dishes.

30. ***Atriplex crassifolia*** C. A. Mey.

Family : Chenopodiaceae

Distribution : From Kashmir to Punjab, Himachal Pradesh & Pakistan.

Habitat : Open dry slopes along roadside and wastelands, Changthang and Spiti valleys around 3,910 m.

Plant Characters : An erect or diffuse, annual herb. Stem erect or branched, branches white. Leaves small, oblong or ovate-oblong, petioled with obtuse sides. Male flowers in clusters borne in leaflets interrupted spike. Fruiting bracts rhombic ovate with a smooth disc at the base.

Flowering & fruiting : June to September.

Uses : The young leaves and shoots are washed and boiled properly and cooked as vegetable.

31. ***Atriplex hortensis*** Linn.

Family : Chenopodiaceae

Name : Giangsar (L)

Distribution : From Assam to Meghalaya, Punjab and Maharashtra.

Habitat : Rocky places, Nubra valley around 3,025 m.

Plant Characters : An erect, annual herb. Stem 30-40 cm long, branched. Leaves triangular-cordate upper ovate-lanceolate. Flowers monoecious in axillary and terminal panicles. Female without perianth. Fruiting bracts free, obtuse.

Flowering & fruiting : July to September.

Uses : The tender leaves are eaten raw or cooked as green vegetable.

32. *Barbarea vulgaris* R. Br.

Family : Brassicaceae

Name : Yellow rocket (E)

Distribution : From Kashmir to Kumaun & Sikkim, North Europe, Australia.

Habitat : Along roadside & rocky slopes, Changthang and Spiti valleys between 3,560-3,680 m.

Plant Characters : A biennial or perennial herb. Stem erect or branched, smooth, upto 15 inch. Lower leaves lyrate, pinnatifid and upper leaves sessile, obovate, entire or pinnatifid. Flowers yellow in racemes. Calyx sub-equal at the base, petals with a broad claw. Pod long, linear, compressed.

Flowering & fruiting : June to August.

Uses : The tender shoots are cooked as vegetable.

33. *Berberis jaeschkeana* Schneid.

See Sr. No. 53, Chapter VI.

Uses : The ripen fruits are eaten.

34. *Berberis ulicina* Hk. f. & T.

See Sr. No. 56, Chapter VI.

Uses : The small fruits are edible when ripen.

35. *Bidens pilosa* Linn.

Family : Asteraceae

Name : Local or Gurgur Chai (L)

Distribution : Western Himalayas from Kashmir to Pakistan, Nepal and Tibet.

Habitat : Cultivated on farmers field, Leh and Suru valleys around 2,870-3,515 m.

Plant Characters : An erect, annual, pilose herb. Stem, erect, branched upto 80 cm tall. Leaves extensively variable. Flower - heads on long stout peduncles, with or without white or yellow rays. Achene black.

Flowering & fruiting : June to September.

Uses : The leaves are used for making local tea called Gurgur Tea / Namkeen Tea.

36. *Bunium persicum* (Boiss.) Fedtsch. *Carum bulbocastanum* Clarke.

Family : Apiaceae

Name : Sahi Zeera or Kala Zeera (H), Caraway black (E), Zeur (L)

Distribution : Western Himalayas from Kashmir to Kumaun.

Habitat : Open dry slopes, Suru and Lahaul valleys around 3,120 m.

Plant Characters : An erect, branched perennial herb. Root round, stout, fleshy. Stem slender, branched. Leaves 2-3 pinnatiscet; segment filiform; lower petiolate, upper sessile. Flowers white, numenous in umbel. Fruits elongated, black at maturity woth smelling oil glands.

Flowering & fruiting : Mid June - mid September.

Uses : The seeds are high valued spice, used as condiment to add flavour in local dishes and the bulbous roots are edible and being uprooted at the time of scarcity of vegetables in early summer.

37. *Capparis spinosa* Linn.

Family : Capparidaceae

Name : Ber (H), Caper bush (E), Kabra (L)

Distribution : Europe, North Africa, Australia, North Western India, throughout Punjab, Rajasthan.

Habitat : Dry rocks & stony slopes along roadside, Leh, Nubra and Spiti valleys between 3,020-3,890 m.

Plant Characters : A diffuse and prostrate, perennial shrub or climber with yellow hooked spines. Leaves variable, ovate to elliptic with a minute sharp point at the tip. Flowers large, white to purple with numerous long stamens, long-stalked, borne on axils of leaves. Fruits fleshy, oblong-ellipsoid, with red flesh and many brown seeds.

Flowering & fruiting : June to August.

Uses : Young and delicate leaves are used as vegetable. Flower buds are pickled as a flavour and condiment especially for fish. Ripen fruits are either eaten raw or made into pickles. The leaves are considered highly nutritious.

38. *Capsella bursa-pastoris* (Linn.) Medik.

Family : Brassicaceae

Name : Shepherd's purse (E); Shamso or Sog-ka (L)

Distribution : Kashmir to Kumaun and Sikkim.

Habitat : Cultivated fields, Leh and Lahaul valleys around 3,510 m.

Plant Characters : A small, annual herb. Stem 15-20 cm long. Leaves shortly stalked, pinnatipartite, cauline leaves linear, sessile, small. Flowers white, pedicilled on long corymbs. Fruit silique glabrous. Seeds numerous, ellipsoid-oblong, punctate.

Flowering & fruiting : July to September.

Uses : The shoots or leaves are boiled properly, then cooked as delicious leafy vegetable.

39. *Capsella elliptica* C. A. Mey

Family : Brassicaceae

Name : Samsho (L)

Distribution : Kashmir to Kumaun.

Habitat : Cultivated fields, Nubra valley around 3,420 m.

Plant Characters : A small glabrous annual herb. Stem slender, branched from the base. Leaves entire or pinnatifid. Flowers corymbose, generally white or pink. Fruits ovate-oblong or truncate. Valves boat shaped. Seeds many, minute.

Flowering & fruiting : Mid June to September.

Uses : The leaves are used as vegetable during early summer.

40. *Capsella thomsonii* Hk. f. & T.

Family : Brassicaceae

Name : Samsho (L)

Distribution : Kashmir to Kumaun, Sikkim and Tibet.

Habitat : Cultivated fields, Zanskar valley around 3,450 m.

Plant Characters : A tomentose, hoary perennial herb. Stem branched from the base. Leaves pinnatisect, linear-oblong; segments lanceolate. Flowers white, crowded in the axils of leafy bracts. Pod broadly oblong, truncate at both ends. Seeds minute, many.

Flowering & fruiting : July - mid September.

Uses : The delicate leaves are good source of vegetable.

41. *Cardamine flexuosa* With.

Family : Brassicaceae

Distribution : Temperate regions to Kashmir to Tibet.

Habitat : Occasional on moist slopes, Rothang pass, Lahaul-Spiti valley, 4,890 m.

Plant Characters : A procumbent- ascending, annual herb. radical leaves few, hardly rosulate, broadly ovate or orbicular with few large lobe like teeth. Flowers white, many-flowered racemes. Silique linear, glabrous, compressed.

Flowering & fruiting : July to August.

Uses : Edible; shoots and tender plants are cooked as green vegetable.

42. *Carum carvi* Linn. var. ***gracile***

Family : Apiaceae

Name : Kala Jeera (E) Umbu, Kosnyot (Go- sNyod) (L)

Distribution : From Himachal Pradesh to Kumaun, Kinnaur, Chamba and Ladakh in J&K.

Habitat : Cultivated fields and grassy wastelands, Leh, Nubra, Suru and Spiti valleys between 2,920-3,529 m.

Plant Characters : An annual or biennial glabrous herb with procumbent or erect stem, 30-60 cm. Leaves bi-pinnately dissected, linear, bracteate with 5-15 white pink flowers; leaflets ovate, dentate, lobes. Flowers white or rosy, in dense umbels. Fruits upto 2.5 cm, elliptic, oblong, yellowish-brown, aromatic, seeds light to dark brown, with five prominent ribs and smelling oil glands.

Flowering & fruiting : July to September.

Uses : Seeds are used as condiment to add flavour for the local dishes.

43. *Chaerophyllum acuminatum* Lindl.

Family : Apiaceae

Name : Chikmo (L)

Distribution : Western Tibet, Himachal Pradesh to Kumaun, Kinnaur and Ladakh in J&K.

Habitat : Found in moist places; Zanskar and Lahaul-Spiti valleys, 3,510 ft.

Plant Characters : A perennial herb. Stem delicate, branched, glabrous, 15-30 cm tall. Leaves 1-2 pinnate, ultimate segment oblong, entire. Bracts usually absent, rays 3-6. Bracteoles 2-5, linear. Fruits oblong-lanceolate.

Flowering & fruiting : July to September.

Uses : The plant is eaten raw or used as pot herb. The root alongwith mint leaves is used as flavouring agent.

44. *Chaerophyllum villosum* Wall. ex DC.

Family : Apiaceae

Name : Mesoso (L)

Distribution : Afganistan to Bhutan, Himalayas from Kashmir to Sikkim, Ladakh and Kumaun.

Habitat : On moist & shady hillsides, Zanskar & Changthang valleys between 3,328-3,605 m.

Plant Characters : A slender erect, perennial herb. Stem branched, hairy, 30-100 cm tall. Leaves 2-3 pinnate, lanceolate lobes further divided into ovate lobed segments; leaf seath usually inflated. Flowers white or pale pink in compound umbels. Bracts absent. Fruits cylindrical narrowed.

Flowering & fruiting : July to September.

Uses : The roots are cooked and eaten as vegetable in Ladakh.

45. *Chenopodium album* Linn.

Family : Chenopodiaceae

Name : Bathua (H), Wild Spinach (E), Neung or Em (L)

Distribution : North temperate Himalayas and cultivated throughout India.

Habitat : Found as a weed in cultivated fields of Cabbage, cauliflower & wheat throughout Ladakh and Lahaul valleys between 2,780-3,940 m.

Plant Characters : An erect, annual or biennial herb. Stem upto 100 cm, branched or unbranched, red streaked. Leaves simple, scentless, extremely variable, more or less sinuate, sometimes toothed. Flower clusters in axillary spikes, pale green often tinged with purple. Fruit depressed globose. Seeds smooth, shiny, solitary, keeled.

Flowering & fruiting : July to September.

Uses : The whole herb is properly boiled and used as vegetable, for making paratha etc either fresh or dehydrated for prolonged winter use by locals.

46. *Chenopodium botrys* Linn.

Family : Chenopodiaceae

Name : Sagani or Shanik (L)

Distribution : A weed distributed from Ladakh to Sikkim.

Habitat : Grown as a weed in cultivated field of Cabbage, cauliflower and wheat, around Ladakh upto 4,230 m.

Plant Characters : An erect, strongly aromatic, annual or biennial herb. Stem upto 45 cm, much branched. Lower leaves petioled, ovate, toothed; upper leaves oblanceolate, entire. Flowers numerous in clusters or axillary panicles, greenish. Seeds solitary and shining.

Flowering & fruiting : July to September.

Uses : The tender leaves are eaten as vegetable after properly boiling.

47. *Chenopodium foliosum* (Moench.) Asch.

Family : Chenopodiaceae

Name : Sangsti (L)

Distribution : From Pakistan to Central Nepal, temperate Eurasia, N. Africa, Kashmir and Kumaun.

Habitat : Abundant on cultivated field & wastelands, Zanskar and Spiti valleys between 2,990-3,650 m.

Plant Characters : A scentless, glabrous, perennial herb. Stem branched, 20-40 cm long. Leaves 5-7 cm long, petioled, triangular-hastate, unequally toothed. Flowers clustered in axillary and terminal spikes, bearing red fruits at maturity, sessile.

Flowering & fruiting : June to August.

Uses : It is used as vegetable after boiling and also dried for winter use. The tender leaves are boiled and then washed thoroughly, mixed with curd and used for a local delicious dish "Tantur".

48. *Chenopodium glaucum* Linn.

Family : Chenopodiaceae

Name : Kupald (L)

Distribution : North temperate zones, common in Ladakh and Kashmir.

Habitat : Wastelands areas, Changthang and Spiti valleys around 4,210 m.

Plant Characters : A much branched, prostrate, succulent herb. Stem 15-20 cm high, shining. Branches decumbent. Leaves small, oblong-ovate, obtuse, lanceolate, toothed. Flowers in dense axillary spikes, simple or compound. Sepals keeled. Seeds large, horizontal.

Flowering & fruiting : July to mid September.

Uses : The tender shoots are boiled properly and eaten as vegetable.

49. *Chorispora sabulosa* Camb.

See Sr. No. 19, Chapter X.

Uses : Young leaves and tender shoots are used as vegetable.

50. *Christolea crassifolia* Camb.

Family : Brassicaceae

Name : Sanak (L)

Distribution : From Afganistan to Central Nepal, Tibet & Central Asia.

Habitat : Dry slopes, Leh, Changthang & Spiti valleys between 3,020-4,220m.

Plant Characters : A spreading to erect perennial herb. Stem with terminal spike-like clusters of white to mauve flowers; thick ovate-oblong, toothed, greyish, often densely hairy leaves. Seeds dark brown to black. Pod long, straight.

Flowering & fruiting : June to August.

Uses : The tender shoots are consumed as vegetable by the locals.

51. *Cichorium intybus* Linn.

Family : Asteraceae

Name : Cichory or wild endive (E), Kasni (L)

Distribution : From Nepal to Kashmir & Himachal Pradesh.

Habitat : Cultivated fields, Lahaul and Suru valleys between 3,290-3,750 m.

Plant Characters : An annual-perennial herb with thick taproots. Stem often many from the base, ribbed, erect, often decumbent below, flesuous upwards. Radical and lower stem leaves short petioled, lyrate pinnatifid, obtuse or acute remotely dentate or lobed, hairy or glabrous; upper ones sessile, sagittate with a semi-amplexicaul base. Heads blue, sessile, 4-10 together in axillary and terminal cluster, combined into a widely branched leafy rigid panicle. Achenes turbinate, 5 angular.

Flowering & fruiting : July to September.

Uses : The tender herb is edible and cooked as fresh vegetable.

52. *Cicer microphyllum* Benth. *C. soongaricum* Auct. non Steph.

See Sr. No. 49, Chapter IX.

Uses : The pods and seeds are edible and collected after maturity also.

53. *Codonopsis clematidea* (Schrenk) Clarke.

See Sr. No. 78, Chapter VI.

Uses : Roots are consumed by locals either as raw or cooked.

54. *Codonopsis ovata* Benth.

See Sr. No. 79, Chapter VI.

Uses : Same as in *Codonopsis clematidea.*

55. *Crambe kotschyana* Boiss.

Family : Brassicaceae

Name : Bemma (L)

Distribution : Kashmir to Tibet and Persia.

Habitat : Rocky slopes, Zanskar valley between 3,540-3,890 m.

Plant Characters : An erect, stout, perennial, herb. Stem branched, 60-100 cm tall. Basal leaves petioled with wavy margins; cauline small rhomboid or elliptic, toothed. Flowers white, numerous in lax branched clusters. Pod globose.

Flowering & fruiting : June to September.

Uses : The young leaves are used as potherb.

56. *Draba alpina* (Linn.) Hk. f. & T.; *D. oreades* Schrenk;

Family : Brassicaceae

Name : Piang (L)

Distribution : From Pakistan to South West China and Central Asia.

Habitat : Rocky slopes, Zanskar valley around 3,540 m.

Plant Characters : A tiny, often mat-forming perennial herb. Stem tufted, glabrous, short or 6-10 inches. Leaves variable, lanceolate to obovate, covered with mixed or simple hairs. Flowers yellow to yellowish-white, in dense terminal cluster. Pod ovate, 4-10 seeded. Siliquae flattened.

Flowering & fruiting : July to September.

Uses : The small delicate herb is used as pot-herb during scarcity in early summers.

57. *Draba cachemirica* Gand.

Family : Brassicaceae

Name : Mulanku (L)

Distribution : Pakistan to Nepal, Tibet, Kashmir and Himachal Pradesh.

Habitat : Damp & moist places, Leh, Zanskar and Spiti valleys between 3,019-3,602 m.

Plant Characters : A prostrate, perennial, tufted herb. Stem branched, 10-20 cm. Roots slender, woody. Leaves rosulate, entire margin, rather rigid, entire, rough and ciliated. Flowers pedicelled, light yellow in terminal cluster. Fruits ovate-lanceolate, hairless. Seeds 2-4 in each locule.

Flowering & fruiting : July to September.

Uses : The herb is used as pot-herb.

58. *Draba stenocarpa* Hk. f. & T.

Family : Brassicaceae

Name : Lepcha (L)

Distribution : Pakistan to Western China, Tibet and Asia.

Habitat : Moist places, Suru and Lahaul valleys around 3,029 m.

Plant Characters : A small, tufted annual herb. Stem simple, erect. Leaves linear-oblong, thickly covered with hairs, entire, sessile. Flowers yellow, in lax racemes. Filaments simple. Pod compressed, elliptic oblong, tapering at both ends. Siliquae compressed. 14-20 seeded.

Flowering & fruiting : July to mid September.

Uses : The tender leaves are cooked as green vegetable.

59. *Draba tibetica* Hk. f. & T.

Family : Brassicaceae

Name : Mulanku (L)

Distribution : Afganistan to South East Asia, Ladakh of J&K, Himachal Pradesh.

Habitat : Stony & dry alpine slopes, Changthang valley around 4,847 m.

Plant Characters : A tufted, perennial herb. Stem branched, woolly. Leaves spathulate-lanceolate, entire. Scape slender usually leafless. Flowers subcapitose, pedicelled, white. Pod hoary, short.

Flowering & fruiting : June to mid August.

Uses : The young shoots and leaves are used as vegetable.

60. *Elsholtzia densa* Benth.

See Sr. No. 27, Chapter VII.

Uses : The tender leaves are eaten raw or in making chatani or used for flavouring local dishes in Ladakh.

61. *Elsholtzia eriostachya* (Benth.) Benth.

See Sr. No. 28, Chapter VII.

Uses : Leaves and flowering parts are used as condiment.

62. *Ephedra gerardiana* Wallich ex Stapf.

See Sr. No. 116, Chapter VI.

Uses : Small ripen and sweet berries are edible.

63. *Eremurus himalaicus* Baker.

Family : Liliaceae

Name : Desert Candle (E), Nyamgtso (L)

Distribution : Afganistan to Himachal Pradesh & Kashmir.

Habitat : Rocky areas, Lahaul and Spiti valleys around 3,850 m.

Plant Characters : A glabrous, scapigerous herb. leaves upto 50 cm long. Flowers white borne in cluster on long raceme. Bracts subulate, filiform. Petals oblong blunt. Anthers exserted, orange. Capsule 1-2 cm. Seeds triquetrous.

Flowering & fruiting : July to September.

Uses : The young flower buds are edible.

64. *Erigeron multiradiatus* (DC.) Benth. ex Clarke.

Family : Asteraceae

Name : Durang-phang (L)

Distribution : Pakistan to Bhutan, Kashmir.

Habitat : Open slopes, Changthang valley around 4,530 m.

Plant Characters : An erect, hairy, tufted, annual herb. Stem stout or slender simple or branched. Leaves all cauline, obovate or lanceolate, obtuse or acute, entire 1-2 toothed. Flower-heads 2-5 cm, dark purple, solitary with long pedicel. Pappus red, numerous, hairy. Achenes glabrous.

Flowering & fruiting : July to August.

Uses : The leaves and tender shoots are eaten as potherb.

65. *Ermania himalayensis* (Camb.) Schelz; *Cherianthus himalayensis* Camb.

Family : Brassicaceae

Name : Fiam (L)

Distribution : Pakistan to Sikkim, Tibet and Kashmir.

Habitat : Alpine pastures, Zanskar valley between 3,320-3,540 m.

Plant Characters : A suberect, small, densely grey haired, perennial herb. Stem tufted. Basal leaves 1-1.5 cm, stem leaves smaller, narrower, the uppermost subtending flowers. Flowers violet-purple borne in short spike like clusters. Pod hairy, linear-oblong.

Flowering & fruiting : July to mid September.

Uses : Young leaves are used as green leafy vegetable.

66. *Ermania stewartii* (T. Ander) Scheultz.; *Cherianthus stewartii* T. Ander.

Family : Brassicaceae

Name : Tschalu (L)

Distribution : Kashmir to Himachal Pradesh and Pakistan.

Habitat : Higher passes, Zanskar valley around 4,134 m.

Plant Characters : A small hoary, suberect, perennial herb. Stem branched, slender. Lower leaves subcoriaceous woolly; upper leaves spathulate toothed at the tip. Flowers purple or lilac in terminal racemes. Pods glabrous, tapering at both ends.

Flowering & fruiting : Mid June to mid September.

Uses : The young leaves are used as vegetable in Ladakh.

67. *Erodium stephanianum* Willd.

Family : Geraniaceae

Name : Tikziang (L)

Distribution : Pakistan to Western Tibet, China, Siberia and Ladakh.

Habitat : On sandy places, Zanskar valley around 3,560 m.

Plant Characters : A diffuse, branched, hairy, annual herb. Leaves opposite, mostly radical, pinnately divided. Stipules cuneate, acute, ciliate. Flowers purple, in peduncled axillary umbels. Sepals ovate, awned. Petals as long as the sepals, broadly ovate. Capsule hairy.

Flowering & fruiting : Mid June to September.

Uses : The fresh leaves are used as green vegetable.

68. *Erodium tibetanum* Edgew. & Hk f.

Family : Geraniaceae

Name : Tikziang (L)

Distribution : From Kashmir to Western Tibet and Ladakh.

Habitat : On sandy places, Changthang valley around 4,760 m.

Plant Characters : A hoary, stemless, annual herb. Leaves oblong-ovate, more or less inciso-crenate. Stipules cuneate, scarious, ciliate. Flowers 2 or more. Bracts ciliate. Sepals with a small knob, ciliate. Petals elliptic. Fruit a capsule.

Flowering & fruiting : July to August.

Uses : The young leaves are used as vegetable.

69. *Eruca sativa* Mill.

Family : Brassicaceae

Distribution : Pakistan to India.

Habitat : Cultivated for fodder and seed oil, often met with as an escape on waste lands along roadside, Lahaul valley, 4,210 m.

Plant Characters : An erect, annual herbs, 30-100 cm long. Leaves lyrate-pinnatifid, lower sub-rosulate, petioled, upper ones sessile. Flowers yellow, turning to while after anthesis, 1.5 to 2.0 cm long. Sepals oblong, often pale-violet. Petals obovate, long clawed, pink veined.

Flowering & fruiting : May to August.

Uses : The young herb is eaten as vegetable.

70. *Erysimum hieracifolium* Linn.

Family : Brassicaceae

Name : Gong-thog-pa (L)

Distribution : Pakistan to Western Tibet and Kashmir.

Habitat : Moist shady slopes, Leh valley around 3,605 m.

Plant Characters : An erect, perennial herb. Stem branched, erect, covered with rough two branched hairs. Leaves elliptic to oblanceolate, sinuate, toothed. Flower many, pedicelled, bright yellow in terminal clusters. Sepals finely hairy. Fruit narrow, tetragonal. Pod strict, erect, narrow, tetragonal, scaberulous.

Flowering & fruiting : Mid June to August.

Uses : The tender leaves are used as vegetables.

71. *Fagopyrum esculentum* Moench.

Family : Polygonaceae

Name : Buckwheat (E), Dayat (L)

Distribution : From Kashmir to Sikkim, Khasi hill, Manipur, Nilgiris.

Habitat : Cultivated as well as growing as a weed in the cultivated fields & wasteland, Leh valley & throughout Ladakh and Lahaul valleys between 2,780-3,890 m.

Plant Characters : An erect, slender, glabrous, annual herb with smaller leaves and dense inflorescence. Leaves stalked, triangular, cordate, acute. Flowers in axillary and terminal branched clusters, pink coloured. Fruit 3-cornered achene. Nut ovate, angles acute.

Flowering & fruiting : July to September.

Uses : It is an important source of vegetable for local tribes. The nutritive value showed it is good source of protein 25.1%, minerals 21.3%, phosphorus 574 mg/100g and rich in energy. Seeds are also used for making flour to prepare different dishes.

72. *Fagopyrum tataricum* (Linn.) Gaertn.; *Polygonum tataricum* Linn.

Family : Polygonaceae

Name : Duckwheat (E), Dayat (L)

Distribution : From Pakistan to Bhutan.

Habitat : Cultivated as well as grown in fields & wasteland, Nubra valley and Lahaul valleys between 2,850-3,980 m.

Plant Characters : An erect, glabrous, annual herb. Stem usually simple. Leaves triangular-cordate, petioled, light green. Flowers in axillary and terminal clusters with greenish or yellowish flowers. Nuts in 3-deep grooves.

Flowering & fruiting : Mid June to August.

Uses : The delicate herb and leaves are used to prepare delicious leafy vegetable.

73. *Fragaria vesca* Linn.

Family : Rosaceae

Name : Gooseberry (E)

Distribution : Temperate Eurasia, Pakistan to Kashmir.

Habitat : Moist slopes, Lahaul valley between 10,000-12,000 ft.

Plant Characters : An erect or ascending, perennial, stoloniferous herb. leaves 3-foliolate; leaflets sessile or petiolate or petiolulate, sparsely hairy or glabrescent above, pubescent beneath, toothed; petioles 3-12 cm long with spreading or adpressed hairs. Flowers white, solitary or 1-3 together; pedicels with spreading or adpressed hairs. Bracteoles entire or toothed. Calyx-lobes silky-hairy, entire or toothed. Petals rhomboic-ovate, slightly longer than the calyx. Berries reddish-brown.

Flowering & fruiting : July to September.

Uses : The ripen juicy, sweet and tasty fruits are edible.

74. *Gentiana tianschanica* Ruper.

Family : Gentianaceae

Name : Wanglo (L)

Distribution : Pakistan to Central Asia and Kashmir.

Habitat : On open moist areas, Suru and Spiti valleys around 3,040 m.

Plant Characters : An erect, small, perennial herb. Stem erect or spreading, 10-30 cm high. Basal leaves oblong, acute; cauline leaves distant, linear-lanceolate, acute. Flowers pinkish-white to deep blue in terminal clusters. Corolla tubular-funnel-shaped with triangular, entire or 2-lobed. Capsule oblong-ellipsoid.

Flowering & fruiting : Mid July to September.

Uses : The whole plant is edible and cooked as vegetable.

75. *Hippophae rhamnoides* Linn. spp. ***turkestanica*** Rousi.

See Sr. No. 158, Chapter VI.

Uses : Fruits are rich source of various vitamins and used for making juice, squash and other food products. Vitamin B_1, B_{12}, C, K, E and flavonoides are present in the entire plant. This is a good source of antioxidants. Based on ethnobotanical information, this lab has formulated a multivitamins herbal beverage from its fruits and got patented and commercialized throughout India. Many other health food products are in pipeline.

76. *Hippophae salicifolia* D. Don.

See Sr. No. 159, Chapter VI

Uses : Same as for *Hippophae rhamnoides* Linn. spp. *turkestanica.*

77. *Hippophae tibetana* Schlecht.

See Sr. No. 160, Chapter VI.

Uses : It is also a source of vitamins and used as edible.

78. *Juglans regia* Linn.

See Sr. No. 13, Chapter XIII.

Uses : Walnut is grown at lower altitudes of Leh, Suru and Nubra valleys, and used for its edible nuts.

79. *Lactuca decipiens* Hk. f. & T. ex Clarke, *Cicerbita decipiens* (Hk. f. & T. ex Clarke) Beauv.

Family : Asteraceae

Name : Khala (L)

Distribution : From Afganistan to Kashmir and Kumaun.

Habitat : Wastelands and open slopes, Suru & Lahaul valleys around 2,750m.

Plant Characters : An erect, perennial herb. Stem branched above, 40-90 cm high. Basal leaves with narrow backward-pointing lobes; upper leaves entire, toothed. Flowers blue, several, borne in terminal clusters. Achenes shortly beaked and dirty white pappus-hairs.

Flowering & fruiting : July to September.

Uses : The tender shoots and leaves are cooked as vegetable.

80. *Lactuca dolichophylla* Kitam.; *L. logifolia* DC.

Family : Asteraceae

Name : Khala (L)

Distribution : North temp regions from Afganistan to China & Garhwal.

Habitat : Cultivated fields & wastelands, Leh and Zanskar valleys around 3,350 m.

Plant Characters : An erect, tall, hairless, perennial herb. Stem upto 90 cm branched above. Leaves 7-15 cm long, sessile, lanceolate to linear-lanceolate, entire or sparingly lobed, stalkless. Flower-heads, blue to violet, borne on terminal clusters. Involucral bracts few. Achenes elliptic, ribbed, black, compressed, beaked.

Flowering & fruiting : July to September.

Uses : Young leaves and shoots are cooked as vegetable. The protein and mineral quantity are 27.1% and 17.5% respectively.

81. *Lactuca lahulensis* Mamgain *et* Rao.

Family : Asteraceae

Name : Khala (L)

Distribution : North temperate regions from Pakistan to China.

Habitat : Occasional on moist slopes, Lahaul valley around 3,390 m.

Plant Characters : An erect, glabrous, perennial herb. Stem 50-80 cm high. Upper leaves finely membranous with long and broadly winged petioles, sagittately auricled; lamina broadly cordate to deltoid. Heads on subpandiculate branches, more or less clustered, drooping, 3-4 flowered; florets all ligulate, ligule white or purplish. Achene 8-9 ribbed on each face, raised, yellowish brown, slightly longer than pappus.

Flowering & fruiting : July to September.

Uses : Young leaves, flower buds and shoots are washed, boiled and cooked as vegetable during scarcity.

82. *Lactuca lessertiana* (DC.) Clarke.

Family : Asteraceae

Name : Chumaseo (L)

Distribution : Temperate & Alpine Himalayas from Kashmir to Sikkim.

Habitat : Open slopes, Suru and Lahaul valleys around 3,060 m.

Plant Characters : An erect, short, annual herb. Stem upto 30 cm stout, with many lateral branches. Leaves usually lanceolate, entire or variably toothed or shallow-lobed. Flower-heads blue or purple, erect or drooping, borne on a stout leafless stem. Fruits black, strongly ribbed, pappus white of yellowish.

Flowering & fruiting : July to September.

Uses : The tender leaves are cooked as leafy vegetable.

83. *Lactuca orientalis* (Boiss.) Boiss.

Family : Asteraceae

Name : Khala (L)

Distribution : From Afganistan to Nepal, China, Bhutan and Kashmir.

Habitat : On wastelands, Zanskar valley around 3,850 m.

Plant Characters : An erect, glabrous, perennial herb. Stem branched, branches with white-green wings. Leaves small, decurrent. Lower pinnatifid, triangular, acute; upper leaves linear, entire. Flower-heads, solitary in spikes. Achenes narrowed, pale. Pappus hairs silvery-white.

Flowering & fruiting : Mid June to August.

Uses : The tender leaves are cooked as vegetable.

84. *Lactuca scariola* Linn. var. *sativa* (Linn.) Clarke.

Family : Asteraceae

Name : Khala (L)

Distribution : From Pakistan to Himachal Pradesh, Kashmir and Kumaun.

Habitat : Wastelands, Changthang valley around 3,880 m.

Plant Characters : An erect, leafy, biennial herb. Stem reaching upto 1-1.5 m. somewhat prickly. Leaves 8-12 cm, runcinate pinnatifid, spinescent, cauline, sagittate. Flower-heads 2-3 inches long, yellow. Achenes oblanceolate, ribbed. Seeds gray.

Flowering & fruiting : June to August.

Uses : Young leaves are cooked as vegetable.

85. *Lactuca tatarica* C.A. Mey.

Family : Asteraceae

Name : Khalask (L)

Distribution : North temperate regions from Afganistan to Kashmir & Kumaun.

Habitat : Cultivated fields & wastelands, Nubra & Leh valleys between 3,760-4,380 m.

Plant Characters : A perennial herb. Stem erect, variable in length, dichotomously branched. Leaves radical and cauline, radical leaves runcinate. Cauline leaves lanceolate, finely toothed, sessile, amplexicaule. Flower-heads upto 2 cm long, purplish-violet, solitary or clostered, in terminal panicles. Achenes strongly ribbed beak minute.

Flowering & fruiting : July to mid September.

Uses : The herb is used as vegetable as well as in salad.

86. *Lamium amplexicaule* Linn.

Family : Lamiaceae

Name : Supila (L)

Distribution : Pakistan to Temperate Eurasia, Kashmir, Kumaun & Sikkim

Habitat : Moist places, Zanskar and Spiti valleys, 3,350 m.

Plant Characters : An annual diffuse, hairy herb. Stem branched, glabrous, 10-15 cm tall. Leaves long-petioled below, orbicular inciso-crrenate, whorls, distant, sessile above. Flowers purple red, in axillary clusters. Corolla tube long slender nearly straight. Nutlets brown.

Flowering & fruiting : July to September.

Uses : The whole plant is used as vegetable in Ladakh.

87. *Lathyrus sativus* Linn.

Family : Fabaceae

Name : Local pea (E), Sanma or Sandil (L)

Distribution : From Western Asia to North Africa, Europe and America.

Habitat : Commonly cultivated upto 3,950 m throughout Ladakh and Lahaul-Spiti areas.

Plant Characters : An annual, glabrous, herb. Leaves ending in 3-fid tendrils; stipules broad, entire. Flowers solitary, axillary, reddish purple, blue or white; pods, seeds 4-5 in a pod, grayish brown or yellowish.

Flowering & fruiting : May to September.

Uses : Seeds are cooked as vegetable.

88. *Lavatera kashmiriana* Camb.

Family : Malvaceae

Name : Sazamool (H), Wild Hollyhock (E), Sazposh (L)

Distribution : From Pakistan to Kashmir.

Habitat : Cultivated fields, Suru valley around 3,130 m.

Plant Characters : A tall densely woolly-haired perennial herb. Stem branched. Lower leaves rounded heart shaped, petioled, blade, orbicular, cordate, 5-lobed, upper with 3-5 narrower lobes. Flowers on terminal spike, large bright pink with darker veins. Fruit many, black carpels.

Flowering & fruiting : June - September.

Uses : The tender shoots are used as vegetable.

89. *Lepidium capitatum* Hk. f. & Th.

Family : Brassicaceae

Name : Sangtso (L)

Distribution : From Pakistan to Kashmir, South West Asia, Europe and Africa.

Habitat : Dry & sandy slopes, Suru and Lahaul valleys around 2,850 m.

Plant Characters : A prostrate, perennial herb. Stem glabrous, branched. Radical leaves sessile, pinnatifid, toothed; cauline leaves sessile. Flowers tiny, white borne in dense racemes. Petals exceeding the sepals. Pod orbicular wingless. Seeds pale-brown.

Flowering & fruiting : July to September.

Uses : The tender leaves and shoots are cooked as vegetable.

90. *Lepidium latifolium* Linn.

Family : Brassicaceae

Name : Dittander (E) & Sauson (Seoji) (L)

Distribution : Afganistan to Kashmir, Western Asia, Europe and North Africa.

Habitat : Wastelands & stony slopes, Leh, Suru and Zanskar valleys 3,260-3,650 m.

Plant Characters : A tall erect (30-100 cm) branched perennial herb. Leaves mostly lanceolate, entire or sometimes toothed, usually hairless. Flowers tiny white borne on branched inflorescence in terminal cluster. Sillique small, ovate-orbicular, seeds brown, rounded.

Flowering & fruiting : July to September.

Uses : The delicate leaves and tender shoots are eaten as vegetable. The leaves are nutritive and rich in protein upto 31.7%, phosphorus content upto 500 mg/100g and mineral content is about 11.8%.

91. *Lepidium ruderale* Hk. f. & T.; *L. apetalum* Linn. ex Willd.

Family : Brassicaceae

Name : Saugso or Khrag-khrog (L)

Distribution : Temperate Europe, Pakistan to Bhutan, Kashmir and Kumaun.

Habitat : Wastelands, Zanskar & Leh valleys between 3,550-3,850 m.

Plant Characters : An erect or diffused, pubescent, annual herb. Stem branched from the base. Cauline leaves sessile, sagittate, glabrous, pinnatifid. Flowers minute pedicelled in long lateral and terminal corymbs. Corolla with white-pinkish tips. Pod orbicular, shortly winged.

Flowering & fruiting : June to August.

Uses : The tender shoots and leaves are cooked as vegetable.

92. *Malva verticillata* Linn.

Family : Malvaceae

Name : Chinese Mallow (E) & Cam-pa ma-ning (L)

Distribution : Punjab to Assam, Pakistan, temperate Eurasia and Africa.

Habitat : Cultivated areas, Changthang valley around 3,960 m.

Plant Characters : An erect, pubescent annual herb. Stem branched, reaching upto 70 cm long. Leaves usually downy with heart-shaped base, 5-7 lobed. Flowers pink or mauve, borne in axillary dense clusters. Fruits 10-12 ribbed, 1-seeded parts.

Flowering & fruiting : Mid June to August.

Uses : Young leaves and tender shoots are boiled and cooked as green vegetable.

93. *Mentha longifolia* Linn. Hudson.

See Sr. No. 37, Chapter VII.

Uses : It is commonly used as chatani and as flavouring agent to local dishes.

94. *Microula tibetica* Benth.

See Sr. No. 203, Chapter VI.

Uses : The young leaves are first boiled to get rid of bitterness and are then cooked as vegetables.

95. *Morus alba* Linn.

See Sr. No. 17, Chapter XIII.

Uses : The ripen fruits are edible.

96. *Nepeta floccosa* Benth.

See Sr. No. 43, Chapter VII.

Uses : The dried leaves and shoots are used to add flavour to local dishes.

97. *Nepeta longibracteata* Benth.

See Sr. No. 47, Chapter VII.

Uses : The whole plant is used as flavouring agent.

98. *Orobanche hansii* Ker.

See Sr. No. 221, Chapter VI.

Uses : The tender shoots are used for making soup in Ladakh.

99. *Oxyria digyna* (Linn.) Hill.

Family : Polygonaceae

Name : Mountain Sorrel (E), Lamanchu or Chu-Lchum (L)

Distribution : From Pakistan to South West China & North temperate Zone.

Habitat : Moist places, Leh valley around 3,650 m.

Plant Characters : A fleshy acid flavoured, glabrous, succulent, perennial herb. Stem 15-50 cm high. Basal leaves numerous, long-petioled, round to kidney shaped, reniform; cauline 1-2, petiole sometimes 20 cm. Flowers minute, rose-pink or green, in branched inflorescence. Fruits 2 mm broad, round, winged, pink.

Flowering & fruiting : July to September.

Uses : The leaves contain vitamin C, used raw or cooked. According to nutritive value it is rich in fibre 10.8%, iron 120 mg/100g and mineral content 29.8%.

100. *Pedicularis bicornuta* Klotz. ex. Klotz. & Garcke.

Family : Scrophulariaceae

Name : Piaseng (L)

Distribution : From Pakistan to Himachal Pradesh.

Habitat : Moist alpine slopes & irrigated meadows, Zanskar valley 3,570 m.

Plant Characters : A robust perennial herb. Stem stout, reaching upto 20-50 cm. Leaves many, alternate, linear, pinnately lobed, stalked. Flowers many, pale yellow in dense terminal racemes. Calyx-oblong, hirsute. Corolla with sigmoid upper lip. Capsule oblong, lanceolate, acuminate.

Flowering & fruiting : July to mid September.

Uses : The tender leaves are used as potherb.

101 *Pedicularis mollis* Wall ex Benth.

See Sr. No. 230, Chapter VI.

Uses (T) : The leaves are edible as vegetable.

102. *Pedicularis punctata* Decne.

Family : Scrophulariaceae

Name : Memerang (L)

Distribution : From Pakistan to Kashmir.

Habitat : Damp & alpine slopes, Zanskar & Lahaul valleys around 3,870 m.

Plant Characters : A tufted herb. Stem slender, usually several from the base. Leaves oblong to elliptic with rather broad ovate, finely toothed lobes. Flowers pinkish-red with a white spot borne in dense spikes. Corolla-tube slender with a curved bifid beak. Capsule broadly oblong, oblique, acute. Seeds oblong, obtuse, base aciculate.

Flowering & fruiting : June to August.

Uses : The tender shoots and leaves are used as pot-herb.

103. *Pedicularis pyramidata* Royle. ex. Benth.

Family : Scrophulariaceae

Name : Scantso (Shakku) (L)

Distribution : From Pakistan to Himachal Pradesh.

Habitat : Moist places, Nubra valley around 3,050 m.

Plant Characters : An erect, unbranched, glabrous, perennial herb. Stem one or many, usually 30-50 cm. Leaves opposite, pinnatisect with serrate margins. Flowers pink, in dense spikes. Corolla-tube longer than hairy calyx, upper lip inflated at base and curved inwards. Capsule ovoid, acute, tip exerted. Seeds large, rigid, deeply pitted, pale.

Flowering & fruiting : July to September.

Uses : The leaves are used as vegetable.

104. *Perovskia abrotanoides* Kiril.

See S. No. 51, Chapter VII.

Uses : The leaves are used as spices.

105. *Picris hieracioides* Linn.

Family : Asteraceae

Name : Trumbu (L)

Distribution : From Sikkim to Bhutan, North West Asia, Japan & North Africa.

Habitat : On alpine areas, Suru and Lahaul valleys around 2,795 m.

Plant Characters : An erect, branched, glandular-hairy annual herb. Leaves alternate, narrowly lanceolate; radical leaves petioled, sinuate-toothed; cauline leaves sessile, half amplexicaul or stem-clasping, entire. Heads yellow, involucral bracts narrow, yellowish-green, glabrous with black hairs. Achenes narrowly ellipsoid, reddish-brown.

Flowering & fruiting : July to September.

Uses : The young plant is used as pot-herb.

106. *Podophyllum hexandrum* Royle.

See Sr. No. 240, Chapter VI.

Uses : The ripen red fruits are eaten.

107. *Polygonum affine* D. Don.

Family : Polygonaceae

Name : Samtsong (L)

Distribution : From Afganistan to Eastern Nepal.

Habitat : Open slopes and roadside, Zanskar & Spiti valleys around 3,660 m.

Plant Characters : A densely tufted, perennial herb. Stem many from the base, 15-25 cm long. Leaves mostly radical, shortly petioled, linear-elliptic; cauline leaves few, sessile, reddish. Flowers rose-pink borne in crowded racemes. Nuts trigonous.

Flowering & fruiting : July to September.

Uses : The young leaves are cooked as vegetable.

108. *Polygonum aviculare* Linn.

Family : Polygonaceae

Name : Chhiankat (L)

Distribution : From Kashmir to Kumaun.

Habitat : On slopes and dry places, Nubra & Zanskar valleys around 3,200m.

Plant Characters : An annual, glabrous, herb. Stem branched from the base. Leaves elliptic or elliptic-oblong or lanceolate obtuse flat nerveless, stipules shorter than the internodes. Flowers axillary, pedicel short, joint at tip. Perianth ovoid 3-gonous. Nuts ovoid, minutely wrinkled.

Flowering & fruiting : July to September.

Uses : The tender leaves are cooked as potherb.

109. *Polygonum sibiricum* Lam.

Family : Polygonaceae

Distribution : From Afganistan to Kashmir.

Habitat : Stony slopes, Leh valley around 3,365 m.

Plant Characters : A dwarf, glabrous, fleshy herb. Stem branched from the base. Leaves narrowly linear-obtuse. Stipule short, silvery white. Flowers minute, pink in capitate, terminal panicle.

Flowering & fruiting : July to September.

Uses : The leaves are eaten as salad.

110. *Polygonum viviparum* Linn.

Family : Polygonaceae

Distribution : Kashmir to Sikkim, Western Tibet, Europe, Asia & America.

Habitat : Meadows & fields, Suru valley around 2,930 m.

Plant Characters : A highly variable, small, slender, annual herb. Leaves glabrous or pubescent below; radical and lower leaves long petioled, linear or linear-oblong, acute or obtuse; upper ones sessile, erect. Flowers white or pink, in erect, solitary slender spikes. Bracts membranous, ovate, acuminate.

Flowering & fruiting : July to September.

Uses : The leaves and stalks are edible and eaten raw as salad.

111. *Potentilla anserina* Linn.

See Sr. No. 244, Chapter VI.

Uses : The whole plant is edible and used in tea.

112. *Potentilla atrosanguinea* Lodd.; *P. argyrophylla* (Wall. ex Lehm.) Griers. & Long.

See Sr. No. 245, Chapter VI.

Uses : Leaves are edible and cooked as vegetable.

113. *Prunus armeniaca* Linn.

Family : Rosaceae

Name : Khubani (H), Apricot (E); Chuli (L)

Distribution : Cultivated in USA, Spain, France, Italy, Turkey, Iran, Africa, Australia and India. Distributed from Kashmir to Ladakh, Chenab.

Habitat : Widely cultivated and wild found upto 3,750 m Leh, Nubra, Suru, Zanskar and Lahaul regions.

Plant Characters : A moderate-sized tree, about 10 m tall. Bark reddish-brown. Leaves appearing after or with the flowering, broadly ovate to round ovate, acuminate, crenate, petioled, glandular; stipule lanceolate. Flowers pinkish white. Fruits round, pubescent, yellow or yellowish orange, firm, sweet, karnels sweet or bitter.

Flowering & fruiting : Mid April to May.

Uses : Apricot is the only natural fruit crop grown in these areas. Dehydrated apricots are the only fruits, which can be purchased at any time from the market. Seeds of some varieties of apricot are edible and nutritious.

114. *Psychrogeton andryaloides* (DC.) Novopokr. ex Krasch.; *Erigeron andryaloides* (DC.) Benth. ex Clarke.

Family : Asteraceae

Name : Luchchugba (L)

Distribution : Kashmir to Lahaul & Western Tibet.

Habitat : Dry stony slopes, Suru valley around 2,795 m.

Plant Characters : A grey or white densely woolly-haired tufted, perennial herb. Rootstock very stout. Stem 5-10 cm, branched from the base, spathulate. Flowers-heads yellow, solitary. Ray florets short and recurved; disc-florets yellow. Involucral bracts linear woolly-haired.

Flowering & fruiting : June to August.

Uses : Tender leaves are used as vegetable.

115. *Rheum spiciforme* Royle

See Sr. No. 257, Chapter VI.

Uses : The thick leaf-stalks contain Vitamin 'C' and is used as raw or salad or added in custard.

116. *Rheum tibeticum* Maxim ex Hk. f.

Family : Polygonaceae

Name : Rhubarb (E), Lachhu (L)

Distribution : From Pakistan to Kashmir & Western Nepal.

Habitat : Dry slopes, Changthang valley between 2,835-3,960 m.

Plant Characters : A stemless, perennial herb. Roots very short. Leaves large, very coraceous, orbicular, cordate with short thick petiole. Panicle quite glabrous with thick petiole. Flowers dark purple, fruit ovoid-oblong, base cordate apex notched wings narrow. Bracts minute.

Flowering & fruiting : July to August.

Uses : Leaf stalks are eaten raw as well as cooked for green vegetable.

117. *Rheum webbianum* Royle.

See Sr. No. 258, Chapter VI.

Uses : The long and thick leaf stalk contain vitamin 'C' and is used as a raw or salad or added in custard in Ladakh.

118. *Rhodiola heterodonta* (Hk. f. & T.) A. Boiss.

See Sr. No. 260, Chapter VI.

Uses : The young leaves and tender shoots are edible. The tender parts are boiled and then washed thoroughly with water, then mixed with curd by which a local delicious dish "Tantur" is prepared in Ladakh.

119. *Rhodiola imbricata* Edgew.; *Sedum roseum* (Linn.) Scop.

See Sr. No. 261, Chapter VI.

Uses : The young leaves are used as vegetable in Ladakh.

120. *Rhodiola tibetica* (Hk. f. & T.) Fu.

See Sr. No. 262, Chapter VI.

Uses : The tender shoots are used as vegetable.

121. *Ribes alpestre* Decne.

See Sr. No. 67, Chapter XII.

Uses : The ripen fruits are eaten.

122. *Ribes orientale* Desf.

See Sr. No. 69, Chapter XII.

Uses : The berries are edible.

123. *Rosa webbiana* Wallich ex Royle

See Sr. No. 78, Chapter X.

Uses : The fruits are rich source of vitamin 'C' and consumed by locals when ripen.

124. *Rumex acetosa* Linn.

See Sr. No. 269, Chapter VI.

Uses : The leaves are used as fresh relish.

125. *Rumex angulatus* Rech. f.

Family : Polygonaceae

Name : Shommena (L)

Distribution : Afganistan to South West China, Kashmir & Kumaun.

Habitat : Rocky slopes, Suru and Spiti valleys around 3,110 m.

Plant Characters : A tall stout, deeply grooved, perennial herb. Stem branched from the above. Leaves elongated, oblong, obtuse, petioled. Flowers in lax pedicel equaling the fruiting perianth.

Flowering & fruiting : July to September.

Uses : The young leaves and shoots are eaten raw or as pot-herb.

126. *Rumex hastatus* D. Don.

Family : Polygonaceae

Name : Gungantse (L)

Distribution : From Afganistan to South West China.

Habitat : On cultivated fields & wastelands, Leh valley around 3,260 m.

Plant Characters : A perennial herb. Stem tufted, much branched from the base, branches stiff and spreading. Leaves rhombic deltoid or hastately 3-lobed, entire. Flowers in small whorls, pink or green tinged on terminal panicles. Flower stalk lengthening in fruit. Fruits pinkish, wings papery and delicately net-veined.

Flowering & fruiting : August to September.

Uses : The young leaves are used as vegetable in Ladakh.

127. *Rumex nepalensis* Spreng.

Family : Polygonaceae

Name : Wild Palak (L)

Distribution : Himachal Pradesh to Kashmir & Kumaun.

Habitat : Moist places, Lahaul-Spiti valley around 3,620 m.

Plant Characters : An erect, branched, robust, perennial herb upto 1.0 m. tall. Radical leaves long-petioled, oblong, ovate-oblong or triangular-ovate, acute or obtuse, base widely or narrowly cordate upper sessile, lanceolate. Flowers bisexual in whorls, forming almost leafless long panicle. Nut enclosed in the enlarged inner tepals-segments.

Flowering & fruiting : June to September.

Uses : The leaves are used as vegetable.

128. *Rumex patientia* Linn. spp. *tibeticus* Rech. f.

Family : Polygonaceae

Name : Soma or Shomena (L)

Distribution : From Kumaun to Kashmir & South West China.

Habitat : Wet places & near water channels, Leh and Suru valleys around 3,450 m.

Plant Characters : An erect, robust, perennial herb. Stem terete, upto 110 cm. Leaves linear-lanceolate, entire margin, petioled. Flowers small, yellowish green in spike like cluster forming a long compact inflorescence. Fruits enclosed by rounded net veined valves with acute apex.

Flowering & fruiting : June to September.

Uses : The young leaves are used as vegetable in Ladakh.

129. *Sedum ewersii* Ledeb.

Family : Crassulaceae

Name : Stone Crop (E), Gomni (L)

Distribution : From Afganistan to Himachal Pradesh & Central Asia.

Habitat : Moist slopes & along water streams, Leh and Lahaul valleys around 3,840 m.

Plant Characters : A small succulent, glabrous, perennial herb. Stem usually branched at the base, reddish, 10-20 cm long. Leaves mostly opposite, ovate to rounded with a somewhat wavy margin. Flowers small purple borne in cluster at the top. Seeds ellipsoid, somewhat obovoid.

Flowering & fruiting : July to September.

Uses : The leaves are used as a salad and for making sauce in Ladakh.

130. *Silene conoidea* Linn.

Family : Caryophyllaceae

Distribution : Pakistan to Central Nepal, temperate Eurasia & North Africa.

Habitat : Drier slopes & weed of fields, Suru valley around 3,140 m.

Plant Characters : A stout erect glandular- hairy annual herb. Stem reaching upto 40 cm high. Leaves narrow-lanceolate. Flowers pink borne in erect terminal clusters. Calyx tubular, narrowed upwards. Petals small obovate, slightly notched. Capsule hard, shining, ehclosed in the globular inflated calyx.

Flowering & fruiting : July to September.

Uses : The tender shoots are edible and cooked as vegetable.

131. *Silene setisperma* Majumdar.

Family : Caryophyllaceae

Distribution : From Pakistan to Eastern Nepal.

Habitat : Rocky slopes, Suru valley around 3,040 m.

Plant Characters : A small annual, herb. Stem 15-30 cm high. Leaves elliptic to lanceolate, the lower stalked. Flowers usually solitary and pendulous with distinctive inflated, with brownish petals. Calyx with dark brown ribs. Corolla brown with short rounded lobes.

Flowering & fruiting : July to September.

Uses : The tender shoots are edible among local people.

132. *Solanum nigrum* Linn.

Family : Solanaceae

Name : Makoi (H), Tsigma (L)

Distribution : Kumaun to Kashmir, Nepal & throughout India.

Habitat : Wastelands & cultivated fields, Suru and Nubra valleys around 3,050 m.

Plant Characters : An annual erect herb. Stem branched, reaching upto 60 cm high, glabrous, angled or longitudinally ribbed, dark green. Leaves ovate or oblong-obovate with coarse angular teeth, petiole much shorter than the leaf. Flowers white borne on slender long drooping stalks forming umbel like clusters. Calyx persistent. Corolla 5-lobed with a small tube. Stamens with connate anthers. Berry upto 7 mm diam, rounded, green, yellow or finally black.

Flowering & fruiting : July to September.

Uses : The tender shoot, leaves and fruits are edible.

133. *Sonchus oleraceus* Linn.

Family : Asteraceae

Name : Milk Thistle (E), Khala (L)

Distribution : Kashmir to Western Tibet & India.

Habitat : On higher passes & wasteland, Leh & Spiti valleys around 3,650m.

Plant Characters : An annual herb having milky latex, 45-90 cm high. Leaves lanceolate, pinnatifid, amplexicaule. Flower-heads yellow; seeds narrowly egg shaped, flattish, ribbed with a brush of hairs.

Flowering & fruiting : July to September.

Uses : The young leaves are eaten as vegetable or salad. The young leaves contain vitamin 'C'.

134. *Stellaria palustris* Ehrh. ex Retz.

Family : Caryophyllaceae

Name : Khalta (L)

Distribution : Pakistan to South West China & Tibet.

Habitat : Open slopes & rocks, Suru valley around 3,020 m.

Plant Characters : A perennial herb. Stem erect, slender, branched. Leaves 3-6 cm, sessile, with rounded base. Flowers pedicelled, bracteate. Sepals and petals oblong. Capsule ovoid.

Flowering & fruiting : July to September.

Uses : The tender shoots are eaten as potherb.

135. *Taraxacum officinale* Wigg.

See Sr. No. 303, Chapter VI.

Uses : The leaves are used as fresh vegetable and roots are also edible. A very nutritious wild edible plant.

136. *Taraxacum sikkimensis* Wigg.

See Sr. No. 304 Chapter VI.

Uses : The whole herb is used as vegetable and roots are eaten raw.

137. *Thlaspi alpestre* Linn. ex Hk. f. & T.

See Sr. No. 308, Chapter VI.

Uses : The herb is edible as vegetable.

138. *Thlaspi arvense* Linn.

See Sr. No. 309, Chapter VI.

Uses : The plant is used as substitute of *T. alpestre*. The young herb is used for vegetable after proper boiling.

139. *Thymus serpyllum* Linn. *T. linearis* Benth.

See Sr. No. 66, Chapter VII.

Uses : The leaves are used as flavouring agent in fresh as well as in dry form.

140. *Tragopogon pratense* Linn.

See Sr. No. 313, Chapter VI.

Uses : The tender shoots/heads are edible.

141. *Tulipa stellata* Hk. f. var. ***chrysantha*** Boiss.

See Sr. No. 91, Chapter X.

Uses : The roots are edible as raw.

142. *Tribulus terrestris* Linn.

See Sr. No. 314, Chapter VI.

Uses : The oil extracted from fruits is sometimes used for cooking purposes.

143. *Urtica dioica* Linn.

Family : Urticaceae

Name : Bichhu booti (H), Stinging Nettle (E)

Distribution : Pakistan to Bhutan, Nepal & India.

Habitat : Open fields & roadside, Suru, Zanskar and Lahaul valleys between 3,450- 3,950 m.

Plant Characters : A herbaceous, monoecious or dioecious. Stinging hair copious, leaves ovate-cordate, oblongserrate, stipules usually free. Cymes unisexual rarely androgynous, inner fruiting sepals longer than the outer. Perianth white green, hairy.

Uses : The leaves and tender shoots are boiled properly and cooked as green vegetable or dehydrated.

144. *Urtica hyperborea* Jacq. ex Wedd.

Family : Urticaceae

Name : Stinging Nettle (E), Dzatsutt or Zozot (L)

Distribution : Pakistan to Bhutan, Tibet & Mangolia.

Habitat : Rocky slopes along roadside, Leh and Changthang valleys around 4,960 m.

Plant Characters : A densely tufted, perennial undershrub with stinging hairs. Leaves heart-shaped, tooth-edged, long, stalkless with soft hair between the stinging hairs. Flowers green, crowded in clusters. Fruits flat enwrapped in the sepals, 1-seeded.

Flowering & fruiting : July to September.

Uses : The young leaves are cooked as vegetable during summer or solar dehydrated and stored for winter use. It is said to provide warmth. It is generally used for making "Thupka", a local dish in Ladakh. Usually the dried leaves are sold in local market.

9

Fodder Plants

In general the agriculture and allied activities are the major source of economy of throughout the Himalayan region. The fodder plants and their availability in trans Himalayan region play important role in the growth of animal husbandry. In this region the cultivation of fodder is restricted to the lower hills. Beyond this zone, the cultivation of fodder is not practical because of adverse climate and the land constraints. Almost all the available cultivable land is used for the cultivation of cereals, pulses, oil seeds and vegetables. Rangelands and pasturelands, comprising approximately 70 per cent of the total geographical area are the major sources of forage for sheep, goats, cattle and other domestic and wild animals.

Ladakh comprising of two districts namely Kargil and Leh has tentative population of 84, 362 cattle (native & crossbreds), 4,63,661 goats & 2,62,473 sheep of various breeds, 2,055 equines, 36,238 yaks/crossbreds with cattle and about 200 double humped camels. Effective and sustainable livestock development are not possible without a corresponding scientific and organised fodder production. These two issues have to be co-related. To sustain existing the huge population of animals economically it is essential to pay serious attention towards the domestication of potential wild fodder species through development of package and practices for their production.

Nomadic grazers (Khadwals, Gaddis, Gujjars, Khas) use sub-alpine and alpine pastures for rearing their livestock. Due to ever-increasing demand of animal products, the livestock population has increased manifolds, thereby increasing the grazing pressure on these pastures. This has resulted in the deterioration of the grass cover as well as valuable forest species. As a result, a significant area of this natural resource has been covered by noxious plant species like *Anemone, Adonis, Aquilegia, Clematis, Artimisia, Echinops, Ranunclus, Stipa, Sorbinia* and *Sambucus*. All these factors have led to the decrease in carrying capacity of these pastures.

The lower alpine zones are covered with lovely meadows, pasturelands, moraines, screes and marshes. The higher passes, which are covered with snow for more than six months in a year and which has a bright sunshine, (with more quantity of UV rays), is more amicable to vegetation during the brief summer (Polunin *et. al*, 1989). The constraints at high altitudes, which need to be recognized are:

Physical	Short growing season, low plant growth rates, therefore limited potential
Soils	Often shallow, acidic, low P availability and degraded
Ecological	Fragile ecosystems, ingress of weeds, devoid of forest
System	Traditional systems, often migratory, pastorals and farmers conservative and averse to risk, lack of cash flow to fund, improvement and communal grazing of range

The high altitude systems, temperate and alpine systems at altitudes of > 2500 to < 5000 m; include the entirely pastoral migratory goat/ sheep systems. There is a range of temperate grasses with species which are adaptable to the sub-humid and semi-arid zones of the Himalaya and which potentially could be used as fodder crops on arable land or improve alpine and sub alpine range some of which is degrading through overuse. In the lower Himalayan region also the natural grasslands are under heavy grazing pressure and have been badly degraded. Insignificant land area is under fodder cultivation and there is least scope of increasing this.

Status of pasture land

Pasturelands are the characteristic of trans Himalayan cold arid zones. Almost every village has pasture land / area or community land for grazing their livestock. The people of Chanthang, Zanskar, Lahaul and Spiti valleys are mainly dependent upon pasturelands which are natural vast grazing ground for their animals since a long times. The local people are fully dependent on agriculture and rearing livestocks and these are the major source of their income. It seems that since the inception of agri-pastoral system, the people of these areas have developed systematic grazing pattern so that they could feed their animals on natural pasturelands round the year.Usually alpine pastures are grazed in summer, area below it in autumn and low lying plain valleys in winter and spring. Prolonged winters, climatic changes and allied activities adversely affect high pasture lands in Changthang, Zanskar and Spiti areas, in turn manifold effects on livestocks such as reduction in body weight, milching capacity, mortality due to starvation, increased susceptibility due to poor sanitation and low quality fodder, force disposal of part of livestock due to fodder scarcity. Degradation of pastureland is multidimensional problem and simple grazing policy may not be viable solution.

Field studies

However, integration of forage crops in the existing crop sequences, growing food cum fodder crops, intensive forage cultivation, raising improved grasses on terrace-risers, developing suitable cropping systems for rain fed and irrigated areas, improving silvi and horti pastoral systems in addition to natural pasture and wasteland improvement are some of the approaches to solve the forage problem. Introduction of improved grass/legume species in the native vegetation not only helps in increasing total biomass production but also improve the quantity of the produce. However, the introduction of improved species in the existing vegetation particularly with seed is not an easy task, unless initial competition from the native vegetation is checked by some means.

"Ole" (Lucerne/Alfalfa) is the most common and cultivated crop in this region. It is a cross pollinated crop, and as such a mixture of Alfalfa is grown. It produces highly nutritive and excellent quality hay. Its deep root system enables it to tolerate the harsh climatic conditions for many years, making it most economical to grow. There are three species of Alfalfa ; *Medicago sativa, M. falcata* and *M. media,* found in Ladakh. The landraces grown in this region are more resistant to cold and drought.

Field Research Laboratory, Leh has surveyed most of parts of Ladakh and Lahaul-Spiti, to identify the suitable wild fodder plants for exploring possibilities to develop the package and practices of fodder production in these areas. Besides three species of alfalfa being cultivated, many other potential species may be cultivated successfully on commercial scale without disturbing cultivable area by putting very less efforts. This lab has started cultivation practices on *Prangos pabularia, Cicer microphyllum, Aconogonum tortuosum* and *Rheum webbianum* etc during last few years. For developing package and practices of other potential fodder species, germplasm collection, propagation and domestication of fodder plants, the trials have already been initiated.

The potential fodder species have been described in alphabetical order alongwith families, English (E), hindi (H) / local names (L), distribution, habitat, brief botanical notes and uses in the enumeration. (Table- 6). In addition, the nutritional profile of 22 potential fodder plants have been appended in table- 6a covering crude protein, fiber, cellulose etc.

Enumeration of Fodder Flora

1. *Aconogonum alpinum* All.

Family : Polygonaceae

Name : Nyalo (L)

Distribution : From Kashmir to Himachal Pradesh & Afganistan.

Habitat : On wet alpine pastures, Lahaul valley around 3,985 m.

Plant Characters : An erect, dense perennial shrub. Stem smooth, 70-100 cm with nodded spreading branches. Leaves entire, smooth, elliptic to ovate, acute. Flowers light yellow to cream-coloured borne in small dense cylindrical branches.

Flowering & fruiting : July to September.

Uses : It is used as fodder species in the form of fresh and dry both, widely collected by the local inhabitants.

2. ***Aconogonum tortuosum*** (D. Don.) Hara. *Polygonum tortuosum* D. Don.

Family : Polygonaceae

Name : Serpalulu / Itimeng / Nyalo (L)

Distribution : From Kashmir to Sikkim & Afganistan.

Habitat : Dry stony slopes & along roadsides, Zanskar, Changthang and Spiti valleys around 3,875 m.

Plant Characters : A low dense perennial shrub. Stem smooth, shining, 30-50 cm with spreading branches. Leaves entire, leathery, elliptic to broadly ovate, blunt to acute. Flowers cream-coloured borne in short dense cylindrical clusters.

Flowering & fruiting : July to September.

Uses : It is one of the most important wild fodder species. The plant is used green as well as in dry form. It is harvested during August/September from natural localities and the dry mass is finally stored on the house roofs after making loose bundles for prolonged winter uses.

3. ***Acroglochin persicarioides*** (Poir.) Moq. *Amaranthas persicarioides* Poir.

Family : Chenopodiaceae

Name : Thiangchi (L)

Distribution : From Kashmir to Himachal Pradesh.

Habitat : Along roadsides & dry places, Changthang and Nubra vallys between 3,980-4,285 m.

Plant Character : An erect, glabrous, annual herb. Stem branched. Leaves alternate, obtuse, lobulate. Flowers green. minute, sessile, borne on axillary terminal cymes.

Flowering & fruiting : July to September.

Uses : It is used as a fodder.

4. *Agropyron jacquemontii* Hk.

Family : Poaceae

Name : Zamak (L)

Distribution : Western Tibet to Kumaun & Nubra.

Habitat : On cultivated fields, Nubra valley around 3,320 m.

Plant Characters : A rhizomatous, erect, tufted perennial grass. Leaves convolute, glabrous, tip obtuse. Spikelets subterminal, close together, awns 4-6 fid, sometimes strongly recurved, rarely straight; keels ciliolate.

Flowering & fruiting : July to September.

Uses : It is used as a fodder.

5. *Agropyron repens* (Linn.) P. Beauv.

Family : Poaceae

Name : Zamak (L)

Distribution : From Europe to temperate Asia & Kashmir.

Habitat : On cultivated fields, Leh and Lahaul valleys around 3,355 m.

Plant Characters : A tufted perennial grass. Stem erect upto 80 cm, glabrous, fistular. Leaves lanceolate, blade upto 30 cm, smooth, glabrous. Spike upto 20 cm long. Spikelets arranged alternately on zig-zag rachis. Glumes 2 unequal. Lemmas 3 cm long, awaned. Caryopsis oblong.

Flowering & fruiting : Mid June to September.

Uses : It is one of the important wild fodder grasses of cold desert areas. It is harvested after maturity during September and stored for winter uses.

6. *Ajuga bracteosa* Wall. ex Benth.

See S. No. 9, Chapter VI.

Uses : It is used as a fodder.

7. *Alopecurus himalaicus* Hk. f. & T.

Family : Poaceae

Name : Ziam (L)

Distribution : From Kashmir to Tibet & Kumaun.

Habitat : Sandy slopes, Zanskar valley around 3,750 m.

Plant Characters : An erect, annual, grass upto 45 cm with a decumbent base, tufted. Leaves erect to spreading, upper shorter lanceolate, acuminate. Upper leaf sheath inflated. Panicle oblong or subcapitate. Spikelets 5-6 mm, membranous, flaccid. Awn longer.

Flowering & fruiting : July to September.

Uses : It is used for fodder.

8. *Androsace primuloides* Duby

Family : Primulaceae

Name : Pemug (L)

Distribution : Kashmir to Sikkim.

Habitat : Along road sides & dry places, Zanskar & Suru valleys, between 3,510-3,980 m.

Plant Characters : A small perennial herb. Stem usually 10-15 cm tall. Leaves rosette, spathulate, numerous, sessile or petioled. Bracts silky. Flowers in lax or dense umbel. Corolla rose-purple, annulate. Capsule oblong.

Flowering & fruiting : July to August.

Uses : It is usually grazed by sheep and goats.

9. *Arctium lappa* Linn.

See Sr. No. 24, Chapter VI.

Uses : The aerial parts are grazed by sheep and goats.

10. *Astragalus adesmiaefolius* Benth. ex Bunge

Family : Fabaceae

Distribution : From Kashmir to temperate Himalayan belts.

Habitat : On stony & drier areas, Nubra valley around 2,955 m.

Plant Characters : A glabrous perennial plant. Stem woody, branched. Leaves pinnate; leaflets 11-13, green. Flowers peduncled in dense raceme. Stipules large, foliaceous. Pod flat and glabrous.

Flowering & fruiting : July to mid September.

Uses : Used as fodder plant.

11. *Astragalus cicerifolius* Royle ex Fisch.

Family : Fabaceae

Name : Yakpo (L)

Distribution : From Pakistan to Kashmir.

Habitat : Common on drier parts, Suru valley around 4,230 m.

Plant Characters : A bushy, perennial herb. Stem branched, woody, spiny, branched from the base. Leaves pinnate 3-7 cm long, leaflets 21-27, with linear stipules. Flowers yellow in clusters, pedicellate. Bracts setaceous. Pod turgid, oblong many seeded.

Flowering & fruiting : July to September.

Uses : The plant is used for fodder.

12. *Astragalus ciliolatus* Benth ex Bunge.

Family : Fabaceae

Distribution : Temperate parts of India to Central Asia, China.

Habitat : On dry open slopes, Leh valley around 3,575 m.

Plant Characters : A herbaceous perennial. Stem branched, pubescent. Leaves petioled, pinnate, bracteate. Leaflets thin oblong finaly pubescent on the lower surface, 15-17 per leaf. Flowers pale-blue, in dense racemes. Calyx campanulate with setaceous teeth. Pods small, silky.

Flowering & fruiting : July to September.

Uses : Used for fodder.

13. *Astragalus confertus* Benth. ex Bunge.

Family : Fabaceae

Name : Singyalo (L)

Distribution : From Pakistan to Central Nepal.

Habitat : On cultivated fields and dry slopes, Nubra valley 3,645 m.

Plant Characters : An erect, herbaceous, perennial plant. Stem branched, hairy. Leaves pinnate, pedicillate, leaflets 11-13. Flowers peduncled in groups. Calyx toothed, adpressed with black hairs. Pod stalked, unilocular.

Flowering & fruiting : June to August.

Uses : Used for fodder.

14. *Astragalus frigidus* (Linn.) A. Gary.

See Sr. No. 17, Chapter XII.

Uses : Used for fodder.

15. *Astragalus gracilipes* (Linn.) A. Gary.

Family : Fabaceae

Name : Flangu (L)

Distribution : From Pakistan to J & K, Himachal Pradesh and Nepal.

Habitat : Cultivated fields, Zanskar and Suru valleys between 3,505-3,905 m.

Plant Characters : An annual weed. Stem branched, densely clothed with ascending strong white hairs. Leaflets 9-13, oblong, obtuse, stipulate minute. Flower heads dense, pale yellow and peduncled. Calyx tubular. Pod cylindrical, pubescent little recurved 10-12 seeded.

Flowering & fruiting : July to mid September.

Uses : Usually grazed by animals.

16. *Astragalus heydei* Baker

Family : Fabaceae

Name : Yanglang (L)

Distribution : From Pakistan to Kashmir, Himachal Pradesh, Kumaun.

Habitat : Common on rocky slopes, Changthang valley around 4,230 m.

Plant Characters : An erect, perennial herb. Rhizome slender, woody. Leaves long; leaflets 15-19, oblong, crowded. Flowers 2-4, rose-red in umbel head. Bracts minute. Calyx more or less campanulate. Pod oblong, inflated, glabrous, many seeded.

Flowering & fruiting : July to August.

Uses : It is used for fodder.

17. *Astragalus himalayanus* Klotz.

Family : Fabaceae

Distribution : From Kashmir to Kumaun.

Habitat : Open rocky areas, Suru valley around 4,250 m.

Plant Characters : An erect or spreading, perennial herb. Stem 30-35 cm high, branches from the base. Leaves usually 3-5 cm; leaflets 19-35, oblong, blunt or notched, pale green with white hairs. Flowers mauve borne in rather dense rounded clusters. Calyx with black and white hairs. Pod linear, oblong, with black hairs.

Flowering & fruiting : July to September.

Uses : Used as fodder.

18. *Astragalus munroi* Benth. ex Bunge

Family : Fabaceae

Name : Serma (L)

Distribution : From Pakistan to Bhutan, J & K, Himachal Pradesh and Tibet.

Habitat : Drier areas, Suru valley around 3,990 m.

Plant Characters : A prostrate and perennial woolly herb. Stem branched, reaching upto 40 cm tall, branches densely villous. Leaves imparipinnately compound; leaflets 19-23, lateral leaflets opposite to subopposite, oblong, entire, obtuse, villous on both sides. Flowers yellow in axillary racemes. Pod broad, inflated, villous.

Flowering & fruiting : July to September.

Uses : Used for fodder.

19. *Astragalus multiceps* Wall.

See Sr. No. 48, Chapter VI.

Uses : Used for fodder.

20. *Astragalus oxyodon* Baker.

Family : Fabaceae

Name : Sharma (L)

Distribution : From Pakistan to Bhutan, Nepal and Western Himalayan region.

Habitat : On alpine slopes Nubra and Changthang valleys between 3,030-3,460 m.

Plant Characters : A perennial herb. Stem short, branched, slender, ascending, clothed with adpressed silky hair. Leaflets pinnate, 15-17. Flowers few. Calyx adpressed, black & white hairs. Pod hairy, many seeded.

Flowering & fruiting : July to September.

Uses : Used for fodder.

21. *Astragalus peduncularis* Royle ex Benth.

Family : Fabaceae

Name : Gyalo (L)

Distribution : From Kashmir to Himachal Pradesh.

Habitat : On cultivated areas, Zanskar and Spiti valleys around 3,650 m.

Plant Characters : An annual pubescent herb. Stem tall, glabrous, branched. Leaves long. Leaflets obtuse or subacute, glabrous above. Stipule deltoid, free, membranous. Flowers many in racemes. Bracts lanceolate. Calyx tubular. Keels and wings nearly equal. Pod linear, unilocular.

Flowering & fruiting : July to September.

Uses : Used for fodder.

22. ***Astragalus rhizanthus*** Royle ex Benth.

See Sr. No. 49, Chapter VI.

Uses : Used as fodder.

23. ***Astragalus subuliformis*** DC.

Family : Fabaceae

Name : Rungentso (L)

Distribution : From Pakistan to Himachal Pradesh, J & K and Kumaun.

Habitat : Open rocky slopes, Leh and Changthang valleys between 3,560-3,910 m.

Plant Characters : A prostrate perennial herb with densely adpressed white - pubescent stems. Leaves imparipinnately compound; leaflets 5-15, linear to subuliform, subobtuse, glabrous above. Flowers yellow in pedunculate racemes. Pod cylindrical, densely pubescent, hairs white, appressed.

Flowering & fruiting : July to September.

Uses : Used for fodder.

24. ***Astragalus webbianus*** Grahm. ex Benth.

Family : Fabaceae

Name : Khangskar (L)

Distribution : From Afganistan to Kashmir, Himachal Pradesh and temperate Alpine zones.

Habitat : Alpine pastures, Zanskar and Lahaul valleys around 4,310 m.

Plant Characters : A densely tufted perennial herb. Stem branched from the base. Tufts densely caespitose. Stipule linear or lanceolate. Leaves imparipinnately compound; leaflets oblong, glabrous. Flowers yellow, subsessile. Pods oblong, glabrous, 15-20-seeded.

Flowering & fruiting : July to September.

Uses : Used for fodder.

25. ***Astragalus zanskarensis*** Benth ex Bunge.

Family : Fabaceae

Name : Nanchu (L)

Distribution : From Pakistan to Himachal Pradesh, J&K and Uttranchal.

Habitat : On grazing grounds & pasturelands, Zanskar valley around 3,640m.

Plant Characters : A low perennial under-shrub. Rootstocks very stout, creeping, stem simple. Leaves with ovate or rounded blade, 20 cm, glabrous with crenate margins. Flower pink or white in a drooping cluster. Calyx hairy with marginal bristles. Corolla yellow. Fruits ovate-lanceolate.

Flowering & fruiting : July to September.

Uses : Used for fodder.

26. *Avena fatua* Linn.

Family : Poaceae

Name : Jangali Jai (H), Wild Oat (E), Yupo (L)

Distribution : Cultivated fields in Punjab, West Bengal, Assam, Maharashtra, J&K, Himachal Pradesh and Uttranchal.

Habitat : Cultivated fields, Leh and Lahaul valleys around 3,075 m.

Plant Characters : An erect, annual grass. Stem upto 70 cm, fistular, glabrous, many, arising from the base. Leaves linear, lanceolate with scabric margins. Panicles pyramidal, much-branched. Spikelets oblong-lanceolate, 2-3 flowered, pendulous. Glumes 2, equal, oblong. Lemmas oblong finely hairy with longer awn. Stamens with long anthers and ovary with feathery style. Caryopsis oblong, slightly hairy at the top.

Flowering & fruiting : July to September.

Uses : It is the most common wild fodder crop mixed with wheat straw and fed to the domestic animals as green as well as dry. The crude protein content is about 6.73% which is quite useful as fodder. It is harvested during September and stored for winter use.

27. *Avena sativa* Linn

Family : Poaceae

Name : Jai (H), Common Oat (E), Yupo (L)

Distribution : Originated near Eastern Centre comprising Asia minor.

Habitat : Cultivated in all the valleys since ages, 4,350 m Ladakh.

Plant Characters : An annual or perennial grass. Stem stout, hollow, upto 90 cm long. Leaves long, linear, lanceolate. Glumes very long-acuminate. Rhachilla disarticulating below glume III.

Flowering & fruiting : July to September.

Uses : It is cultivated mainly for food, feed and fodder.

28. *Barbarea vulgaris* R. Br.

See Sr. No. 32, Chapter VIII.

Uses : The plant is used as fodder.

29. *Bothriochloa ischaemum* (Linn.) Keng.

Family : Poaceae

Name : Zamak (L)

Distribution : From temperate Western Himalayas to almost throughout India.

Habitat : Cultivated fields, Leh and Lahaul valleys around 3,200 m.

Plant Characters : A perennial tufted grass. Stem erect, 30-70 cm, unbranched, glabrous. Leaves crowded at the base, linear-lanceolate, margins fringed with hairs. Panicle long. Spikelets paired, sessile, usually fertile, awned. Pedicelled spikelets hairy, awnless. Caryopsis oblong, small, glabrous.

Flowering & fruiting : July to September.

Uses : The plant is used as fodder.

30. *Bothriochloa pertusa* (Linn.) Cams.

Family : Poaceae

Name : Zamak (L)

Distribution : Temp Western Himalayas from Kashmir to Garhwal.

Habitat : Alpine slopes, Zanskar valley around 4,450 m.

Plant Characters : A tufted perennial grass. Stem erect or decumbent, slender. Leaves narrowly linear-lanceolate. Sheaths compressed. Spikes 3-20, slender, silky, rachis glabrous. Upper glume of the sessile spikelets obtuse, awnless. Grain glabrous, shining.

Flowering & fruiting : July to September.

Uses : It is used as fodder grass.

31. *Brachyactis pubescens* (DC.) Aitch. *Conyza pubescens* DC.

Family : Asteraceae

Distribution : Kashmir to Kumaun and Tibet.

Habitat : Moist sandy slopes, Changthang valley around 4,215 m.

Plant Characters : An erect or ascending, stout, glandular annual herb. Leaves obovate, toothed, with a short petiole. Involucral bracts small. Heads many, sessile or peduncled, axillary and terminal. Receptacle broad. Achenes obovate.

Flowering & fruiting : June to August.

Uses : Used for fodder.

32. *Bromus gracillimus* Bunge.

Family : Poaceae

Name : Piantso (L)

Distribution : From Kashmir to other parts of India, Bhutan.

Habitat : Abundant on cultivated fields, Suru valley around 3,125 m.

Plant Characters : An erect, annual grass. Stem slender, smooth. Leaves linear, flaccid, leaf sheath glabrous. Ligule ovate-lanceolate. Panicle oblong, erect. Branches 3-5; spikelets 3-6 flowered. Grain linear-oblong.

Flowering & fruiting : June to July.

Uses : Used as fodder plant.

33. *Bromus inermis* Leyss.

Family : Poaceae

Name : Piantso (L)

Distribution : Abundant from Nepal to Bhutan, Pakistan, china & North Europe.

Habitat : Moist places, Leh and Spiti valleys around 3,150 m.

Plant Characters : A tufted perennial grass. Stems many, arising from the rootstock, 25-75 cm long. Leaves linear, pubescent, margin hairy. Ligule truncate, very short, toothed. Panicles erect compact. Spikelets linear, long, flattened, 9-13 flowered. Glumes 2, equal, hairy. Lemmas 5-9 nerved, covered with stiff hair. Palea transparent, loosely ciliate, lodicules absent. Caryopsis oblong-lanceolate, glabrous.

Flowering & fruiting : July to September.

Uses : It is used for fodder.

34. *Bromus japonicus* Thunb. ex Murr.

Family : Poaceae

Name : Piantso (L)

Distribution : From Himalaya to Bhutan, Europe, Asia, China & Japan.

Habitat : Common on alpine rocks, Suru, Changthang and Spiti valleys around 3,250 m.

Plant Characters : An erect, annual or biennial grass. Leaf-blades upto 15 cm long, glabrous or hairy, sheaths pubescent. Panicle 6-20 cm long, very lax and spreading, the branches and pedicles very long, filiform and flexuous. Spikelets ovate to oblong lanceolate, 7-15 flowered. Lemma 2-toothed at apex, awned below the tip.

Flowering & fruiting : July to September.

Uses : It is used for fodder.

35. *Bromus oxyodon* Schrenk.

Family : Poaceae

Name : Piantso (L)

Distribution : From Kashmir to Kumaun, Bhutan, Pakistan & china.

Habitat : Cultivated fields, Nubra & Zanskar valleys around 3,150 m.

Plant Characters : An erect, annual grass. Stem erect, usually unbranched many clums arising from the base. Leaves linear, narrow, subacute, hairy. Young sheath villous. Ligule ovate, lanceolate. Panicle usually contracted, erect. Spikelets pale-green, densely 5-10 flowered. Glumes 2, nearly equal. Grains oblong, glabrous.

Flowering & fruiting : July to September.

Uses : The plant is used as fodder.

36. *Bromus scoparius* Linn.

Family : Poaceae

Name : Piantso (L)

Distribution : Throughout the world.

Habitat : Cultivated fields & moist places, Suru valley around 3,125 m.

Plant Characters : An annual grass. Stem erect, 25-60 cm high, usually unbranched, several clums arising from the base. Leaves linear covered with dense stiff hairs. Panicles erect, compact. Spikelets linear, flattened, 9-13 flowered. Glumes 2, nearly equal, boat shaped, acute, stamens 3 with linear anthers longer than the slender filaments. Caryopsis oblong lanceolate, glabrous.

Flowering & fruiting : July to September.

Uses : It is used as fodder grass.

37. *Bromus tectorum* Linn.

Family : Poaceae

Name : Zamak (L)

Distribution : From Kashmir to Kumaun & Arunchal Pradesh, Europe, North Africa.

Habitat : Rocky & sandy slopes, Leh and Lahaul valleys around 3,595 m.

Plant Characters : A tufted, perennial grass. Stem erect, glabrous. Leaves long, linear, acute, hairy. Sheath hairy, rarely glabrous. Panicle long faccid, nodding. Spikelets green, 4-10 flowered. Flowering glumes long, narrow. Grain linear-oblong, dorsally compressed.

Flowering & fruiting : July to September.

Uses : It is used as fodder.

38. *Bupleurum falcatum* Linn.

Family : Apiaceae

Name : Sickle-leaved Hae's ear (E), Zeerakarpu (L)

Distribution : Pakistan to Bhutan, temperate Eurasia, Kashmir and Bihar.

Habitat : Cultivated fields, Zanskar and Lahaul valleys around 3,580 m.

Plant Characters : An erect, perennial herb. Stem slender, branched, reaching upto 1 m high. Stem leaves linear-lanceolate and usually curved. Flowers yellow in compound umbels. Branchlets lanceolate. Primary rays 5-8. Fruits oblong, 4-ribs, distinct.

Flowering & fruiting : July to September.

Uses : Usually grazed by the animals.

39. *Bupleurum longicaule* Wall. ex DC.

Family : Apiaceae

Name : Zeerakarpu (L)

Distribution : Pakistan to Bhutan.

Habitat : Moist & shady places, Suru and Lahaul valleys 2,890 m.

Plant Characters : A perennial herb. Stem 15-40 cm high, with several branches, erect. Leaves 8-10 cm, narrow lanceolate, long-pointed. Flowers blackish-brown, 2-8 secondary umbels and subtended by an involucre of 6-8 oval green bractioles. Fruits oblong, ribs winged.

Flowering & fruiting : July to September.

Uses : It is used as fodder.

40. *Bupleurum thomsonii* Clarke ex Hk. f. & T.

Family : Apiaceae

Name : Zeerakarpu (L)

Distribution : Kashmir to Bhutan and temperate Eurasia.

Habitat : Moist & cultivated fields, Zanskar valley around 3,670 m.

Plant Characters : An erect, glabrous herb. Stem 30-40 cm, with many lateral branches. Lower leaves linear to lanceolate, upper leaves linear oblong, subobtuse, mucronate small, obovate-lanceolate. Fruits oblong, yellowish brown with prominent ridges.

Flowering & fruiting : June to September.

Uses : It is used as fodder.

41. *Calamagrostis emodensis* Griseb.

Family : Poaceae

Name : Yanglo (L)

Distribution : From Kashmir to Bhutan & temperate Himalayas.

Habitat : Alpine passes, Nubra and Suru valleys around 3,730 m.

Plant Characters : A perennial grass. Stem many arising from the base. Leaves long broad, flat, glaucous, scaberulous. Panicle dense or diffuse much branched. Spikelets crowded, awn exserted. Glume 2, boat shaped. Lemmas ovate-oblong. Grain oblong.

Flowering & fruiting : July to September.

Uses : Used as fodder.

42. *Calamagrostis garhwalensis* C. E. Hubb. *et* Bor.

Family : Poaceae

Name : Yanglo (L)

Distribution : Kashmir to Bhutan & Tibet.

Habitat : Open slopes, Suru and Lahaul valleys around 3,150 m.

Plant Characters : A slender, stout perennial, grass. Branches arising from the creeping stem. Leaves acute, flat, narrow. Ligule acute. Panicle dense. Spikelets 4-8 mm long, narrow. Glumes 3, unequal. Palea often minute.

Flowering & fruiting : July to September.

Uses : Used for fodder.

43. *Calamagrostis stolizkai* Hk. f. & T.

Family : Poaceae

Name : Singshi (L)

Distribution : From Kashmir to Tibet & Zanskar.

Habitat : Higher passes, Zanskar valley around 4,350 m.

Plant Characters : A stout, perennial, grass. Stem short, stout. Leaves erect, convolute. Spikelets in branched panicle. Panicle short, oblong, contracted. Spikelets shortly pedicelled. Awn about twice as long as the spikelets, shortly exerted.

Flowering & fruiting : July to September.

Uses : Used as fodder.

44. ***Carex infuscata*** Nees.

Family : Cyperaceae

Distribution : From Kashmir to Sikkim & Kumaun.

Habitat : On alpine meadows, Changthang & Lahaul valleys around 4,870m.

Plant Characters : An erect, tufted, perennial herbaceous plant. Stem short, drooping above. Leaves subequal, equal or slightly exceeding the stems, flat, apices shortly acuminate. Inflorescence consisting of 3-4 ovate, spikelets. Bracts foliaceous.

Flowering & fruiting : July to September.

Uses : Used for fodder.

45. ***Carex obscura*** Nees.

Family : Cyperaceae

Distribution : From Kashmir to Kumaun & in Western Himalayas.

Habitat : On moist alpine places, Changthang valley around 4,860 m.

Plant Characters : An erect, densely tufted sedge, perennial herbaceous plant. Rhizome elongated, woody, shortly stoloniferous. Stem erect, compressed, trigonous. Leaves equal or much exceeding the stem, flat, apices long-attenuated, sheathing. Inflorescence consisting of 3-8 oblong or cylindric, spikelets. Utricle obovate or elliptic, trigonous, inflated.

Flowering & fruiting : June to September.

Uses : Used as fodder.

46. ***Chaerophyllum acuminatum*** Lindl.

See Sr. No. 43, Chapter VIII.

Uses : The whole plant is used as fodder.

47. ***Chesneya cuneata*** (Benth.) Ali. *Caragana cuneata* (Benth.) Baker.

Family : Fabaceae

Name : Leemopa (L)

Distribution : From Pakistan to Kashmir.

Habitat : Wastelands & stony slopes, Suru and Lahaul valleys between 3,019-3,604 m.

Plant Characters : A tufted, tomentose, silky perennial, herb with thick tap roots. Stem branched, 15-25 cm high. Leaves pinnate, leaflets oblong to obovate, with blunt or notched apex velvetly haired. Flowers pink-purple, 1-4 in clusters. Calyx cylindric, velvetly haired.

Flowering & fruiting : July to September.

Uses : Usually grazed by wild animals.

48. *Chloris virgata* Sw.

Family : Poaceae

Distribution : Pakistan to Kashmir & North temperate Himalayas, China, Mongolia, South Africa.

Habitat : Wastelands, Zanskar valley around 3,750 m.

Plant Characters : A perennial tufted, grass. Stem rather stout, flattened, decumbent, usually leafy. Leaves broad, flat, acute, upper sheaths more or less inflated. Spikelets in branched panicle.

Flowering & fruiting : July to September.

Uses : Used as fodder.

49. *Cicer microphyllum* Benth. *C. soongaricum* Auct. non Steph.

Family : Fabaceae

Name : Wild Gram (E), Seri or Srad-Kar (L)

Distribution : From Afganistan to Western Nepal, Kashmir & Kumaun.

Habitat : On pastures and dry slopes, Changthang, Leh, Zanskar and Lahaul-Spiti valleys between 4,230-4,870 m.

Plant Characters : A very diffuse, erect, spreading shrublet, glandular, herb. Stem upto 50 cm, much branched. Leaves pinnate, ending in a coiled tendril, sessile, cuneate. Leaflets many in widely spaced pairs, ovate-wedge shaped, toothed apex. Flowers solitary or two, in the axils of leaves, stalked, axillary 2-3 cm long. Calyx hairy, lobed. Corolla bluish-purple. Pods linear-oblong, beaked, and silky-hairy.

Flowering & fruiting : Mid June to September.

Uses : The whole plant is used as fodder. The green plant is generally grazed by wild animals especially sheep and goats and the whole plant is harvested in the month of September and dried properly for prolonged winter use. It has good protein content about 15.30% which is very much useful for milch and young animals.

50. *Clematis orientalis* Linn.

Family : Ranunculaceae

Name : Clematis (E), Tiktikma or dByi-mong dkar-po (L)

Distribution : From Pakistan to Uttranchal, Central Nepal, China and Eurasia.

Habitat : Wastelands, Zanskar valley between 3,370-4,210 m.

Plant Characters : A prostrate, woody perennial climber. Stem branched; branchlets sometimes puberulous. Leaves opposite, bipinnate, lanceolate,-ovate; leaflets ovate, deeply cleft. Flowers creamish-yellow, axillary or terminal. Filaments membranous, tapering from a broad base, silky. Achenes oblong, silky, hairy.

Flowering & fruiting : Mid June to September.

Uses : The whole plant is used for fodder.

51. *Clematis tibetana* Kuntze.

See Sr. No. 77, Chapter VI.

Uses : The leaves are used as fodder.

52. *Convolvulus arvensis* Linn.

Family : Convolvulaceae

Name : Tiktikma (L)

Distribution : From Pakistan to Central Nepal, North & South temperate Zones.

Habitat : Cultivated fields & wastelands, Leh, Nubra and Lahaul valleys 3,540 m.

Plant Characters : A slender climbing, prostrate, creeping or twining, perennial herb. Leaves ovate-lanceolate, with hastate or sagittate base, basal lobes petioled. Flowers usually solitary, axillary, borne on stalks, white or pink or lilac, funnel shaped. Capsule opening by 4-valve, hairless, ovoid-globose.

Flowering & fruiting : Mid June to September.

Uses : It is used for fodder.

53. *Crepis stoliczkai* Clarke.

Family : Asteraceae

Name : Richkut (L)

Distribution : From Pakistan to Nepal and Tibet.

Habitat : Dry rocky slopes, Suru valley around 3,016 m.

Plant Characters : A perennial herb. Leaves few, narrow, usually radial, petioled, pinnitifid. Flowering stem erect, sparingly branched. Flower-heads 2-3 cm long, cylindric, puberulus, cylindric. Involcral bracts linear-obtuse. Achenes ribbed, red-brown, pappus white.

Flowering & fruiting : July September.

Uses : Used for fodder.

54. *Crepis tenuifolia* Willd.

Family : Asteraceae

Name : Remang (L)

Distribution : Pakistan to Himachal Pradesh and Kumaun.

Habitat : Stony slopes, Leh and Suru valleys between 3,020-3,601 m.

Plant Characters : A much-branched perennial herb. Upper leaves linear-lanceolate, entire margin. Basal leaves broader, long stalked, partially lobed. Flower-heads corymbose, pale or bright-yellow. Involucral bracts hoary. Achenes black or brown. Pappus-hairs white or yellowish-white.

Flowering & fruiting : Mid June to August.

Uses : It is generally grazed by sheep and goats.

55. *Cymbopogon jwarancusa* (Johns) Schult.

Family : Poaceae

Distribution : Pakistan to India, Nepal & North Africa.

Habitat : Stony slopes, Suru valleys between 3,050-3,800 m.

Plant Characters : A perennial herb upto 1.5 m., pale or glaucous green; clums often tufted, erect. Leaves linear-lanceolate; blades flat or convulate, tip attenuate to filliform. Inflorescence a false, decompound, narrow panicle, 15-40 cm long. Pedicelled spikelets purplish, slifhtly longer than sessile lower glume 7-9 vained, keel scaberulus; upper glume ciliate on margins, lower lemma oblong, hyaline, fimbriate, upper lemma deeply bifid with an awn 7-10 mm long.

Flowering & fruiting : May to August.

Uses : It is widely grazed by the animals.

56. *Cynoglossum petiolatum* (Hk. f.) DC. *Anchusa petiolata* Hk. f.

See Sr. No. 31, Chapter XII.

Uses : The whole plant is useful as fodder.

57. *Cynoglossum wallichii* G. Don.

See Sr. No. 101, Chapter VI.

Uses : The whole plant is useful as fodder.

58. *Dactylis glomerata* Linn.

Family : Poaceae

Name : Tahaung (L)

Distribution : From Pakistan to Central Nepal, Tibet, Europe & North Africa.

Habitat : Dry areas, Leh & Suru valley between 2,980-3,500 m.

Plant Characters : A perennial tufted herb. Stem tall upto 90 cm, rough, fistular. Leaves lanceolate, acuminate, margin scabrid. Panicle erect, 10-15 cm long. Spikelets 1 cm long, 3-5 flowered, crowded in one side. Glumes 2, lanceolate, boat shaped, slightly unequal, persistent. Lemmas larger than glumes, stiff, 5 nerved, shortly awned. Stamens 3 with purple anthers.

Flowering & fruiting : June to August.

Uses : The grass is used as fodder.

59. *Deschampsia caespitosa* (Linn.) Beauv.

Family : Poaceae

Distribution : Kashmir to Bhutan, Tibet & cold regions.

Habitat : Sandy places, Changthang valley around 4,760 m.

Plant Characters : A perennial herb. Stem short, shining, slender. Leaves long, flat. Sheath smooth, long. Panicle contracted, sparingly branched. Spikelets silvery. Awn equaling to glume. Lodicules ovate fleshy.

Flowering & fruiting : July to September.

Uses : Used as fodder.

60. *Deschampsia koelerioides* Regel.

Family : Poaceae

Distribution : Kashmir to Baltistan & Turkestan.

Habitat : On alpine slopes, Zanskar valley around 3,780 m.

Plant Characters : A perennial, grass. Stem short, 20-40 cm tall, leafy. Leaves basal, linear, erect; ligule oblong. Panicle erect, ovoid or subspiciform Spikelets brown-yellow, shortly pedicelled.

Flowering & fruiting : June to September.

Uses : Used for fodder.

61. *Deyeuxia holciformis* (Jaub. & Spach.) Bor.; *Calamogrostis holciformis* Jaub. & Spach.

Family : Poaceae

Distribution : Kashmir to Western Tibet, Nubra & Kumaun.

Habitat : Cultivated fields, Suru valley around 3,080 m.

Plant Characters : A slender, grass. Stem 20-40 cm high. Leaves flat; ligule short. Panicle branched. Spikelets purple. Glumes 4, nerved, narrow lanceolate.

Flowering & fruiting : July to September.

Uses : Used for fodder.

62. *Digitaria ischaemum* (Schreb) Schreb ex Mubl.; *Panicum ischaemum* Schreb.

Family : Poaceae

Distribution : Kashmir to Tibet & Kumaun.

Habitat : Roadside, Changthang valley around 4,355 m.

Plant Characters : A tufted, grass. Stem 15-30 cm, diffusely decumbent. Spikes 2-4, subdigitately spreading. Pedicel glabrous. Spikelets imbricate, elliptic, subacute. Glume 5-nerved.

Flowering & fruiting : July to September.

Uses : Used as fodder.

63. *Dilophia salsa* T. Thoms.

Family : Brassicaceae

Distribution : Kashmir to Tibet and Kumaun.

Habitat : Alpine pastures, Zanskar valley between 4,215-4,545 m.

Plant Characters : A tufted, perennial, herb. Stem fleshy, glabrous. Leaves spathulate, linear, toothed. Flower numerous, white in racemes. Sepals and petals persisting round the opening pod. Pod compressed, obcordate.

Flowering & fruiting : July to mid September.

Uses : Used for fodder.

64. *Echinochloa colonum* (Linn.) Link.; *Panicum colonum* Linn.

Family : Poaceae

Name : Yeseshin (L)

Distribution : Afganistan to Bhutan, Nepal & Tibet.

Habitat : Sandy slopes, Leh valley around 3,995 m.

Plant Characters : An annual erect grass. Stems many, upto 70 cm long, branches arising from the base, glabrous. Leaves linear-lanceolate, glabrous, margins scaberulous, sharp. Panicle long. Spikelets hispid, in cluster of 3 glumes 2, unequal. Lemmas 2, awn less, slightly hairy at the base.

Flowering & fruiting : Mid June to September.

Uses : The plant is used for fodder.

65. *Echinospermum redowskii* Lehm.

Family : Boraginaceae

Name : Buckma (L)

Distribution : Pakistan to China, Central Nepal, Tibet and Kashmir.

Habitat : Sandy places, Zanskar valley around 3,620 m.

Plant Characters : An annual or perennial, hispid herb. Stem branched, tomentose, 6-20 cm long. Leaves lanceolate, and alternate. Flowers sub-sessile, bracteate, and yellow-pink. Calyx 4-lobed, oblong-sessile. Nutlets separable, small.

Flowering & fruiting : June to August.

Uses : Used for fodder.

66. *Ephedra gerardiana* Wallich ex Stapf.

See Sr. No. 116, Chapter VI.

Uses : The whole plant is used as fodder at the time of scarcity of green fodder during winters. It is sometimes grazed by wild animals in summer also. 9.27% crude protein content is quite useful for animals.

67. *Eragrostis persica* P. Beauv.; *Poa pilosa* Linn.

Family : Poaceae

Distribution : From Pakistan to Bhutan & Kashmir.

Habitat : Stony places, Leh and Spiti valleys around 3,780 m.

Plant Characters : An annual, tufted grass. Stem suberect or spreading, slender, branched from the base. Leaves lanceolate, glabrous, margin smooth. Ligule oblong, membranous. Panicle upto 10 cm long, erect, inclind or nodding. Spikelets oblong or linear flattened, 5-12 flowers, green with purple tinged. Glumes 2, unequal, 1-nerved, apex acute, persistent. Caryopsis grooved, glabrous, shining.

Flowering & fruiting : July to September.

Uses : The plant is used for fodder.

68. *Eragrostis pilosa* P. Beauv.; *Poa pilosa* Linn.

Family : Poaceae

Distribution : From Pakistan to Bhutan & Kashmir, South Europe.

Habitat : Stony places, Leh and Spiti valleys around 3,780 m.

Plant Characters : An annual, tufted herb upto 80 cm. clums erect or ascending usually slender and simple. Leaves linear, flat, flaccid or strict, sometimes convulate, erect. Spikelets usually grey, tipped with purple, linear, 4-6 mm long, 5-10 flowered. Glumes hyaline, unequal, ovate, lemma ovate, acute, palea-keel scaberulus.

Flowering & fruiting : July to September.

Uses : The plant is used for fodder.

69. *Eragrostis poaeoides* P. Beauv.

Family : Poaceae

Name : Gieng (L)

Distribution : From Afganistan to Nepal, China & Bhutan.

Habitat : Rocky places, Zanskar valley around 3,705 m.

Plant Characters : An annual tufted grass. Stem 25-35 cm long, slender. Leaves rather short, flat; leaf margins glandular. Panicle ovate, stiff. Branches solitary capillary. Spikelets linear to ovate upto 12 flowered. Glumes 1-nerved. Flowering glumes broadly ovate, obtuse. Grains globose or ellipsoid.

Flowering & fruiting : July to September.

Uses : Widely grazed by domesticated and wild animals.

70. *Eremopoa persica* (Trin.) Rozhev.; *Poa persica* Trin.

Family : Poaceae

Distribution : Kashmir to Western Tibet & Asia minor.

Habitat : Moist places, Suru valley around 3,105 m.

Plant Characters : A slender, flaccid, annual grass. Stem upto 40 cm high. Leaves narrowly linear, quite smooth, flat. Ligule lanceolate. Panicle branched. Spikelets 2-15 flowered, green. Lodicules ovate-oblong. Anthers minute. Caryopsis narrow linear.

Flowering & fruiting : July to September.

Uses : Used for fodder.

71. *Fagopyrum esculentum* Moench.

See Sr. No. 71, Chapter VIII.

Uses : The plant is quite useful as fodder.

72. *Ferula jaeschkeana* Vatke.

See Sr. No. 29, Chapter VII.

Uses : Leaves are used as fodder.

73. *Festuca altaica* Trin. ex Ledeb.

Family : Poaceae

Name : Kongola (L)

Distribution : Kashmir to Nepal, Bhutan & Kumaun, Central Asia, Siberia.

Habitat : Moist places, Suru valley between 2,850-3,200 m.

Plant Characters : A perennial grass. Stem tufted, branches arising from the creeping stem. Leaves long, subsetaceous, sheaths crowded open to the middle, ligule very short. Panicle very loose. Spikelets 3-6 flowered, variegated. Glumes membranous.

Flowering & fruiting : July to September.

Uses : A common fodder grass.

74. *Festuca cumminsii* Stapf.

Family : Poaceae

Distribution : Afganistan to Nepal, Bhutan & Kashmir.

Habitat : Dry places, Zanskar valley around 3,850 m.

Plant Characters : A perennial grass. Stem tufted, slender, branched. Leaves setaceous, smooth, nerve 5 green, strong above, obscure beneath. Panicle with scabrid rachis. Spikelets pale-green or purple. Awn half to the glumes. Grains linear oblong.

Flowering & fruiting : July to September.

Uses : Used as fodder.

75. *Festuca kashmiriana* Stapf. ex Hk. f.

Family : Poaceae

Distribution : Pakistan to Bhutan & Kashmir.

Habitat : Common on slopes, Nubra and Spiti valleys around 3,110 m.

Plant Characters : A perennial grass. Stem tufted, slender. Leaves setaceous, rarely flat, smooth, nerved. Panicle loose. Spikelets elliptic-oblong, 3-5 flowered. Glumes 2 thinly membranous, flowering glumes oblong-lanceolate. Grains glabrous and shining.

Flowering & fruiting : July to September.

Uses : This is used for fodder.

76. *Festuca lucida* Stapf. ex Hk. f.

Family : Poaceae

Name : Kongola (L)

Distribution : From Pakistan to Central Nepal & Bhutan.

Habitat : On open slopes, Leh & Zanskar valley around 3,150 m.

Plant Characters : A perennial tall grass. Leaves, linear, glabrous, sheaths long, loose white. Panicle broadly ovate; branchlets very scabrid. Spikelets glistening. Glumes oblong, acute or sub-obtuse, mucronate.

Flowering & fruiting : Mid June to August.

Uses : Used for fodder.

77. *Festuca nitidula* Stapf.

Family : Poaceae

Distribution : Pakistan to Himachal Pradesh & Kashmir.

Habitat : Sandy slopes, Changthang and Zanskar valleys around 4,650 m.

Plant Characters : A perennial grass. Stem tufted, branched, node solitary at the base. Leaves very narrow, 5-7 nerved, smooth. Panicle flexous, ovate. Spikelets 2-5 flowered, elliptic-lanceolate. Margins of glumes narrowly hyaline. Awn much shorter than its glume. Grain linear-oblong.

Flowering & fruiting : Mid June to August.

Uses : Used for fodder.

78. *Festuca ovina* Linn.

Family : Poaceae

Name : Niupang (L)

Distribution : From Afganistan to Kashmir.

Habitat : Dry slopes, Leh and Suru valleys around 2,980 m.

Plant Characters : An erect, slender, perennial grass. Leaves, setiform Panicle narrowly stiff. Branches very soft. Spikelets 3-5 flowered, awn shorter than its glumes.

Flowering & fruiting : Mid June to August.

Uses : Used as fodder.

79. *Filago arvensis* Linn.

Family : Asteraceae

Name : Panksa (L)

Distribution : Pakistan to South West China, Tibet and Kashmir.

Habitat : Sandy places, Lahaul valley between 3,010-3,529 m.

Plant Characters : A small tufted annual herb. Stem branched from the base, densely haired. Leaves linear-lanceolate. Flower-heads clustered, yellow, axillary terminal. Involucral bracts with scarious tip. Achenes minute. Pappus hair white.

Flowering & fruiting : Mid June to August.

Uses : Used for fodder.

80. *Galinsoga parviflora* Cav.

Family : Asteraceae

Distribution : Throughout the Himalayas.

Habitat : Found as a weed in the cultivated fields & wasteland, Leh, Nubra, Changthang, Suru & Zanskar valleys between 2,740-3,625 m.

Plant Characters : An erect, annual, glabrous, herb. Stem delicate, green succulent upto 15-45 cm long. Leaves opposite, entire or toothed. Flower-heads small, sub terminal and axillary, heterogamous. Ray flowers few, female 1-serrate, fertile, ligules yellow. Achene angled.

Flowering & fruiting : June to September.

Uses : The whole plant is used as green fodder.

81. *Geranium meeboldii* Brig.

Family : Geraniaceae

Distribution : Pakistan to Western Nepal, South West Tibet.

Habitat : Dry sandy slopes, Zanskar valley around 3,548 m.

Plant Characters : A tall pubescent, annual herb. Stem branched from the base. Leaves palmately 5-lobed; lobes rhomboid, acute, incised. Stipules subulate-lanceolate. Peduncles long, glabrous-hairy. Sepals awned. Petals obovate, rounded at the tip.

Flowering & fruiting : July to mid September.

Uses : Used for fodder.

82. *Hackelia uncinata* (Royle ex Benth.) Fischer

Family : Boraginaceae

Name : Sukmei (L)

Distribution : Pakistan to Nepal.

Habitat : Cultivated fields, Lahaul valley around 3,450 m.

Plant Characters : A perennial herb. Stem 40-60 cm. Leaves 8-15 cm, linear, rigid, glaucous. Flowers lilac, 1-2 with narrow-elliptic blade. Spathes narrow, green, papery marginated. Nutlets ovoid, broadly winged.

Flowering & fruiting : June to August.

Uses : Commonly used as fodder.

83. *Halocharis violacea* Bunge.

Family : Chenopodiaceae

Name : Sheing (L)

Distribution : Afganistan to Kashmir and South Asia.

Habitat : Rare along moist places, Changthang valley between 4,560-4,639m.

Plant Characters : An annual, hairy, diffuse, herb. Stem 20-30 cm high, woolly, branched, prostrate. Leaves linear, subtriquetrous, obtuse. Spikes subcapitate. Flowers axillary, solitary, 2-sexual, 2-bracteolate. Perianth squarely hispid.

Flowering & fruiting : Mid June to August.

Uses : Used for fodder.

84. *Halogeton glomeratus* (M. Bieb.) C. A.•Mey.; *H. tibeticus* Bunge.

See Sr. No. 43, Chapter XII.

Uses : The whole plant is used for fodder and widely grazed by sheep and goats.

85. *Haloxylon thomsonii* Bunge.

See Sr. No. 44, Chapter XII.

Uses : The whole plant is used for fodder.

86. *Heracleum candicans* Wall. ex DC.

Family : Apiaceae

Name : sPru-dkar (L)

Distribution : From Pakistan to South West China.

Habitat : On slopes & shrubberies, Lahaul and Spiti valleys around 4,270 m.

Plant Characters : A tall, slender, erect, robust, perennial herb. Stem pubescent or closely pubescent. Leaves 1-2 pinnate, radical pinnatifid; upper with large boat-shaped. Flowers white in large compound umbels. Fruits obovate, flattened obconic.

Flowering & fruiting : July to September.

Uses : This plant is mixed with grasses and used as fodder.

87. *Heracleum pinnatum* C. B. Clarke.

See Sr. No. 156, Chapter VI.

Uses : The whole plant is used for fodder.

88. *Heracleum thomsonii* Clarke.

Family : Apiaceae

Distribution : From Pakistan to Himachal Pradesh & Kashmir.

Habitat : Dry slopes, Lahaul valley between 3,480-4,390 m.

Plant Characters : An erect, glabrous or pubescent herb, upto 70 cm high. Stems numerous from the base. Leaves usually basal, pinnate to bipinnate; linear to oval or bipartite, sessile or shortly petioled, pinnatifid to pinnatisect; upper leaves few, reduced. Umbels long pedunculate. Bracts 3-5, linear-pubescent. Rays 3-12. flowers white to yellowish-white. Fruits oblong with narrow wings.

Flowering & fruiting : July to October.

Uses : Used for fodder.

89. *Hippophae rhamnoides* Linn. spp. ***turkestanica*** Rousi.

See S. No. 158, Chapter VI.

Uses : Seabuckthorn leaves and twigs are commonly grazed by sheep, goats and other animals. An analysis report shows that Seabuckthorn leaves are the best source of fodder with crude protein content 23.93% considered very much useful for young and milking animals.

90. *Hippophae salicifolia* D. Don.

See Sr. No. 159, Chapter VI.

Uses : Same as in case of *H. rhamnoides* ssp *turkestanica.*

91. *Hippophae tibetana* Schlecht.

See Sr. No. 160, Chapter VI.

Uses : As in case of *H. rhamnoides* ssp *turkestanica.*

92. *Indigofera heterantha* Wall. ex Brandis

See Sr. No. 48, Chapter XII.

Uses : It is considered a very useful leguminous fodder.

93. *Iris kumaonensis* D. Don ex Royle *I. hookeriana* Foster.

Family : Iridaceae

Name : Wild Iris (E), Tesmamentok (L)

Distribution : Pakistan to Kashmir.

Habitat : Moist alpine pastures, Lahaul and Spiti valleys around 3,980 m.

Plant Characters : A perennial herb with stout rootstock. Stem 10-30 cm long, crowded, outer basal sheaths fibrous. Leaves 20-25 cm, linear, 5 mm broad. Perianth tube 4-8 cm long, lilac. Capsule 3-5 cm long, trigonous, oblong, narrow at both ends.

Flowering & fruiting : June to August.

Uses : The whole plant is used for fodder. Sometimes it is harvested and is stored after drying for winter uses.

94. *Iris lactea* Pallas. var. ***ensata*** Thunb.

Family : Iridaceae

Name : Wild Iris (E), Tesmamentok (L)

Distribution : From Pakistan to Himachal Pradesh, Tibet & Asia.

Habitat : Cultivated fields, Leh valley around 3,450 m.

Plant Characters : A perennial herb with stout rootstock. Stem 20-40 cm. Leaves 8-15 cm, linear, rigid, glaucous. Flowers lilac, 1-2 with narrow-elliptic blade. Spathes narrow, green, papery marginated. Capsule 6 cm long, ribbed.

Flowering & fruiting : June to August.

Uses : It is considered the best fodder plant in Ladakh. The total crude protein content is about 10.62% and considered suitable for milking animals. It is commonly harvested in green stage and sometimes dried for winter use also.

95. *Isopyrum grandiflorum* Fisch ex DC.

Family : Ranunculaceae

Distribution : From Central Nepal to Kashmir, Burma and Eastern India.

Habitat : Moist places & grassy land, Changthang and Lahaul valleys around 4,250 m.

Plant Characters : An annual herb. Stem densely tufted. Leaves petioled, auricled, pinnate. Leaflets cuneate, 3-fid, obovate. Scapes long, bracteate. Flowers solitary, axillary, terminal, white. Carpels 3-7.

Flowering & fruiting : June to July.

Uses : Used for fodder.

96. *Juncus himalensis* Klotz.

Family : Juncaceae

Name : Gimnalo (L)

Distribution : From Pakistan to South East Tibet.

Habitat : Open slopes, Nubra valley around 3,150 m.

Plant Characters : A perennial herb with stoloniferous rootstock, slender green. Stem tufted, erect, leafy below arising from the base. Leaves linear, flattish, tapering towards the apex somewhat tubular. Flowers small in cymes, branched, bracteate. Involucral bracts leafy, ovate-lanceolate. Flowers shortly pedicelled, brown. Capsule obovate. Seeds elongated, black to brown.

Flowering & fruiting : July to September.

Uses : Used as fodder.

97. *Juncus membranaceus* Royle ex D. Don.

Family : Juncaceae

Distribution : Pakistan to South East Tibet, Bhutan & India.

Habitat : Marshy slopes, Suru valley around 2,950 m.

Plant Characters : A perennial herb with a stoloniferous rootstock. Stem tufted, with 1-3 cauline leaves, tall slender, many from the base, fistular and glabrous. Leaves filliform, sheath long auricled, linaer, flattish, tapering towards apex. Flowers small in cymes forming a hemispherical 8-24 flowered head. Involucral bracts leafy, ovate-lanceolate, hyaline. Capsule long-beaked, seeds elongated, black to brown.

Flowering & fruiting : July to September.

Uses : Used as fodder.

98. *Juncus sphacelatus* Decne.

Family : Juncaceae

Distribution : From Afganistan to China, Tibet & Sikkim.

Habitat : Moist places, Leh and Spiti valleys around 4,950 m.

Plant Characters : A perennial herb with stoloniferous rootstock. Stems many from the base, 20-35 cm long. Leaves linear, upto 15 cm. Flowers small in cymes forming a hemispherical head. Bracts lanceolate. Capsule dark brown.

Flowering & fruiting : June to August.

Uses : A very useful fodder grass.

99. *Juncus triglumis* Linn.

Family : Juncaceae

Distribution : Afganistan to Burma, China & Sikkim.

Habitat : Open places, Zanskar valley around 4,680 m.

Plant Characters : A villous, pubescent low perennial herb with a woody rootstock. Stem tufted, slender, 10-20 cm long. Leaves short, linear, tapering at the end. Flowers in terminal cymes. Cymes 3-4 flowered, dark brown. Sepals oblong-lanceolate, acute. Capsule exserted obtuse or mucronate. Seed tailed.

Flowering & fruiting : July to September.

Uses : Used for fodder.

100. *Kochia indica* Wight.

See Sr. No. 52, Chapter X.

Uses : The whole plant is used for fodder.

101. *Kochia prostrata* (Linn.) Schrad. *Salsola prostrata* Linn.

Family : Chenopodiaceae

Distribution : From Pakistan to Himachal Pradesh and Kashmir.

Habitat : Dry areas, Suru & Zanskar valleys between 2,750-3,150 m.

Plant Characters : A villous, pubescent low perennial herb with a woody rootstock. Rootstock as thick as the thumb. Stem with erect and spreading branches and flat, linear leaves, branches numerous, 12-20 cm long. Leaves flat, linear, subacute, clusters in long leafy spikes. Flowers densely villous. Spikes slender, wings of the fruiting perianth hard and scarious.

Flowering & fruiting : June to September.

Uses : Used as fodder.

102. *Kochia stellaris* Moq.

Family : Chenopodiaceae

Distribution : From Pakistan to Kashmir, Kumaun and China.

Habitat : Sandy & dry areas, Suru valley around 2,750-3,120 m.

Plant Characters : A diffusely branched, annual herb. Roots as thick as finger and possibly perennial. Stem densely tomentose. Leaves small, linear, acute, clusters in long or short simple or branched leafy spikes. Flowers greenish-yellow, fruiting tapels winged, strongly nerved.

Flowering & fruiting : July to Mid September.

Uses : Used for fodder.

103. *Koeleria argentea* Griseb.

Family : Poaceae

Distribution : From Kashmir to Kumaun, Garhwal & India.

Habitat : Alpine meadows, Changthang valley around 4,300 m.

Plant Characters : A perennial tufted grass. Leaves flat, glabrous, or sparsely hairy. Panicle oblong. Spikelets 2-3 flowered, silvery white, glabrous, shining, tip of glume 1-2 toothed, awn of flowering glume very short. Glumes very short.

Flowering & fruiting : Mid June to August.

Uses : It is used for fodder.

104. *Krascheninnikovia ceratoides* (Linn.) Guelden.; *Axyris ceratoides* Linn.

Family : Chenopodiaceae

Name : Gausan (L)

Distribution : From Kashmir to Kumaun and Nepal.

Habitat : Sandy areas, Changthang and Zanskar valleys upto 3,870 m.

Plant Characters : A shrubby annual. Stem branched, with stellate hairs. Leaves linear, oblong, obtuse with short petiole. Fruiting bracts crowded along the upper parts of the branches with silky, reddish-brown hairs.

Flowering & fruiting : June to August.

Uses : The whole plant is used as fodder in fresh as well as dry form.

105. *Lathyrus humilis* (Ser.) Fisher. ex Sprengel. *L. altaicus* Ledeb.

Family : Fabaceae

Name : Kaown (L)

Distribution : From Pakistan to Uttranchal and Central Asia.

Habitat : Cultivated fields & wastelands, Leh valley around 3,180 m.

Plant Characters : A more or less erect, perennial herb. Stem weak, 20-45 cm Leaves pinnate, narrowly elliptic, leaflets 6-10 in pairs, oblong, hairless, margin entire. Tendril mostly simple, coiled. Flowers reddish or bluish-purple borne on axillary long peduncles, subsessile. Calyx lobes unequal. Corolla with a broad standard. Pod 3-5 cm long hairless.

Flowering & fruiting : Mid June to September.

Uses : A very useful leguminous fodder plant.

106. *Lathyrus sativus* Linn.

See Sr. No. 87, Chapter VIII.

Uses : Used as fodder.

107. *Lepidium latifolium* Linn.

See Sr. No. 90, Chapter VIII.

Uses : The whole plant is used as fodder and grazed by animals. Analysis report shows that dittander is very rich in crude protein (24.62%) which is quite suitable for milch and young animals. This is commonly harvested in the month of September and stored on the house roof and used as dry fodder during winter period.

108. ***Ligusticum thomsoni*** Clarke. ex Hk. f. & T.

Family : Apiaceae

Distribution : Afganistan to Kashmir and Western Tibet.

Habitat : Dry places along roadside, Zanskar & Spiti valleys around 3,650m.

Plant Characters : A tall, perennial herb. Stem branched from the base reaching upto 1 m. Leaves pinnate, pinnae oblong, lobed or pinnatifid. Bracts 5-8 inches, deciduous. Rays 10-20, bracteoles like the bracts smaller. Fruits ellipsoid.

Flowering & fruiting : Mid June to mid September.

Uses : Used for fodder.

109. ***Lindelofia anchusoides*** (Lindl.) Lehm.; *Cynoglossum anchusoides* Lindl.

Family : Boraginaceae

Name : Makpen (L)

Distribution : From Afganistan to Himachal Pradesh.

Habitat : On stony slopes, Nubra and Spiti valleys around 4,860 m.

Plant Characters : An erect, perennial herb upto 40 cm long. Stem thick, hollow with adpressed hairs. Leaves cauline, grayish hairy, lanceolate, and acute sessile. Flowers pink or bluish or bright blue, funnel shaped, clustered at the end of branched stem. Nutlets with large hooked bristles. .

Flowering & fruiting : July to mid September.

Uses : It is used as fodder.

110. ***Lindelofia longiflora*** (Royle ex Benth.) Baill.; *Cynoglossum longiflorum* Royle ex Benth.

Family : Boraginaceae

Name : Makpen (L)

Distribution : From Pakistan to West Nepal.

Habitat : On open slopes, Nubra, Suru and Spiti valleys between 3,260-4,860 m.

Plant Characters : An erect, perennial herb. Stem hollow, hairy, and thick. Leaves cauline, sessile, lanceolate, hairy. Flowers pedicelled, hairy borne in terminal or sub-terminal raceme. Corolla bluish-black tipped to deep blue or purple. Nutlets ovate, smooth.

Flowering & fruiting : July to August.

Uses : Used as fodder.

111. *Lindelofia stylosa* (Kar. & Kir.) Brand.; *Cynoglossum stylosum* Kar. & Kir.

Family : Boraginaceae

Name : Makpen (L)

Distribution : Pakistan to Himachal Pradesh and Kashmir.

Habitat : Open slopes, Leh, Zanskar and Suru valleys around 4,890 m.

Plant Characters : An erect, hairy, perennial herb. Stem many 40-70 cm long. Radical leaves long petioled, lanceolate, the upper stalkless. Flowers claret-coloured red or dark purple with protruding styles, borne in lax nodding clusters or terminal racemes. Corolla tube cylindrical. Nutlets forming a pyramid densely shortly glochidiate.

Flowering & fruiting : June to September.

Uses : This plant is used as fodder.

112. *Lonicera asperifolia* (Decne.) Hk. f. & T.; *Xylosteon asperifolium* Decne.

See S. No. 52, Chapter XII.

Uses : Used as a fodder.

113. *Lonicera heterophylla* Decne.

Family : Caprifoliaceae

Name : Fumarpe (L)

Distribution : From Pakistan to Kashmir, Uttranchal & West Asia.

Habitat : Rocky slopes, Changthang valley around 4,655 m.

Plant Characters : An erect, perennial shrub. Stem branched, glabrous, reaching upto 1-2 m tall. Leaves 3-5 cm, oblong narrow at both ends with sinuate margins. Peduncles rigid, compressed, thickened upwards, scarcely pubescent. Corolla short. Bracts linear. Drupes never connate.

Flowering & fruiting : July to September.

Uses : The shrub is grazed by wild animals.

114. *Lonicera rupicola* Hk. f. & T.

Family : Caprifoliaceae

Name : Tarno or Lamora (L)

Distribution : From Uttranchal to South West China, Kumaun & Sikkim.

Habitat : Stony slopes, Zanskar valley around 3,860 m.

Plant Characters : A perennial shrub. Stem much-branched. Leaves oblong, tomentose beneath, subobtuse at both ends. Flowers on stalkless clusters, pale-pink to purple. Calyx glabrous. Bracts linear-obovate. Berries ellipsoid.

Flowering & fruiting : July to September.

Uses : The plant is used for fodder.

115. *Lotus corniculatus* Linn.

Family : Fabaceae

Name : Bird's foot Trefoil (E)

Distribution : From Pakistan to Central Nepal & Kashmir.

Habitat : Cultivated fields, Lahaul and Spiti valley around 3,850 m.

Plant Characters : A prostrate, perennial herb. Stem slender, glabrous or obscurely silky. Leaves 5-foliolate, lower 3, upper –2. Leaflets variable, broadly, ovobate-wedge-shaped to lanceolate, the lower pair at the base of the rachis and separated from the terminal three. Flowers yellow on axillary umbels. Calyx teeth lanceolate-cuspidate.

Flowering & fruiting : Mid June to mid September.

Uses : A very useful leguminous fodder plant and the whole plant is grazed or harvested.

116. *Matthiola flavida* Boiss.

Family : Brassicaceae

Distribution : Pakistan to Kashmir.

Habitat : Sandy & dry areas, Changthang and Leh valleys around 4,260 m.

Plant Characters : A perennial herb. Stem branched, whitish-woolly reaching upto 70 cm high. Stem leaves lanceolate, entire, toothed or shallow-lobed. Flowers often fragrant, brownish purple borne in branched inflorescence. Sepals woolly haired. Fruits densely hairy.

Flowering & fruiting : June to August.

Uses : Used for fodder.

117. *Medicago falcata* Linn.

Family : Fabaceae

Name : Yellow flowered Alfalfa / Lucerne (E), Ole (L)

Distribution : Temperate Eurasia, Kashmir to Nepal.

Habitat : Grow as wild & as cultivated upto 4,250 m throughout Ladakh and Lahaul-Spiti areas.

Plant Characters : An erect or semi-spreading perennial herb. Stems much-branched, 40-85 cm. Leaves pinnately trifoliate mostly narrow, oblong, 8-10 mm with toothed margin; leaflets oblanceolate. Flowers small, yellow, in axillary clusters. Pods strongly curved, 1-2.5 cm long.

Flowering & fruiting : Mid June to September.

Uses : It is the only cultivated fodder crop of Ladakh. The plant is usually used as dry in the form of hay. It is considered the most nutritious fodder crop. It is harvested in the month of September. Leaf fall immediately after drying is very common, therefore, baling or bundling is generally practiced for its storage. Bailed or bundled hay can be kept on the house roof for prolonged winter use.

118. *Medicago lupulina* Linn.

Family : Fabaceae

Name : Ole or Bu-su-hang pho (L)

Distribution : Punjab to Kumaun, Bihar, Kashmir and North Bengal.

Habitat : Open slopes, Suru & Leh valleys 3,750 m.

Plant Characters : A biennial, pubescent, herb. Stem much branched from the base 30-50 cm long. Stipule toothed. Leaves pinnately 3-foliolate. Leaflets obovate, faintly inciso-crenate. Flowers in axillary, racemes, densely capitate. Peduncles exceeding the leaves. Corolla yellow. Pods long, smooth, black.

Flowering & fruiting : Mid June to September.

Uses : The plant is used as nutritive fodder.

119. *Medicago media* Linn.

Family : Fabaceae

Name : Alfalfa/Lucerne (E), Ole (L)

Distribution : Western Asia, Mediterrarean regions, Kashmir to Ladakh.

Habitat : Common as wild & cultivated, upto 4,000 m in all the valleys of Ladakh and Lahaul-Spiti.

Plant Characters : An erect, perennial herb. It is believed to be a product of natural hybridization between *M falcata* and *M. sativa*. Leaves pinnately trifoliate; flowers mixture of yellow and purple; pods linear, sickle-shaped, with 5-10 seeded.

Flowering & fruiting : Mid June to September.

Uses : As in case of *Medicago falcata*.

120. *Medicago sativa* Linn.

Family : Fabaceae

Name : Alfalfa / Lucerne (E), Ole (L)

Distribution : Native to South-West Asia. Found in Afganistan, Iran and cultivated in West Persia.

Habitat : Widely cultivated upto 4,450 m all over Ladakh and Lahaul-Spiti region.

Plant Characters : An erect, much branched, perennial herb. Leaves pinnately trifoliate; leaflets, dentate towards apex, obovate-linear. Flowers purple or violet or mauve; pods 6-8 seeded, yellow or brown, kidney shaped.

Flowering & fruiting : Mid June to September.

Uses : As in case of *Medicago falcata*

121. *Melilotus alba* Medik ex Desr.

Family : Fabaceae

Name : White sweet clover (E), Jellek or Ole (L)

Distribution : Temperate Eurasia, throughout India, chiefly in north parts. Cultivated in North India.

Habitat : Moist alpine slopes, Suru and Lahaul valleys around 2,960-3,180 m.

Plant Characters : An erect, rather slender, annual or biennial herb. Stem branched from the base, 30-120 cm high. Leaves trifoliate, elliptic to obovate, blunt with acute teeth. Flowers numerous white borne in long erect axillary spikes. Pod ovoid with a network of raised veins, 2-seeded. Corolla white.

Flowering & fruiting : July to September.

Uses : The whole plant is used as fodder in Ladakh.

122. *Melilotus officinalis* Linn.

Family : Fabaceae

Name : Yellow sweet clover (E), Jellek or Ole or rGya-spos (L)

Distribution : Temperate Eurasia, Pakistan to Kashmir.

Habitat : Found abundantly in Ladakh and Lahaul-Spiti areas on irrigated land, upto 4,000 m.

Plant Characters : An erect, biennial herb. Stem slender, 30-100 cm long, branched, glabrous. Leaves 3 pinnate, toothed, trifoliate; leaflets obovate, oblong; flowers yellowish in lax racemes, 5-20 cm long, dense, shortly stalked, bracteate. Corolla yellow, odorous, usually 3 times long as calyx. Pods brown when ripe; smooth.

Flowering & fruiting : Mid June to August.

Uses : It is a leguminous wild fodder species and also cultivated in many parts of Ladakh and Lahaul areas alongwith *Medicago sp*. It has maximum protein content about 24.01% crude protein which is highly useful for milch and young animals. It is also harvested in the month of September and stored for winter use.

123. *Myricaria squamosa* Desv.; *M. hoffmeisteri* Klotz.

See Sr. No. 20, Chapter XIII.

Uses : The twigs or leaves are chopped for fodder in Ladakh.

124. *Oryzopsis brachyclada* Pilger.

Family : Poaceae

Name : Pianiutan (L)

Distribution : From Pakistan to Kashmir & Himachal Pradesh.

Habitat : Dry & open slopes, Leh valley around 3,550 m.

Plant Characters : An erect, perennial grass. Stem many, tall. Leaves narrow, flat, smooth margin. Branches of panicle short, erect, not widely spread. Spikelets 7-9.5 mm long. Glumes acuminate, almost tailed. Lemmas 5.5 mm long, shortly pilose, gray hair.

Flowering & fruiting : June to August.

Uses : Used as fodder.

125. *Oryzopsis gracilis* (Mez). Pilger.

Family : Poaceae

Name : Fulungkan or Chipkiang (L)

Distribution : From Pakistan to Kashmir & Himachal Pradesh.

Habitat : On open pastures, Leh valley around 3,860 m.

Plant Characters : An erect, perennial grass. Leaves filiform, very glabrous, setaceous, crowded at the base of the plant. Panicle ovate-oblong. Glumes 5-6 mm long. Lemmas 3-5 mm long, covered all over with a gray pubescence.

Flowering & fruiting : June to August.

Uses : Used for fodder.

126. *Oxytropis cachemiriana* Camb.

Family : Fabaceae

Name : Yamchu (L)

Distribution : Central Asia, Pakistan to Kashmir.

Habitat : Stony & rocky slopes, Nubra valley around 3,120 m.

Plant Characters : An erect, perennial herb. Stem several, spreading from woody rootstock. Leaves imparipinnately compound, long; leaflets sessile, pilose, densely haired. Leaflets elliptic, entire. Stipules lanceolate encircling the stem. Flowers pinkish-purple. Pods black and yellow.

Flowering and fruiting : July- September.

Uses : Used as fodder.

127. *Oxytropis lapponica* (Wahl.) Gay.

Family : Fabaceae

Name : Rushkum (L)

Distribution : From Pakistan to Kashmir & temperate Eurasia.

Habitat : On rocky slopes, Leh and Spiti valleys around 4,650 m.

Plant Characters : A semiprostrate, perennial herb covered with silky hair. Leaves 5-15 cm with many pairs of ovate-lanceolate leaflets. Flowers long peduncled, mauve in ovoid clusters. Calyx with dark hair and bristle like teeth. Pods in cluster with persistent beak.

Flowering & fruiting : July to September.

Uses : The plant is grazed by sheep and goats usually in dry form during winters in the scarcity of fodder.

128. *Oxytropis tatarica* Camb. ex Bunge.

Family : Fabaceae

Name : Zingula or Sarkash (L)

Distribution : From Pakistan to Kashmir & North West Himalayas.

Habitat : Dry, rocky slopes, Nubra valley around 3,920 m.

Plant Characters : A dwarf, stemless, tufted, silky-hairy, perennial herb. Stem many arising from the base. Leaves narrowly lanceolate, scaberulous, ligule ovate, 13-21 leaflets, compound. Flowers blue to purple, borne in ovoid peduncled heads. Pod hairy.

Flowering & fruiting : July to September.

Uses : Usually grazed by wild animals.

129. *Parrya exscapa* C. A. Mey.

Family : Brassicaceae

Name : Sumu (L)

Distribution : From Pakistan to Kashmir.

Habitat : Rocky places & dry slopes, Nubra valley around 3,120 m.

Plant Characters : A small tufted herb. Leaves 4-10 cm, oblanceolate, entire or toothed margins. Scapes long, 1-flowered. Flowers usually lilac, purple in terminal cluster. Pod linear. Seeds 1-2 serrate, winged.

Flowering & fruiting : July to September.

Uses : The plant is grazed by the goats and sheep.

130. *Phragmites australis* (Cav.) ex Steud. *P. communis* Trin.

Family : Poaceae

Name : Piachnig (L)

Distribution : From Kashmir to Nepal & Pakistan.

Habitat : On field areas, Leh and Lahaul valleys around 3,160 m.

Plant Characters : An erect, perennial herb. Stem tall upto 2 m many arising from a creeping rootstock. Leaves long, linear to broadly lanceolate. Spikelets spreading. Sheaths much shorter than the blades. Panicles usually brownish purple. Lemmas oblong, lanceolate. Grains small, smooth, shining.

Flowering & fruiting : July to September.

Uses : The whole plant is used as fodder.

131. *Phragmites karka* Trin.

Family : Poaceae

Name : Piachnig (L)

Distribution : From Kashmir to Nepal, Pakistan, Africa & Australia.

Habitat : On field areas, Leh and Lahaul valleys around 3,160 m.

Plant Characters : An erect and tall perennial grass reaching upto 2 m. Leaves long, linear, broader at the base, margins sharp. Sheaths much shorter than the blades. Panicles more spreading. Spikelets long, flattened. Glumes 2, much shorter than spikelet Lemmas oblong, lanceolate. Grains small, smooth, shining.

Flowering & fruiting : July to September.

Uses : The whole plant is used as fodder.

132. *Pleurospermum candollii* (DC.) Benth. ex Clarke.

Family : Apiaceae

Name : rTsad-rgod (L)

Distribution : From Pakistan to Uttranchal.

Habitat : On alpine passes & moist slopes, Suru valley between 2,920-3,105m.

Plant Characters : A stout, erect, perennial plant. Stem hollow reaching upto 10-30 cm high, usually very thick. Base of stem covered with persistent old leaf-bases. Leaves petiolate with paired broad wedge-shaped to broadly ovate leaflets. Flower white borne in a solitary terminal compound umbel. Fruits ellipsoid, oblong, ridges narrowly winged. Seeds grooved on the inner face.

Flowering & fruiting : July to Mid September.

Uses : It is used as fodder.

133. *Pleurospermum stellatum* (D. Don.) Clarke

Family : Apiaceae

Name : rTsad (L)

Distribution : Kashmir to Himachal Pradesh.

Habitat : On alpine meadows & along roadside, Changthang and Lahaul valleys around 4,290 m.

Plant Characters : A perennial herb. Stem branched, 5-15 cm. Leaves 3-5 cm, oblong, ovate with serrate margin. Bracts pinnatifid. Rays 3-5 long. Fruits dorsally compressed.

Flowering & fruiting : July to August.

Uses : Used as fodder.

134. *Poa alpina* Linn.

Family : Poaceae

Name : Humulu (L)

Distribution : Pakistan to Kashmir, Nepal & North temperate & arctic regions.

Habitat : Open slopes, Leh, Changthang & Lahaul valleys around 3,860 m.

Plant Characters : A densely tufted grass. Stem erect, many, arising from the base. Leaves linear, acute, leaf-blade flat. Panicle ovate or oblong. Each spikelet with many flowers. Lemmas longer than glumes. Stamens with linear exerted anthers.

Flowering & fruiting : July to September.

Uses : As fodder.

135. *Poa annua* Linn.

Family : Poaceae

Name : Wukung (L)

Distribution : Throughout temperate and sub alpine Himalaya from Kashmir to North India, Nepal, Europe & temperate Asia.

Habitat : On open slopes, Leh and Zanskar valleys around 4,160 m.

Plant Characters : A tufted, biennial or perennial grass. Stem shortly creeping, 20-30 cm tall. Leaves flat, flaccid, acute, sharp margin. Panicle 3-7 cm long, spreading. Spikelets oblong, 3-7 flowered. Glumes 2, shorter one upto 3 mm long, persistent. Lemmas longer than glumes. Stamens with oblong extruded anthers. Grains oblong, glabrous and shining.

Flowering & fruiting : June to August.

Uses : Usually grazed by the animals.

136. ***Poa bulbosa*** Linn.

Family : Poaceae

Name : Wukung (L)

Distribution : From Pakistan to UP, Nepal and temperate Asia & Europe.

Habitat : Rocky slopes, Leh and Spiti valleys around 3,860 m.

Plant Characters : An annual tufted grass. Stem many erect, 30-40 cm. Leaves narrow, lanceolate, acute, mostly crowded at the base. Sheaths small, loose or adpressed to the stem. Panicles oblong, terminal. Spikelets sub-sessile or sessile, ovate-oblong. Glume 2, outer one shorter, acute, persistent. Lemmas oblong, acute, hyaline. Palea short, membranous.

Flowering & fruiting : June to August.

Uses : The grass is useful as fodder.

137. ***Poa falconeri*** Hk. f.

Family : Poaceae

Name : Zamak (L)

Distribution : From Pakistan to Nepal & Kashmir.

Habitat : Moist alpine pastures, Nubra valley around 4,730 m.

Plant Characters : A perennial tufted grass. Stem many arising from the base. Leaves narrowly lanceolate, scaberulous, ligule ovate. Panicles narow, very lax. Spikelets 2-3 flowered. Glumes equal. Lemmas longer than the glumes. Grains oblong, glabrous.

Flowering & fruiting : June to August.

Uses : The grass is useful as fodder.

138. ***Poa polycolea*** Stapf.

Family : Poaceae

Name : Zamak (L)

Distribution : Western Himalayas from Kashmir to Kumaun.

Habitat : Open fields, Zanskar valley around 4,250 m.

Plant Characters : A tufted perennial grass. Leaves lower, 8 inch long, acuminate, flat, more or less convulate, subflaccid, glabrous, smooth or margins. Panicle pyramidal, rigid or flaccid. Spikelets loosely disposed, oblong, bright green or variegated; glumes thin, lanceolate, keel glabrous or sparsely silky; palea rather shorter.

Flowering & fruiting : June to August.

Uses : The grass is used as fodder

139. *Poa stapfiana* Bor.

Family : Poaceae

Distribution : From Pakistan to Kashmir, Himachal Pradesh & Nepal.

Habitat : Moist places, Nubra and Spiti valleys around 3,160 m.

Plant Characters : A perennial grass. Stem rooting at the base. Leaves flat, ligule oblong; margin smooth. Sheaths small, glabrous. Panicle pyramidal. Spikelets oblong, lax, 3-6 flowered. Glume silky. Keels of the palea long-ciliate.

Flowering & fruiting : June to August.

Uses : The grass is used as fodder.

140. *Poa stewartiana* Bor.

Family : Poaceae

Distribution : Pakistan to Kashmir & Himachal Pradesh.

Habitat : Open slopes, Changthang valley around 3,860 m.

Plant Characters : A perennial tufted grass. Stem branched, smooth. Leaves long, linear. Basal sheaths without any thickening at the base. Spikelets oblong, elliptic. Inflorescence dense, contracted. Glumes 4, obtuse. Ligule over 1 mm long. Keel of palea ciliate. Grain oblong.

Flowering & fruiting : July to August.

Uses : A very useful grass fodder.

141. *Polygonum cognatum* Messin.

Family : Polygonaceae

Distribution : From Kashmir to Ladakh.

Habitat : On dry slopes, Suru and Spiti valley around 3,050 m.

Plant Characters : A diffuse prostrate annual herb. Stem branched, glabrous, angled. Leaves petioled, elliptic, obtuse. Stipules ovate, hyaline. Perianth with pinkish white margin. Nut ovate, black, shining.

Flowering & fruiting : July to September.

Uses : Used for fodder.

142. *Polygonum glaciale* (Meissn). Hk. f. & T.; *P. perforatum* Aitch. var. *gracile* Meissn.

Family : Polygonaceae

Distribution : From Afganistan to China & E. Nepal.

Habitat : Open dry slopes, Zanskar valley around 4,260 m.

Plant Characters : A dwarf, glabrous, diffuse, annual herb. Stem branched from the base, branches diffuse. Leaves long ptioled, broadly ovate, obtuse; stipules short. Flower-heads small, sessile or peduncled, greenish-white. Perianth hyaline, obtuse. Nuts minute, punctate, trigonous, black.

Flowering & fruiting : July to September.

Uses : The whole plant is used as fodder.

143. *Polygonum nepalense* Meissn.

Family : Polygonaceae

Distribution : From Sikkim to Kashmir, Afganistan, Java & Japan.

Habitat : Moist alpine slopes, Suru and Lahaul valleys around 3,120 m.

Plant Characters : An erect-ascending, annual herb. Leaves ovate-triangular to rhomboid, glabrous except hairy on nerve beneath. Flowers red-purple in solitary heads, with sessile semi-amplexicaule, ovate-lanceolate involucral leaf at base. Nut trigonous, red-brown, pitted.

Flowering & fruiting : July to September.

Uses : It is used as fodder and grazed by the animals.

144. *Polygonum paronychioides* C. A. Mey.

Family : Polygonaceae

Name : Sewagawan (L)

Distribution : From Afganistan to Himachal Pradesh.

Habitat : Wastelands places, Changthang valley around 3,365 m.

Plant Characters : A perennial herb. Stem branched from the base, tufted. Leaves linear, fleshy with recurved margins. Stipule large, hyaline lanceolate. Flowers reddish-brown. Fruiting perianth urceolate. Nut broad, black, smooth.

Flowering & fruiting : July to mid September.

Uses : Used for fodder.

145. *Polygonum vacciniifolium* Wall ex Meissn.

Family : Polygonaceae

Distribution : From Afganistan to Kashmir& Himachal Pradesh.

Habitat : Wastelands & moist places, Changthang valley around 4,360 m.

Plant Characters : A tufted, perennial herb. Stem glabrous, branched, creeping. Leaves shortly petioled, orbicular, small; stipules rigid, brown, laciniate. Flowers rose and red with exerted stamens.

Flowering & fruiting : July to August.

Uses : The plant is used as fodder.

146. *Populus ciliata* Wall. ex Royle.

See Sr. No. 24, Chapter XIII.

Uses : The leaves are commonly used as fodder.

147. *Populus euphratica* Oliv.

See Sr. No. 26, Chapter XIII.

Uses : The leaves are used as fodder during summer.

148. *Populus nigra* Linn.

See Sr. No. 27, Chapter XIII.

Uses : One of the most important tree and the leaves are grazed or collected for fodder.

149. *Potentilla bifurca* Linn.

Family : Rosaceae

Distribution : From Kashmir to Sikkim.

Habitat : On rocky & alpine slopes, Nubra valley around 3,160 m.

Plant Characters : A perennial plant. Rootstock woody, long. Stem woody, 3-7 cm branched with white crown leaves. Leaves long, petioled, pinnate. Leaflets silky beneath, 5-7 lobed. Flowers solitary in long pedicelled, yellow. Achenes triangular, smooth, with hairy receptacle.

Flowering & fruiting : June to September.

Uses : The whole plant is used as a source of fodder.

150. *Potentilla microphylla* D. Don.

Family : Rosaceae

Name : Thakto (L)

Distribution : From Pakistan to Kashmir.

Habitat : Dry rocky slopes, Suru valley around 3,160 m.

Plant Characters : A densely tufted, glabrescent to sparsely silky, perennial herb. Stem branched. Leaves pinnate, leaflets broadly oblong, pinnatifid to the middle, glabrous above. Flowers yellow, solitary or occasionally two together on flowering scape. Petals orbicular-obovate.

Flowering & fruiting : July to September.

Uses : Used for fodder.

151. *Potentilla sericea* Linn.

Family : Rosaceae

Name : Pemma (L)

Distribution : Afganistan, Kashmir to Kumaun & Sikkim.

Habitat : On stony slopes & grazing grounds, Zanskar and Lahaul valleys around 3,780 m.

Plant Characters : A veriable perennial herb-covered densely with soft shining silky hairs. Leaves pinnate, crowded, oblong. Leaflets 5-11 silky oblong. Flowers yellow in corymbose. Calyx lobes triangular-ovate or lanceolate. Petals orbicular-obovate. Achenes smooth.

Flowering & fruiting : July to September.

Uses : Used as fodder.

152. *Prangos pabularia* Lindl.

Family : Apiaceae

Name : Prangos (E); Prangos or Plans (L)

Distribution : From Kashmir to Himachal Pradesh.

Habitat : Stony slopes & amongst boulders, Suru valley around 3,105 m.

Plant Characters : An erect, aromatic, tall, perennial herb. Stem hollow, simple or branched, reaching upto 100-150 cm high. Leaves decompound being many times cut into linear segments. Flowers golden yellow in large umbels. Primary umbels with 5-6 linear bracts and 10-15 stout primary rays; secondary umbels each with 5-6 linear bractioles. Fruiting umbels 10-15 cm across; fruit oblong 1-1.5 cm by 1 cm broad.

Flowering & fruiting : July to August.

Uses : Prangos is considered the best nutritious and widely used fodder plant having 12.66% crude protein in Dras areas. It is believed to increase yield in early lactation. Locals harvest this wild plant during September and store on the house roofs for winter use. It is believed to give warmth to animals under extreme winters.

153. *Pseudomertensia echioides* (Royle ex Benth.); *Lithospermum echioides* Royle ex Benth.; *Mertensia echoides* (Royle ex Benth.) Benth. & Hk. f.

Family : Boraginaceae

Distribution : From Pakistan to Himachal Pradesh.

Habitat : Cultivated fields & rocks, Suru, Zanskar & Lahaul valleys 2,910 m.

Plant Characters : A softy haired, perennial herb. Stem usually slender unbranched, 15-40 cm high. Leaves oblong to lanceolate, upto 20 cm long. The lower ones narrowed to a long stalk, the upper ones few, smaller. Flowers blue-purple borne in dense clusters. Corolla-tube slender, without scales in throat. Styles long, protruding. Inflorescence remaining dense in fruit. Nutlets white shining.

Flowering & fruiting : July to September.

Uses : Used for fodder.

154. *Ribes alpestre* Wall. ex Decne.

See Sr. No. 67, Chapter XII.

Uses : It is used for fodder.

155. *Ribes orientale* Desf.

See Sr. No. 69, Chapter XII

Uses : The whole plant is used as fodder.

156. *Robinia pseudacacia* Linn.

See Sr. No. 33, Chapter XIII

Uses : The twigs and leaves are considered as a very useful fodder.

157. *Rosa webbiana* Wall. ex Royle.

See Sr. No. 78, Chapter X.

Uses : The leaves are usually grazed by sheep and goats.

158. *Salix alba* Linn.

See Sr. No. 35, Chapter XIII

Uses : The leaves / twigs provide the excellent reserve of feed for livestock in summer months. 18.12% crude protein content really makes the fodder valuable for animals.

159. *Salix daphnoides* Vill.

See Sr. No. 39, Chapter XIII

Uses : The leaves are chopped for fodder.

160. *Setaria glauca* (Linn.) P. Beauv. (*Panicum glaucum* Linn.)

Family : Poaceae

Name : Yangjuk (L)

Distribution : Pakistan from Kashmir & Himalaya, throughout India.

Habitat : Cultivated fields, Leh valley around 3,060 m.

Plant Characters : An erect, annual grass. Stem 30-50 cm, branched, glabrous. Leaves narrow or broad in middle, lanceolate, acute. Panicle spiciform, solitary at the top of each branch. Spikelets nearly sessile, ovoid. Glumes unequal. Lower florets usually male with 3 stamens. Upper florets bisexual, fertile. Grains ovoid, shiny.

Flowering & fruiting : July to September.

Uses : The plant is used as fodder.

161. *Setaria viridis* (Linn.) P. Beauv. (*Panicum viride* Linn.)

Family : Poaceae

Name : Cheehe (L)

Distribution : Pakistan to Kashmir & temperate Himalayas.

Habitat : Stony slopes, Leh valley around 3,160 m.

Plant Characters : An annual erect herb. Stem 30-35 cm long, branched or unbranched, glabrous. Leaves lanceolate, acute with hairy margins. Panicle spiciform, cylindric, often interrupted. Bristle on involucre 3-6 with erect or spreading barbs. Grains small, ovoid, smooth.

Flowering & fruiting : July to September.

Uses : The plant is used as fodder.

162. *Sisymbrium brassiciforme* C.A. Mey.

Family : Brassicaceae

Name : Pelaung (L)

Distribution : From Kashmir to Kumaun & Central Nepal.

Habitat : On cultivated fields & wastelands, Nubra, Changthang and Lahaul valleys around 3,105 m.

Plant Characters : An annual or perennial herb. Stem branched from the base, glabrous. Branches leafy. Leaves petioled, pinnatifid, hairy. Flowers yellow borne in terminal or axillary corymbs, pedicelled. Pod cylindrical, linear, 4-6 cm long.

Flowering & fruiting : June to September.

Uses : It is used as fodder.

163. *Sisymbrium loeselii* Linn.

Family : Brassicaceae

Distribution : Kashmir to Kumaun & Central Nepal.

Habitat : Moist slopes, cultivated fields & wastelands, Leh and Spiti valleys around 3,105 m.

Plant Characters : An annual tall herb. Stem upto 100 cm, branched from the base. Leaves much dissected, glabrous. Flowers small, yellow in long branched racemes. Pod upto 5 cm long, linear.

Flowering & fruiting : June to September.

Uses : Used as fodder.

164. *Sisymbrium sophia* Linn.

Family : Brassicaceae

Distribution : From Kashmir to Kumaun & Tibet.

Habitat : Wastelands, Changthang valley around 3,555 m.

Plant Characters : An annual or biennial, erect plant. Stem tomentose, branched. Leaves numerous, segmented, linear. Flowers small, yellow. Pod cylindric, pedicilled.

Flowering & fruiting : July to September.

Uses : The plant is used as fodder.

165. *Stachys tibetica* Vatke.

Family : Lamiaceae

Name : Yakjak (L)

Distribution : From Pakistan to Kashmir & Central Asia.

Habitat : Dry stony slopes, Leh, Changthang and Lahaul valleys between 2,860-4,980 m.

Plant Characters : A low, much-branched perennial shrub. Branches slender upto 60 cm. Leaves lanceolate to narrow, elliptic, entire or shallowly lobed, short stalk. Flowers sessile, pink in widely spaced whorls. Calyx campanulate, glabrous. Nutlets oblong.

Flowering & fruiting : Mid June to September.

Uses : The whole plant is used as fodder.

166. *Stellaria media* (Linn.) Vill. Syn. *Alsine media* Linn.

Family : Caryophyllaceae

Name : Chickweed (E), Khalta (L)

Distribution : Himalayas & other temperate parts of the world.

Habitat : As a weed of cultivated fields & near water sources, Leh valley around 3,450 m.

Plant Characters : A flaccid, green, delicate, annual, herb. Stem procumbent, 15-60 cm long, much branched; lower leaves petioled, upper sessile; flowers white, small, star shaped, in leafy cymes; capsules ovoid; seeds brown, acutely tubered.

Flowering & fruiting : June - September.

Uses : It is used as fodder.

167. *Stipa orientalis* Trin.

Family : Poaceae

Name : Pilli (L)

Distribution : From Pakistan to Kashmir & Kumaun.

Habitat : Open slopes & dry alpine areas, Leh & Spiti valleys around 4,760 m.

Plant Characters : A tufted, perennial grass. Stems branched, glabrous, with shiny sheath. Leaves linear-lanceolate, flattened. Panicle elongated with short pedicel. Glumes 2, oblong, with acute apex. Lemmas as long as glumes, membranous. Grains oblong, smooth, shiny.

Flowering & fruiting : July to September.

Uses : It is the most important species of grass which is harvested either raw or mature for domestic fodder from higher alpine passes.

168. *Stipa sibirica* (Linn.) Lamk.

Family : Poaceae

Name : Pilli (L)

Distribution : From Pakistan to Kashmir, Kumaun, Siberia to Korea.

Habitat : Dry alpine slopes, Leh valley around 3,360 m.

Plant Characters : A tufted, perennial grass. Stems upto 80 cm jointed fistular, branched, arising from the rootstock. Leaves linear-lanceolate, glabrous, flat. Panicle much branched upto 30 cm long. Spikelets oblong, lanceolate, 1-flowered. Glumes 2, unequal, lanceolate, hyaline, 3-7 nerved, glabrous. Lemmas oblong as long as glumes, membranous. Grains oblong, smooth, shining.

Flowering & fruiting : July to September.

Uses : As in case of *Stipa orientalis.*

169. *Tamarix gallica* Pallas. II. (Desv.) Thist. Dyer.

See Sr. No. 47, Chapter XIII

Uses : The plant is used as fodder.

170. *Trifolium pratense* Linn.

Family : Fabaceae

Name : Purple or red clover (E)

Distribution : Kashmir to Garhwal.

Habitat : Moist places & cultivated fields, Suru and Lahaul valleys around 2,960 m.

Plant Characters : An erect to decumbent, hairy, perennial herb. Stipules very broad. Leaves obovate, obscurely dentate, 3 foliate, leaflets oblong, toothing obscure. Flowers reddish-purple, sessile. Pods 1-seeded.

Flowering & fruiting : July to September.

Uses : The whole plant is grazed by animals alongwith other grasses. It can also be used as dry fodder. The crude protein content of this species is about 17.56% on dry matter basis which is the best for milking animals.

171. *Trifolium repens* Linn.

Family : Fabaceae

Name : White clover (E)

Distribution : Temperate, alpine Himalayas & temperate Eurasia.

Habitat : Moist cultivated fields, Zanskar & Lahaul valleys around 3,360 m.

Plant Characters : A creeping, perennial herb. Stem slender, glabrous, running along the surface of the soil, branched, roothing at the nodes. Stipules long stalked; stalk upto 20 cm, slender, glabrous, trifoliate. Leaflets obovate to obcordate with a small notch at the apex. Flowers white or pink in dense globular heads. Peduncles as long or longer than leaf petioles. Calyx 5-toothed, campanulate. Corolla double the length of calyx. Pod small, ovate, 1-3 seeded.

Flowering & fruiting : July to September.

Uses : The herb is having same fodder properties as in case of *Trifolium pretense*.

172. *Triglochin palustris* Linn.

Family : Juncaceae

Name : Na-ram (L)

Distribution : Kashmir to Western Tibet, South America & temperate Himalayas.

Habitat : Fresh water marshes, Leh valley between 3,537-3,658 m.

Plant Characters : An erect, scapigarous, fleshy, perennial herb. Stem 15-70 cm, high, tuberous and stoloniferous. Leaves 2-12 in flaccid. Scape slender. Flowers in racemes, pedicels short. Fruit long, appressed to the rachis.

Flowering & fruiting : July to September.

Uses : Used as fodder.

173. *Trigonella corniculata* Linn.

Family : Fabaceae

Name : Watpo (L)

Distribution : From Afganistan to Bhutan & Nepal.

Habitat : Open slopes, Suru valley around 2,990 m.

Plant Characters : A diffuse branched, annual herb. Stipule deeply of finely toothed, points setaceous. Petiole as long as the exceeding leaflets. Leaves pinnately 3-foliate, toothed. Flowers lemon yellow in racemes. Pod linear.

Flowering & fruiting : July to September.

Uses : A very useful leguminous fodder plant.

174. *Trigonella emodi* Benth.

Family : Fabaceae

Name : Watpo (L)

Distribution : From Afganistan to Bhutan.

Habitat : Moist places, Leh and Lahaul valleys around 2,980 m.

Plant Characters : A spreading perennial herb. stems several, branched, 20-25 cm. Leaves trifoliate. Leaflets obovate, finely toothed, hairy. Flowers yellow in axillary cluster. Corolla twice as long as calyx. Pods flattened, veined, hairless.

Flowering & fruiting : July to September.

Uses : A very useful leguminous fodder plant.

175. *Tripogon filiformis* Nees ex Steud.

Family : Poaceae

Distribution : Temperate Himalayas from Kashmir to Nepal.

Habitat : Frequent on dry slopes, Lahaul valley around 3,440 m.

Plant Characters : A tufted perennial plant with erect, slender upto 50 cm high. Leaf-blades filiform, convolute 5-15 cm long. Spikes 5-22 cm long, straight or flexuous. Spikelets 5-8 flowered, 3-10 mm long. Lower glume 1 nerved, lanceolate; upper glume 1 nerved , narrowly lanceolate, straight on the back. Sometimes emerginate or minutely 2-toothed, mucronate or very shortly awned.

Flowering & fruiting : July to September.

Uses : Used as fodder.

176. *Ulmus wallichiana* Palchon.

See Sr. No. 48, Chapter XIII

Uses : The twigs and leaves are chopped as fodder for animals.

□□□

10

Ornamental Plants

Flowers symbolize purity, beauty, love, passion and tranquility such as Rose for love, Pansy for thoughts, Carnation (white) for women's love, French marigold for jealousy / sorrow, African marigold for vulgar mind, Narcissus for self esteem, Daffodil for regard, amaryllis for pride, iris for message, snapdragon for presumption, jasmine for amiability, Lily for purity, Stock for luxury and Sweet pea for departure.

Flowers have been closely linked with Indian culture from the Vedic times (3000 - 2000 BC). A study of ancient literature, mythological epics, paintings, cave murals, sculpture, architecture and tribal arts, reveals the importance of flowers during those periods. The great mythological epics Ramayana and Mahabharata written during 500 BC describe some beautiful flowering trees such as *Ficus benghalensis* (Banyan), *Ficus religiosa* (Pipal), *Saraca indica* (Ashoka), *Terminalia arjuna* (Arjun tree) and *Nerium oleander*. The great sanscrit writers Sudreka, Ashwagosh and Kalidas had mentioned the importance of flowers in their literatures. The lotus was a very popular flower and was regarded by both the Hindus and the Buddhists as a symbol of purity. Similarly Hiuen Tsang, the famous Chinese pilgrim, traveled widely in India and had given a detailed account of life in India at that time. About flowers he wrote that the Hindus needed flowers every morning for worshipping and they also used flowers to adorn their hair.

During Mughal period several beautiful gardens had been constructed which are still famous for their architecture and style. Similarly many famous gardens like the one in Rashtrapati Bhawan, Brindawan and Pinjore have their own importance even today. An important development in Indian gardens, earlier during the Mughal periods and British periods, has been the introduction of several exotic flowers such as chenar, cypress, rose, carnation, gladiolus, daffodil, iris, narcissus, tuberose, and a number of flowering trees, shrubs, climbers etc. With the view to preserve exotic species and for compilation of local flora certain botanical gardens have been established. These are the Botanical Garden of Culcutta and Lloyd Botanical garden of Darjeeling which have much educative and scientific value. The Agri-Horticultural Society of India, established in 1820 in Calcutta, has served a very useful purpose in the development of agriculture and horticulture including floriculture in India.

Importance

Our country is endowed with natural wealth of plant materials, which are of great floricultural importance and significance. Besides beauty and aesthetic value, flowers are important for their economic value also. Sale of flowers (cut and loose), extraction of essential oils and making of economic products from these, are highly profitable business in almost all parts of the world. Now- a-days the cut and loose flowers play an important role during social and religious occasions.

Cut flowers

Now-a-days the cut flowers play a major role and are having great aesthetic importance. The flowers valued as cut flowers, are several species of orchids, rose, carnation, chrysanthemum and gladiolus etc. The uses of cut flowers in home decoration, has become an integral part of living in human society, particularly in advanced countries. Majority of such flowers are being cultivated under protected and controlled conditions such as glass / fibre / polycarbonate / UV stabilized polythene structures, mostly in European and other countries. In India majority of cut flowers are grown in open during winters and have a great potential for their export. Besides their great potential for export, the requirement of cut flowers has increased tremendously within the country itself in recent years. The major growing areas for different ornamental crops are Delhi, Pune, Banglore, Chandigarh for rose; Kalimpong, Kolaghat, Panshkura for gladiolus, tuberose, rose, jasmine, and other important annuals; Coimbatore, Pune, Banglore for tuberose and chrysanthemum. The market of cut flowers are growing rapidly throughout within country and outside also.

Foliage plants

Similarly foliage plants are also becoming very popular in our country as well as in other countries. In metropolitan cities, the foliage plants are in great demand as home decoration, metro and corporate offices as aesthetic value. In addition, with the growing population, lack of open spaces and development of multistoreyed housing systems, people have to depend largely on indoor plants for decorating their surroundings.

Perfumes

The other and very important aspect of commercial floriculture are the byproduction of flowers for extraction of floral perfumes which has great demand in domestic and in international markets. The need of floral perfumes in perfumery trade, in cosmetic industry, food industry and allied industries are growing tremendously.The two major flowers which find their uses in perfumes are rose and jasmine which are being extracted for essential oils for a very long time.

Field study

Since ages, the Himalayan flora is considered as a treasure of medicinal, edible, aromatic and aesthetic plants of repute. Many of cold desert flora of Ladakh and Lahaul-Spiti Himalayas have been known for their attractive colors and aesthetic value. High alpine plant species like *Aquilegia sp, Delphinium sp, Juniperus sp, Meconopsis sp, Rosa sp, Saussurea sp* etc have tremendous importance in the religious places, gompas and are used for various purposes on several occasions.

Keeping the above facts in view, this chapter attempts to enlist and popularise potential wild ornamental flora of Ladakh and Lahaul-Spiti, which are being used by local tribals or have potential to be used as ornamental flora, which can be utilized in future. Plant species of potential wild flora have been described in alphabetical order alongwith families, English (E), hindi (H)/local names (L), distribution, habitat, brief botanical notes and uses in the enumeration. (Table 7).

Enumeration of Ornamental Flora

1. ***Acantholimon lycopodioides*** (Girard) Boiss. *Statice lycopodioides* Girard.

   See Sr. No. 1, Chapter VI.

   Uses : It is used for dry decoration.

2. ***Androsace aizoon*** Duby.

   See Sr. No. 2, Chapter XII

   Uses : This plant is used by local tribes for ornamental purposes in Ladakh.

3. ***Allium przewalskianum*** Regel.; *A. jacquemontii* Regel, *A. stoliczkii* Regel.

   See Sr. No. 7, Chapter VIII

   Uses : The whole plant is the best suited for flower beds due to having very attractive flowers.

4. ***Aquilegia fragrans*** Benth.

   Family : Ranunculaceae

   Name : Fragrans Columbine (E)

   Distribution : Himalayas from Kashmir to Kumaun, temperate Europe.

   Habitat : Along water streams & moist places, Nubra & Suru valleys between 2,950-3,205 m.

Plant Characters : A sweet scented perennial herb. Stem 40-80 cm, much branched, covered with glandular hairs. Leaves compound, leaflets 3-lobed, lobes with shallow teeth, more or less glaucous above. Flowers cream-coloured, large 3-5 cm with mild fragrance and hooked spurs. Spur 15-20 cm. Follicles 6-8, densely hairy with long styles.

Flowering & fruiting : July to mid September.

Uses : It is used as ornamental plant because of attractive and fragrant flowers in spike.

5. *Aquilegia pubiflora* Wall ex Royal.

Family : Ranunculaceae

Name : Fragrans Columbine (E)

Distribution : Himalayas from Kashmir to Himachal Pradesh.

Habitat : Forest areas, Lahaul valley between 3,490-3,760 m.

Plant Characters : An erect or decumbent, glabrous to sparsely hairy, branched perennial herb upto 50 cm, tall. Basal leaves long-petioled, pinnate; lobes linear-lanceolate, obtuse or acute; upper stem leaves comparatively short-petioled and smaller in size and shape. Flowers white with purplish or bluish tinge in loose drooping panicles. Sepals 5, long acuminate. Petals 5, spur short, strongly curved. Follicles black-hairy.

Flowering & fruiting : July to September.

Uses : It is used as ornamental plant because of attractive and fragrant flowers in spike.

6. *Aster diplostephioides* Benth. *Heterochaeta diplostephioides* DC.

Family : Asteraceae

Name : Wild aster (E), Kiamentok / Me-tog Lug-mig rigs-dang-po (L)

Distribution : From Kashmir to South West China.

Habitat : On open stony slopes, Zanskar & Suru valleys between 3,595-3860 m.

Plant Characters : An erect, robust, annual herb. Stem simple, leafy reaching upto 20-30 cm. Leaves sessile, lanceolate, oblong lanceolate, entire and narrow at the base. Flower-heads large, solitary with long narrow somewhat reflexed lilac ray-florets. Achene silky, with reddish pappus.

Flowering & fruiting : July to mid September.

Uses : The handsome flowers are used as ornamental purposes in Ladakh.

7. ***Aster flaccidus*** Bunge.

Family : Asteraceae

Name : Wild aster (E), Niamentok or Me-tog lug-mig rigs-gnyyis-pa (L)

Distribution : From Pakistan to South West China and Central Asia.

Habitat : Open slopes, Changthang & Suru valleys between 3,460-3,520 m.

Plant Characters : An annual herb. Stem small woolly, branched, leafy. Radical leaves obovate, entire, acute. Flower-head solitary, large. Outer involucral bracts linear-lanceolate, acuminate. Achenes 4-5 mm, pappus reddish, two seriate.

Flowering & fruiting : July to mid August.

Uses : This is commonly used for flowers as ornamental plant.

8. ***Aster tibeticus*** Hk. f.

Family : Asteraceae

Name : Wild Aster (E), Skalzangmentok (L)

Distribution : From Pakistan to Nepal, China and Tibet.

Habitat : Open slopes, Leh & Changthang valleys between 4,680-4,890 m.

Plant Characters : A perennial herb. Stem 10-35 cm, pubescent. Leaves small, radical, 2-3.5 cm, broad. Ligules many, long, bright, blue. Flower -heads solitary, 2-3 cm broad. Achenes small, flat, pappus double white.

Flowering & fruiting : July to August.

Uses : Flowers are used ornamentally by locals.

9. ***Braya alpina*** Hk. f. & T. Anders.

Family : Brassicaceae

Name :Lingantso (L)

Distribution : Kashmir to Kumaun, Nepal, Tibet & Pakistan, temperate Europe.

Habitat : Higher alpine passes, Changthang valley around 4,995 m.

Plant Characters : A tufted perennial herb. Root stout, fleshy, long. Stem small branched. Leaves tufted, linear, spathulate. Flowers capitate, white, pedicelled. Pods erect, linear or elliptic-lanceolate, acute. Seeds variable.

Flowering & fruiting : July to September.

Uses : It is used as ornamental plant.

10. ***Braya tibetica*** Hk. f. & T.

Family : Brassicaceae

Name : Lingantso (L)

Distribution : Western Himalayas from Kashmir to Kumaun and Tibet.

Habitat : Alpine passes, Nubra valley around 3,645 m.

Plant Characters : A small, glabrous, perennial herb. Root stout, fleshy. Leaves small, linear-oblong. Flowers tiny, attractive, borne on dense cluster. Fruits short, corymbose.

Flowering & fruiting : June to August.

Uses : Used as an ornamental plant.

11. *Bupleurum tenue* Ham ex Don.

Family : Apiaceae

Name : Kimantso (L)

Distribution : Kashmir to Nepal, Bhutan and temperate Eurasia.

Habitat : Barren rocky slopes, Changthang valley around 4,560 m.

Plant Characters : An erect, glabrous, perennial herb. Stem reaching upto 1 m tall, branched above. Lower leaves linear; upper leaves linear-oblong, subobtuse, mucronate narrowed at the base. Bract and bracteolate small, obovate-lanceolate. Primary rays of the umbel slender. Fruits oblong, yellowish brown, ridges prominent.

Flowering & fruiting : July to August.

Uses : Used as ornamental.

12. *Caltha palustris* Linn.

See Sr. No. 62, Chapter VI

Uses : The plant is having aesthetic value due to attractive flowers.

13. *Campanula argyrotricha* Wall. ex DC.

Family : Campanulaceae

Distribution : Kashmir to Bhutan.

Habitat : Dry stony slopes, Zanskar valley 3,540 m.

Plant Characters : A sub-erect, delicate, tufted, perennial herb. Stem many, spreading from the rootstock. Leaves sessile, ovate, toothed silky-white beneath. Flowers small blue or purple or purplish-white, solitary or few. Corolla hairy outside. Capsule obconical, nodding, rounded.

Flowering & fruiting : July to September.

Uses : The plant is having ornamental value.

14. *Campanula aristata* Wall.

Family : Campanulaceae

Name : Piseng (L)

Distribution : Afganistan to Bhutan, Kashmir and Himachal Pradesh.

Habitat : Alpine slopes and meadows, Changthang valley between 4,870-4,990 m.

Plant Characters : A low, erect, annual herb. Basal leaves elliptic, entire, petioled; cauline leaves linear, entire. Flowers solitary, deep blue or violet, terminal. Capsule narrowly oblong sublinear. Seed very small, ellipsoid.

Flowering & fruiting : June to September.

Uses : The plant is used ornamentally.

15. *Campanula cashmeriana* Royle.

Family : Campanulaceae

Name : Nipeng (L)

Distribution : Afganistan to Uttranchal, Kashmir and Himachal Pradesh.

Habitat : Higher passes, Nubra valley around 3,650 m.

Plant Characters : An attractive, villous or tomentose, annual herb. Stem rigid zigzag, many, hoary, villous or tomentose. Leaves thick, woolly, elliptic blunt, hair soft. Flowers large, bright blue, in terminal clusters. Calyx-lobes broadly lanceolate. Capsule broader than long.

Flowering & fruiting : July - September.

Uses : The plant is used as ornamental for its flowers.

16. *Campanula latifolia* Linn.

Family : Campanulaceae

Name : Large bellflower (E), Nipeng (L)

Distribution : Pakistan to Central Nepal & Kashmir to Kumaun.

Habitat : Moist alpine slopes, Suru valley around 3,035 m.

Plant Characters : An erect, robust, perennial, herb. Stem long, branched. Leaves 5-12 cm, lanceolate, cuneate at base, subsessile; lower pedicelled upper sessile, margin toothed. Flowers dark blue-purple, bell-shaped, borne in terminal clusters. Corolla lobes triangular-lanceolate. Capsule glabrous.

Flowering & fruiting : June to September.

Uses : The plant is used ornamentally.

17. *Campanula pallida* Wall.

Family : Campanulaceae

Name : Nipeng (L)

Distribution : Afganistan to South West China, Kashmir to Himachal Pradesh and Kumaun.

Habitat : Cultivated areas and stony slopes, Zanskar valley between 3,395-3,560 m.

Plant Characters : An erect perennial, herb. Stem branched, hairy, 15-50 cm long. Leaves petioled, ovate-lanceolate. Flowers pale lilac or purple, borne in lax terminal clusters. Calyx teeth lanceolate; corolla hairy. Capsule obconical, nodding.

Flowering & fruiting : July to September.

Uses : Used as ornamental.

18. *Cardus edelbergii* Rech. f.

Family : Asteraceae

Name : Scotch thistle (L)

Distribution : Kashmir to Sikkim.

Habitat : Common on dry sandy riverbeds and waste places and along agriculture fields, Sissu, Lahaul-Spiti valley 3,850 m.

Plant Characters : An erect, branched, spinescent herb upto 60 cm tall. Stems grooved, winged. Leaves sessile, continued down the stem as interrupted spinous wings, pinnately lobed. Heads pink or purple, globose to hemispheric, erect to drooping. Achenes pale-brown, glabrous, obtusely 4-angled.

Flowering & fruiting : June to September.

Uses : Having ornamental value.

19. *Chorispora sabulosa* Camb.

Family : Brassicaceae

Distribution : Afganistan, Kashmir to Kumaun.

Habitat : Frequent on dry sandy slopes, Leh & Changthang valleys 4,290 m.

Plant Characters : A suberect or ascending, branched, sparsely hairy with glandular hairs, 5-25 cm tall perennial herb. Radical leaves spathulate, sinuate toothed, petioled; cauline leaves almost similar to the radical leaves. Flowers pink or violet in 10-25 flowered racemes. Siliquae fruits.

Flowering & fruiting : June to September.

Uses : The plant is used for ornamental purposes due to having attractive and mild fragrant flowers.

20. *Clematis orientalis* Linn.

See Sr. No. 50, Chapter IX.

Uses : The whole plant is used as ornamental.

21. *Clematis tibetana* Kuntze.

See Sr. No. 77, Chapter VI.

Uses : The flowers are used ornamentally.

22. *Codonopsis clematidea* (Schrenk) Clarke.

See Sr. No. 78, Chapter VI

Uses : Due to attractive flowers the plant is having ornamental value.

23. *Codonopsis ovata* Benth.

See Sr. No. 79, Chapter VI

Uses : The plant is used as *C. clematidea.*

24. *Cremanthodium ellisii* (Hk. f.) Kitam. Var. *C. plantagineum* Maxim.

See Sr. No. 92, Chapter VI

Uses : It is suitable as pot flowers.

25. *Cremanthodium reniforme*

Family : Asteraceae

Name : Lukumentok (L)

Distribution : Kashmir to Tibet.

Habitat : Open fields, Lahaul valley around 4,430 m.

Plant Characters : An erect, perennial herb. Stem 15-30 cm. Basal leaves with an oblong blunt blade, 5-12 cm, widely spaced teeth, petioled. Flower-heads solitary, terminal. Rays florets yellow, elliptic; disk florets greenish to brownish. Involucral bracts narrow, lanceolate, woolly haired.

Flowering & fruiting : July to August.

Uses : Used as ornamentally.

26. *Dactylorhiza hatagirea* (D. Don.) Soo.; *O. latifolia* Linn.

See Sr. No. 102, Chapter VI

Uses : The plant is having aesthetic value due to attractive flowers and being a member of orchidaceae.

27. *Delphinium brunonianum* Royle.

See Sr. No. 103, Chapter VI

Uses : Flowers are used as ornamental in fresh as well as in dry form.

28. *Delphinium cashmerianum* Royle.

See Sr. No. 104, Chapter VI

Uses : Flowers are used as ornamental in fresh as well as in dry form.

29. *Delphinium vestitum* Wall. ex Royle.

See Sr. No. 105, Chapter VI

Uses : Having ornamental potential.

30. *Dianthus crinitus* Smith ex Boiss.

Family : Caryophyllaceae

Name : Liamtnang (L)

Distribution : Kashmir to Pakistan, Punjab, Turkestan and Beluchistan.

Habitat : On dry rocks, Changthang valley around 4,805 m.

Plant Characters : A small, perennial herb. Stem with rigid branches, 15-30 cm long. Leaves rigid, acute and cauline shorter than the joints. Bracts 4-8 lanceolate. Calyx toothed. Petals fimbricate white, sweet-scented. Flowers white. Capsule long with many seeds.

Flowering & fruiting : Mid June to August.

Uses : Used as ornamental plant.

31. *Dianthus incana* Hk. f. & T. Anders. *D. lanceolata* Royle

Family : Caryophyllaceae

Name : Mulanku (L)

Distribution : Afganistan to Kashmir, Tibet, Ladakh and Himachal Pradesh.

Habitat : On dry alpine slopes, Nubra Valley around 3,001 m.

Plant Characters : A small often tufted, perennial herb. Stem branched, leafy, 5-20 cm long. Radical leaves tufted, elliptic or oblong-lanceolate or narrow-spathulate, cauline, distant, ovate sessile. Flowers small white borne in terminal clusters. Fruits twisted, pubescent.

Flowering & fruiting : July to mid September.

Uses : Used as ornamental plant.

32. *Dracocephalum heterophyllum* Benth.

See Sr. No. 23, Chapter VII

Uses : The plant is suitable for pot flowers and bears very attractive and fragrant flowers.

33. *Echinops cornigerus* DC.

See Sr. No. 114, Chapter VI

Uses : The flowers are used for dry decoration.

34. *Echinops niveus* Wall. ex DC.

See Sr. No. 115, Chapter VI

Uses : The flowers are used in dry decoration.

35. *Elaeagnus angustifolia* Linn. *E. hortensis,* M Bieb.

See Sr. No. 7, Chapter XIII.

Uses : Flowers are used for fragrance in monastry or houses by the local people.

36. *Epilobium angustifolium* Linn.

Family : Onagraceae

Name : Rose way willow herb (E), Utpal or Wanbo (L)

Distribution : Afganistan to South West China, North Temperate Zones.

Habitat : Along water sources and slopes, Zanskar & Spiti valleys around 3,420 m.

Plant Characters : An erect, leafy, glabrous to glabrescent, perennial herb. Stem branched upto 2 m tall. Leaves 6-12 cm, sessile, lanceolate, denticulate, glabrous beneath, almost sessile, scattered. Flowers rose-purple in long terminal clusters. Stamens conspicuous, curved outwards. Calyx segments lanceolate, acuminate, subpetaloid, free to the base. Petals obovate. Capsule pubescent, hairy.

Flowering & fruiting : July to September.

Uses : The flowers are used ornamentally and the best for making hedge and cut flowers.

37. *Epilobium latifolium* Linn.

Family : Onagraceae

Name : Willow herb (E)

Distribution : Afganistan to Central Nepal, North Asia.

Habitat : Damp places, Zanskar valley around 3,568 m.

Plant Characters : An erect, leafy, perennial herb. Stem branched, glabrous reaching upto 1 m tall. Leaves narrow-elliptic, narrowed above and below, oblong, elliptic, pubescent. Flowers rose-purple, borne on axils of leafy bracts. Capsule long, pubescent.

Flowering & fruiting : Mid June to mid September.

Uses : A beautiful small flowering plant.

38. *Epilobium roseum* Schreb.

Family : Onagraceae

Name : Willow herb (E)

Distribution : From Afganistan to South West China.

Habitat : On pasturelands, Zanskar & Suru valleys between 3,905-4,530 m.

Plant Characters : An erect, perennial herb. Stem branched, uniformly hairy, 15-30 cm. Leaves scattered, ovate-narrow, linear-lanceolate, short stalked. Flowers in axillary spikes. Petals 4, obovate. Capsule linear, 4-celled.

Flowering & fruiting : July to September.

Uses : Used as ornamental plant.

39. *Euphorbia stracheyi* Boiss.

See Sr. No. 125, Chapter VI

Uses : It is used as ornamental plant.

40. *Gagea dschungarica* Regel. G. *persica* Boiss.

Family : Amaryllidaceae

Name : Yellow star of Bethlehem (E)

Distribution : Pakistan to Central Nepal & Kashmir.

Habitat : Pasturelands, Zanskar valley around 4,650 m.

Plant Characters : A small bulbous herb. Stem 4-8 cm long, glabrous, slender. Radical leaves linear-lanceolate. Flowers 1-3 in forked cymes, bracteate, perianth yellow, segment linear, blunt. Capsule long, seeds compressed.

Flowering & fruiting : July to September.

Uses : Uses as ornamental plant.

41. *Gagea lutea* (Linn.) Ker-Gawl.

Family : Amaryllidaceae

Name : Yellow star of Bethlehem (E)

Distribution : Pakistan to Central Nepal & Kashmir.

Habitat : Rare amidst boulders and in rock-crevices, Lahaul valley around 3,905 m.

Plant Characters : A small bulbous herb. Stem 4-8 cm long, glabrous, slender. Radical leaves linear-lanceolate. Flowers 1-3 in forked cymes, bracteate, perianth yellow, segment linear, blunt. Capsule long, seeds compressed.

Flowering & fruiting : July to September.

Uses : Uses as ornamental plant.

42. *Geranium pratense* Linn.

See Sr. No. 145, Chapter VI.

Uses : The flowers are of ornamental value.

43. *Glechoma nivalis* (Benth.) Press

Family : Lamiaceae

Name : Gaumej (L)

Distribution : From Kashmir to South West China.

Habitat : On screes of high altitudes, Changthang and Spiti valleys around 4,450 m.

Plant Characters : A dwarf hairy, spreading, prostrate, mild aromatic herb. Stem arising from spreading fibrous underground stems. Leaves crowed with rounded blades, grey- green, wrinkled, obscurely toothed. Flowers pale mauve in axillary whorls.

Flowering & fruiting : July to August.

Uses : The whole plant is of ornamental value.

44. *Impatiens amplexicaulis* Edgew.

Family : Balsaminaceae

Distribution : From Kashmir to Himachal Pradesh.

Habitat : Along glacial streams, Lahaul valley around 3,940 m.

Plant Characters : An erect, glabrous herb, upto 30-75 cm tall. Leaves sessile, amplexicaul, oblong or oblong-lanceolate, long acuminate, crenate. Flowers purple in axillary umbel-like clusters or racemes. Lip funnel-shaped, abruptly narrowed into a cylindric usually straight spur. Capsule oblong, terete, acute.

Flowering & fruiting : July to September.

Uses : The flowers are used ornamentally.

45. *Impatiens glandulifera* Royle.

Family : Balsaminaceae

Distribution : From Kashmir to Nepal.

Habitat : Along perennial streams, Lahaul valley around 3,610 m.

Plant Characters : An erect, branched, glabrous, robust herb, upto 2.0 m tall. Leaves opposite-verticillate, elliptic ovate, serrate, base glandular. Flowers pink to pinkish-red, in compact terminal racemes or umbel-like clusters, occasionally paniculate. Lip spotted with yellow, broadly bell-shaped, abruptly contracted into a cylindric curved rather thick yellow spur. Capsule clavate.

Flowering & fruiting : July to September.

Uses : The flowers are of ornamental value.

46. *Inula obtusifolia* Ker.

See Sr. No. 169, Chapter VI.

Uses : It is used as ornamental plant.

47. *Iris kumaonensis* D. Don ex Royle *I. hookeriana* Foster.

See Sr. No. 93, Chapter IX.

Uses : It is used as ornamental plant.

48. *Iris lactea* Pallas. var. ***ensata*** Thunb.

See Sr. No. 94, Chapter IX.

Uses : It is used for ornamental purposes and very much suitable as bed flowers.

49. *Juniperus communis* Linn. var. ***saxatilis*** Pallas, *communis* non. Linn. Hk. f.

See Sr. No. 14, Chapter XIII.

Uses : It is suitable as ornamental shrub and commonly planted near Monastry due to religious importance.

50. *Juniperus macropoda* Boiss.

See Sr. No. 15, Chapter XIII.

Uses : Same as in case of *J. communis.*

51. *Juniperus recurva* Buch,-Ham

See Sr. No. 16, Chapter XIII.

Uses : Same as in case of *J. communis.*

52. *Kochia indica* Wight.

Family : Chenopodiaceae

Distribution : From North West India to Coimbatore & Kashmir.

Habitat : Wastelands, Leh valley between 2,850-3,000 m.

Plant Characters : A softly villous, annual herb. Stem diffusely branched from the base, branches divoricate, long. Leaves small, elliptic or linear-oblong, acute. Wings of fruiting perianth short, broadly triangular, ovate, much shorter than the diameter of the disk.

Flowering & fruiting : July to September.

Uses : Having ornamental potential.

53. *Leontopodium leontopodinum* (DC.) Hand. Mazz.; *Antennaria leontopodina* DC. Prodr. syn. *L. alpinum* Cass. DC.

Family : Asteraceae

Name : Himalayan Edelweiss (E), Palu (L)

Distribution : From Afganistan to North Nepal, Bhutan and Kashmir.

Habitat : Alpine passes, Zanskar valley around 4,660 m.

Plant Characters : A small densely woolly perennial herb. Leaves alternate, small lanceolate, woolly on both surface. Flower-heads small, monoecious. Involucral bracts scarious, oblong-lanceolate, acuminate. Achenes pappillose.

Flowering & fruiting : July to August.

Uses : An attractive high alpine ornamental plant.

54. *Leontopodium nanum* (Hk. f. & T.) Hand.; *Antehharia nanna* Hk. f.

Family : Asteraceae

Name : Himalayan Edelweiss (E), Palu (L)

Distribution : From Pakistan to Central Nepal and Kashmir.

Habitat : Higher passes, Leh & Spiti valleys around 5,160 m.

Plant Characters : A grey woolly tufted perennial herb. Leaves linear, many, 1.5-2.5 cm long in lax rosette. Flower-heads sub sessile, solitary. Involucral bracts linear lanceolate. Pappus white.

Flowering & fruiting : July to August.

Uses : An attractive high alpine ornamental plant.

55. *Lomatogonium carinthiacum* (Wulfen). Reichb.

Family : Gentianaceae

Name : Tikta (L)

Distribution : Afganistan to South West China, Europe and North Asia.

Habitat : Open moist slopes, Suru valley around 2,960 m.

Plant Characters : A small annual herb. Stem simple or branched from the base, 15-30 long. Lower leaves, elliptic; cauline leaves elliptic, smaller. Flowers long-stalked, blue in lax cluster. Calyx lobes narrow-elliptic unequal. Corolla lobes green with small pale fringed nectaries at their base. Capsule usually cylindrical.

Flowering & fruiting : July to September.

Uses : The whole plant is used for ornamental purpose.

56. *Malaxis muscifera* (Lindl.) Ktze.

See Sr. No. 195, Chapter VI.

Uses : Used as ornamental.

57. *Malva verticillata* Linn.

See Sr. No. 92, Chapter VIII.

Uses : Used as ornamnetal

58. *Meconopsis aculeata* Royle.

Family : Papaveraceae

Name : Himalayan blue poppy (E), Achatsarmum (L)

Distribution : Western Himalayas from Kumaun to Kashmir.

Habitat : Rare on damp slopes, Zanskar valley & Spiti valley around 4,290 m.

Plant Characters : A slender rather delicate perennial herb. Stem reaching upto 60-80 cm with bristly hairs. Leaves deeply and irregularly pinnately lobed; lobes usually rounded, bristly haired. Flowers usually blue, borne on long stalks. Petals usually 4, rounded to obovate. Capsule oblong, prickly.

Flowering & fruiting : Mid June – August.

Uses : Due to attractive flowers the plant is having aesthetic value and a very potential ornamental plant.

59. *Meconopsis bikramii* Aswal.

Family : Papaveraceae

Name : Himalayan blue poppy (E),

Distribution : Western Himalayas from Kumaun to Kashmir.

Habitat : Rare amongst stones and boulders, Lahaul valley around 3,750 m.

Plant Characters : A decumbent, bristly, perennial herb, 30-50 cm tall. Stems hollow, smooth, except the yellowish straight thin bristles. Radical leaves 15-

20 cm long, 4.5-12.5 cm broad, long petiolate, palmately 3-5 lobed, sparsely throughout. Cauline leaves few 4.0-9.5 cm long, 1.5-3.0 cm broad, oblong to lanceolate, petiolate, irregularly pinnately lobed. Flowering buds globose. Stamens indefinite, filaments 0.5-1.0 cm long, anthers 1.5-2.0 mm long, basifixed, oblong, yellow. Capsule obovate to ovoid, elongated, prickly; seeds numerous, reniform to plano-convex, brown-red.

Flowering & fruiting : July – August.

Uses : Having ornamental potential.

60. ***Oxygraphis polypetala*** (Royle) Hk. f. et T.

Family : Ranunculaceae

Distribution : From Kashmir to South East Tibet & Burma.

Habitat : Snow cleared areas, Suru valley around 3,120 m.

Plant Characters : A small tufted, alpine herb. Roots fibrous. Flowering stem 2.5-10 cm high with tiny early-flowering solitary or several golden yellow or bronze flowers. Leaves all basal, stalked. Sepals 5, elliptic, blunt. Petals oblong-spathulate. Achenes in a globular head surrounded by enlarged calyx.

Flowering & fruiting : July to Mid September.

Uses : It is used as ornamental plant.

61. ***Papaver nudicaule*** Linn.

Family : Papaveraceae

Name : Orange poppy (E)

Distribution : From Pakistan to Western Himalayas from Kumaun to Kashmir.

Habitat : On rocky slopes, Leh valley around 4,890 m.

Plant Characters : An erect, tall, perennial herb. Rootstock scaly. Stem tall, 15-45 cm, hairy. Leaves in a rosette, bristly-haired, glabrous, 3-lobed. Flowers yellow or orange to reddish. Scape several. Sepals densely hairy. Capsule round, hairy.

Flowering & fruiting : June to August.

Uses : It is used as ornamental plant.

62. ***Pedicularis pycnantha*** Boiss.

Family : Scrophulariaceae

Distribution : From Pakistan to Kashmir & Himachal Pradesh.

Habitat : Muddy & moist slopes, Nubra & Lahaul valleys 2,920 m.

Plant Characters : A short, stout sparsely hairy herb. Stems many from the base. Cauline leaves few, opposite, whorled, lanceolate, pinnatifid. Flowers pink, borne in lax or dense spikes. Capsule shortly oblong, acute, shorter than the calyx. Seeds large ellipsoid.

Flowering & fruiting : July to mid September.

Uses : This is used as ornamental plant.

63. *Pedicularis pyramidata* Royle.

See Sr. No. 103, Chapter VIII.

Uses : The plant is used as ornamental.

64. *Physalis alkekengi* Linn.

See Sr. No. 234, Chapter VI.

Uses : The fruits are used as religious ornament.

65. *Polygonum polycnemoides* Jaub. & Spach.

Family : Polygonaceae

Name : Lamtso (L)

Distribution : Himachal Pradesh to Kashmir, Bhutan & Afganistan.

Habitat : Moist & open slopes, Nubra valley around 3,205 m.

Plant Characters : An annual, prostrate herb. Stem branched, pubescent. Leaves small, sessile, linear, brown, glabrous, thick, hairy. Bracts scarious, 3-4 mm. ovate, toothed, enclosing the perianth. Tepals reddish-purple. Nut brown, hairy.

Flowering & fruiting : Mid July to mid September.

Uses : The plant is used as ornamental plant in Ladakh.

66. *Primula denticulata* Smith.

Family : Primulaceae

Name : Sulumentok (L)

Distribution : Afganistan to South East Tibet, Burma, Kashmir & Bhutan.

Habitat : On moist and open slopes, Zanskar & Lahaul valleys around 3,730m.

Plant Characters : A perennial herb. Rootstock stout, short, thick. Leaves erect or spreading, oblong, oblanceolate, toothed, denticulate, midrib broad, never reticulate. Leaves much enlarging after flowering. Scape stout, 10-15 cm long. Bracts short, unequal. Flowers purple to mauvish-blue in compact globular heads. Capsule small.

Flowering & fruiting : June to August.

Uses : It is used as ornamental plant.

67. *Primula elliptica* Royle.

Family : Primulaceae

Name : Sulumentok (L)

Distribution : From Pakistan to Uttranchal.

Habitat : Alpine rocks, Changthang and Spiti valleys around 5,030 m.

Plant Characters : A perennial, scapigerous, herb. Leaves suberect, closely serrated down to the dilated top of the petiole midrib slender, nerves obscure, rounded, petioled. Flowers 3-10 with purple petals. Calyx campanulate, deeply cleft. Corolla lobes toothed. Capsule oblong.

Flowering & fruiting : June to August.

Uses : Used as ornamental plant.

68. *Primula involucrata* Wall. ex Duby.

Family : Primulaceae

Name : Sulumentok (L)

Distribution : Pakistan to South West China.

Habitat : Moist places & marshes, Suru valley around 2,830 m.

Plant Characters : A perennial herb. Stem 10-30 cm high. Leaves oblong, blunt, entire or obscurely toothed margins. Flowers white with a yellow eye borne in a lax umbel. Petals shallowly notched; calyx ribbed with lanceolate lobes.

Flowering & fruiting : July to September.

Uses : It is used as an ornamental plant.

69. *Primula macrophylla* D. Don. (excl. syn. *P. moorcroftiana*). *P. stuartii* Wall. var. *perpurea* (Royle) Hk. f.; *P. nivalis* Pall. var. *macrophylla* (D. Don.) Pax.

Family : Primulaceae

Name : Sulumentok or Kelche-Karpa (L)

Distribution : Afganistan to South East Tibet.

Habitat : Rocky slopes & damp places, Leh valley around 5,330 m.

Plant Characters : A robust, perennial herb. Leaves erect, (10-25 cm), lanceolate or toothed, acute with white farina beneath. Flowers loosely umbelled, purple or lilac with usually a darker eye. Bracts awl-shaped; corolla tube 1-3 cm and calyx cylindrical with linear-lanceolate lobes. Capsule cylindrical, 2-5 cm.

Flowering & fruiting : July to September.

Uses : Having ornamental value.

70. *Primula reptans* Hk. f. ex Watt

Family : Primulaceae

Distribution : Pakistan to Himachal Pradesh & Kashmir.

Habitat : Open places, Suru and Lahaul valleys around 4,530 m.

Plant Characters : A tiny densely tufted herb. Leaves small, ovate to rounded blade, deeply toothed and with recurved margin, leaf stalk short. Flowers solitary, pink or purple. Corolla tube narrow funnel shaped, petals deeply notched.

Flowering & fruiting : July to August.

Uses : It is used for ornamental purposes.

71. *Primula rosea* Royle.

Family : Primulaceae

Name : Prime rose (E), Sulumentok or Brab-lcam (L)

Distribution : Pakistan to Himachal Pradesh, Kashmir, Kullu & Garhwal.

Habitat : Damp places & streams, Leh and Spiti valleys around 5,030 m.

Plant Characters : A glabrous scapigerous perennial herb with rosy fragrance. Leaves many, ovate- lanceolate, finally toothed, rounded. Flowers rose-pink with a yellow. Corolla obcordate, notched. Bracts long pointed, poached at base.

Flowering & fruiting : July to August.

Uses : Ornamental value due to attractive and rosy fragrant flowers.

72. *Rhodiola imbricata* Edgew.; *Sedum roseum* (Linn.) Scop.

See Sr. No. 261, Chapter VI.

Uses : Ornamental potential.

73. *Rhodiola tibetica* (Hk. f. & T.) Fu.

See Sr. No. 262, Chapter VI.

Uses : Ornamental potential.

74. *Rosa brunonii* Lindl.

Family : Rosaceae

Name : Himalayan Musk Rose (E)

Distribution : Pakistan to South West China, India & Tibet.

Habitat : Rare on dry areas, Lahaul valley around 3,690 m.

Plant Characters : A large, climbing shrub with glabrous twigs and recurved prickles. Leaves 3-6 cm long; leaflets elliptic-oblong or oblong-lanceolate, glabrous above, softly pubescent beneath. Flowers in many-flowered corymbs; peduncles and pedicles pubescent, not prickly or bristly. Petals 5, white. Style united. Fruits globose or ovoid, dark brown.

Flowering & fruiting : May to July.

Uses : The flowers are used ornamentally.

75. *Rosa ecae* Aitchinson.

Family : Rosaceae

Name : Wild rose (E), Shayh (L)

Distribution : Native of West Asia, Kashmir to Himachal Pradesh & Ladakh.

Habitat : Common along river belts and along roadside, Suru valley around 2,920 m.

Plant Characters : A stiff erect shrub. Stem branched, prickled, leaflets 5-11 elliptic to narrow – oblong, terminal leaflets slightly large. Flowers large, yellow in terminal clusters. Petals several. Fruits bright red, globular.

Flowering & fruiting : May to August.

Uses : The whole plant is used as ornamental hedge.

76. *Rosa eglanteria* Linn.

Family : Rosaceae

Name : Wild rose (E)

Distribution : Kashmir to Himachal Pradesh & Tibet.

Habitat : Common in forest, Lahaul valley around 3,820 m.

Plant Characters : An armed with large and small prickles, perennial shrub. Leaves 4-8 cm long; leaflets 2-4 pair, elliptic or oblong-obovate, deeply serrate, glandular-pubescent beneath. Flowers solitary or 2-3 together. Calyx-lobes lanceolate, hairy. Petals obcordate, yellow.

Flowering & fruiting : May to August.

Uses : The whole plant is used as ornamental.

77. *Rosa macrophylla* Lindl.

Family : Rosaceae

Name : Jangali Gulab (H), Wild rose (E)

Distribution : Himachal Pradesh to Tibet & South West China and Kashmir.

Habitat : Dry slopes, Lahaul valley around 3,830 m.

Plant Characters : A deciduous shrub with glabrous twigs; prickles variable in amount according to habitat, when present straight or slightly curved with broad or narrow bases. Leaves 4-12 cm long; leaflets 7-11, elliptic, nearly glabrous. Flowers solitary or in terminal corymbs. Calyx-lobes longer than petals. Petals broadly obcordate red. Fruits red, ovoid, glandular-hairy.

Flowering & fruiting : June to September.

Uses : The flowers are used as ornament for decoration.

78. *Rosa webbiana* Wall. ex Royle.

Family : Rosaceae

Name : Jangali Gulab (H), Wild Rose (E), Siah (L)

Distribution : Kashmir to Kumaun, Pakistan & Western Nepal.

Habitat : Dry & rocky slopes, Nubra & Suru valley around 3,240 m.

Plant Characters : A leaf shedding, deciduous, perennial shrub. Stem 1.5-2.5 m with slender branches and straight prickles. Leaves generally 5-9, coarsely toothed, hairless. Flowers pink to deep red, usually solitary with 5 petals. Fruit red, egg shaped to round with persistent calyx.

Flowering & fruiting : July to September.

Uses : Flowers are used for ornamental purposes.

79. *Rosularia alpestris* (Kar. & Kir.) Boiss.; *Sempervivum acuminatum* Decne.; *Sempervivella acuminata* (Decne.) Berger.

Family : Crassulaceae

Distribution : From Afganistan to Himachal Pradesh & Central Asia.

Habitat : Alpine slopes, Zanskar valley around 4,230 m.

Plant Characters : A perennial, succulent herb. Stem 10-20 cm long. Leaves in dense rosette, oblong, acuminate. Flowers pink-purple to white borne in terminal clusters. Petals small with a red mid vein. Filaments of stamens white. Seeds narrow obovoid, striated.

Flowering & fruiting : July to August.

Uses : Used as ornamental.

80. *Salix babylonica* Linn.

See Sr. No. 37, Chapter XIII.

Uses : The plant is used for beautification and as ornamental tree.

81. *Saussurea bracteata* Decne.

See Sr. No. 272, Chapter VI.

Uses : The flowers are used for ornamental purposes.

82. ***Saussurea gossypiphora*** D. Don

Family : Asteraceae

Distribution : From Pakistan to Kashmir.

Habitat : Alpine meadows, Lahaul valley aroud 5,190 m.

Plant Characters : A dwarf, stemless, woolly plant just look like a snowball. Stem covered by fibrous leaf remains at the base, hollow, expanded at the top. Basal leaves many, sessile, pinnatifid or remotely toothed, narrow towards the base; cauline leaves numerous, profusely hairy on the under surface, covered by dense matted wool. Flower-heads deeply embedded in woolly hairs with many purple florets. Involucral bracts purplish. Achenes glabrous.

Flowering & fruiting : July to September.

Uses : Used for ornamental.

83. ***Saussurea obvallata*** (DC.) Edgew.

See Sr. No. 277, Chapter VI.

Uses : The flowers are of ornamental value.

84. ***Sedum ewersii*** Ledeb. *Sedum ewersii* Ledeb. Lcon.

See S. No. 129, Chapter VIII.

Uses : The whole plant is used as an ornamental plant.

85. ***Silene edgeworthii*** Bocquet.

Family : Caryophyllaceae

Name : Globe flower (E)

Distribution : Afganistan to South West China & Central Asia.

Habitat : Stony slopes & moist areas, Suru valley around 3,740 m.

Plant Characters : An erect herbaceous perennial. Stem branched, 50-70 cm high. Leaves ovate-lanceolate to narrowly lanceolate, acute. Flowers erect or nodding in large terminal panicles. Calyx bell shaped. Petals much longer than the sepals, purple or cream white.

Flowering & fruiting : July to September.

Uses : The plant is used for ornamental purposes.

86. ***Silene moorcroftiana*** Wall. ex Benth.

Family : Caryophyllaceae

Name : Khuamenla (L)

Distribution : Afganistan to Central Nepal and Tibet.

Habitat : Dry rocky slopes, Leh valley around 3,340 m.

Plant Characters: A tufted perennial herb. Stem finely pubescent, much branched, 15-30 cm long. Leaves mostly basal, linear-lanceolate, acute. Flowers white solitary or few, terminal. Petals deeply lobed with scales in the throat. Calyx contracted towards the base. Capsule with a stalk

Flowering & fruiting : July to September.

Uses : The plant is used ornamentally.

87. *Syringa emodi* Linn.

See Sr. No. 46, Chapter XIII.

Uses : The plant is used for ornamental purposes.

88. *Thermopsis inflata* Camb.

Family : Fabaceae

Name : Olmango or Gla-ba srad-ma (L)

Distribution : From Kashmir to Himachal Pradesh & Tibet.

Habitat : Alpine slopes, Changthang valley around 5,340 m.

Plant Characters : A perennial herb with creeping rootstock. Stem several, mostly 10-15 cm. Leaves sessile, fleshy, glaucous. Leaflets ovate to obovate. Flowers deep yellow in terminal cluster. Bracts ovate, hairy. Pods inflated broadly oblong.

Flowering & fruiting : July to September.

Uses : The flowers are used ornamentally.

89. *Thylacospermum caespitosum* (Camb.) Schischk.; *Thylacospermum rupifragum* (Kar. & Kir.) Schrenk.

See Sr. No. 81, Chapter XII.

Uses : It can be used as ornamental plant in fresh as well as in dry form.

90. *Trollius acaulis* Lindl.

Family : Ranunculaceae

Name : Globe flower (E)

Distribution : Pakistan to Western Nepal.

Habitat : Moist alpine slopes, Suru valley around 3,060 m.

Plant Characters : An erect, fleshy, perennial herb. Rootstock stout, fibres thick. Stem simple, 12 or more inches long in fruit, base clothed with fibres. Basal leaves long-petioled, petioles. Flowers solitary, terminal, bright golden-yellow. Achenes stiff.

Flowering & fruiting : July to September.

Uses : Used as ornamental.

91. *Tulipa stellata* Hk. f. var. ***chrysantha*** Boiss.

Family : Liliaceae

Name : Wild Tulip (E) & Kapichong (L)

Distribution : Kashmir to Bhutan.

Habitat : On wastelands, Suru valley around 2,860 m.

Plant Characters : A small bulbous perennial herb. Stem reaching upto 15-40 cm high. Leaves linear acute, 20-30 cm, channeled, sheathing below. Flowers solitary, golden yellow on a long stem. Corolla at first bell-shaped, but petals later spreading widely.

Flowering & fruiting : July to September.

Uses : The flowers are of ornamental value.

❑❑❑

11

Magico-Religious Plants

In addition of plant utility as food, medicine and shelter, plant species have been closely associated with many social custums and mythological rituals of man times immemorials. Several flowers, fruits or other parts or whole plants have been used for offering in worship, or some plants are themselves worshipped or considered sacred viz. Sacred basil (*Ocimum sp*), Pipal (*Ficus sp*). Similarly some plants are believed to having magical properties viz. placing a twig of *Achranthus aspera* in hairdo of a woman can facilitate child birth, paste of flowers of *Ageratum sp* on forehead means good luck and success. The most common Dub grass (*Cynodon dactylon*) is a very useful plant for most of religious, social and other ceromonies. Similarly identity of many plants were well documented in our ancient epics Ramayan and Mahabharta for their usefulness.

Cold desert trans Himalayan regions of India are not an exception to it. The tribal communities of Ladakh and Lahaul-Spiti have their own traditional belief about the plants growing in and around their surroundings. Most of the cultivated species are being used for magico-religious purposes such as *Allium spp, Brassica spp, Capsicum annuum, Hordeum vulgare, Linum sp* and different flowers etc. A few species of wild plants are also being used for magico-religious uses. Plants and their parts are commonly used for religious ceremonies, worships in gompas, for incense, ward off of bad evil etc. The widely used magico-religious plants are described in brief. (Table 8).

1. *Aconitum violaceum* Jacq. ex Stapf.

See Sr. No. 6, Chapter VI.

Uses : The roots are used in small quantity to ward off of bad evils.

2. *Arnebia euchroma* (Royle) John. I. M Johnston.

See Sr. No. 28, Chapter VI.

Uses : Fat soluble red dye from roots is used by the local people for colouring prasada in monastry.

3. ***Berberis pachyacantha*** Koehne.

See Sr. No. 55, Chapter VI.

Uses : The priest and Lamas used the twigs in religious ceremonies.

4. ***Carex melanantha*** C.A. Mey.

Family : Cyperaceae

Distribution : From Kashmir to Sikkim & Kumaun.

Habitat : On alpine meadows, Spiti valley around 4,870 m.

Plant Characters : An erect, loosely tufted, perennial herbaceous plant. Rhizome elongated, woody. Stem erect, trigonous. Leaves basal and sub basal, flat, sheathing. Inflorescence consisting of 3-5 spikelets.

Flowering & fruiting : July to September.

Uses : When a male child is born, a feast is given to the village people and on this occasion a small bunch of dried plants is given to the guests.

5. ***Cedrus deodara*** (Roxb. ex D. Don.) G. Don.

See Sr. No. 3, Chapter XIII.

Uses : The wood is used in religious ceremonies.

6. ***Delphinium brunonianum*** Royle.

See Sr. No. 103, Chapter VI

Uses : Flowers are used in religious ceremonies.

7. ***Delphinium cashmerianum*** Royle.

See Sr. No. 104, Chapter VI

Uses : Flowers are used in religious ceremonies.

8. ***Delphinium vestitum*** Wall. ex Royle.

See Sr.No. 105, Chapter VI.

Uses : Flowers are used in religious ceremonies.

9. **Elaeagnus angustifolia** Linn. E. hortensis, M Bieb.

See Sr. No. 7, Chapter XIII.

Uses : Flowering twigs are used for fragrance and religious ornament.

10. ***Epilobium angustifolium*** Linn.

See Sr.No. 36, Chapter X.

Uses : Flowers are used in religious places.

11. ***Epilobium latifolium*** Linn.

See Sr.No. 37, Chapter X.

Uses : Flowers are used in religious places.

12. ***Epilobium roseum*** Schreb.

See Sr.No. 38, Chapter X

Uses : Flowers are used in religious ceremonies.

13. ***Juniperus communis*** Linn. var.

See Sr. No. 14, Chapter XIII.

Uses : The twigs, leaves and bark are used for making incense and dhoop by the tribal people of Ladakh and Lahaul-Spiti.

14. ***Juniperus macropoda*** Boiss.

See Sr. No. 15, Chapter XIII.

Uses : Same as in case of *J. communis.*

15. ***Juniperus recurva*** Buch,-Ham

See Sr. No. 16, Chapter XIII.

Uses : Many old trees of this areas are considered as the resting places of supernatural powers since the ancient times. These plants are kept untouched, as nobody dares to disturb the spirit.

16. ***Peganum harmala*** Linn.

See Sr. No. 231, Chapter VI.

Uses : The seeds are used for child bath and as dhoop for religious ceremonies especially for keeping away of bad evils.

17. ***Physalis alkekengi*** Linn.

See Sr. No. 234, Chapter VI.

Uses : The fruits are used as religious symbol by the Dards of Dah-Hanu village.

18. ***Physochlaina praealta*** (Decne.) Miers.

See Sr. No. 235, Chapter VI.

Uses : Seeds of the plant alongwith *Artimisia brevifolia* are helpful in vermifuge and tooth decay. The whole plants are having halluginosic properties.

19. *Primula macrophylla* D. Don. (excl. syn. *P. moorcroftiana*). *P. stuartii* Wall. var. *perpurea* (Royle) Hk. f.; *P. nivalis* Pall. Var. *macrophylla* (D. Don.) Pax.

See Sr. No. 69, Chapter X.

Uses : The flowers are used in religious ceremonies.

20. *Primula rosea* Royle.

See Sr. No. 70, Chapter X.

Uses : The flowers are used in religious ceremonies.

21. *Rosa brunonii* Lindl.

See Sr. No. 74, Chapter X.

Uses : The flowers are used in religious places.

22. *Rosa ecae* Aitchinson.

See Sr. No. 75, Chapter X.

Uses : The flowers are used in religious ceremonies.

23. *Rosa eglanteria* Linn.

See Sr. No. 76, Chapter X.

Uses : The flowers are used in religious places.

24. *Rosa macrophylla* Lindl.

See Sr. No. 77, Chapter X.

Uses : The flowers are used in religious ceremonies.

25. *Rosa webbiana* Wall. ex Royle.

See Sr. No. 78, Chapter X.

Uses : The flowers are used in religious ceremonies.

26. *Verbascum thapsus* Linn.

See Sr. No. 318, Chapter VI.

Uses : The plant is considered religious and used for keeping away of evils. To protect the house and animal shelters from evil spirits, an axe, sword, a branch of kerpa (*Berberis*) or Pilisti (*Ribes*) along with this plant are put above the main doors and windows.

27. *Waldheimia glabra* (Decne.) Regel.; *Allardia glabra* Decne.

See Sr. No. 68, Chapter VII.

Uses : The whole plant is used for making dhoop in religious places.

28. *Waldheimia stoliczkai* Clarke.

See Sr. No. 69, Chapter VII.

Uses : Flowers are used for incense.

29. *Waldheimia tomentosa* (Decne.). Regel.; *Allardia tomentosa* Decne.

See Sr. No. 70, Chapter VII.

Uses : The whole plant is used for making dhoop in religious places.

12

Fuel Plants

The fuel plays an important role in trans Himalayan cold desert region especially during prolonged winter period. The people of far-flung areas of Ladakh and Lahaul-Spiti utilize the wild plants for fuel. Majority of these plants are being collected during summer and stored on the roof for winter use. It is believed that these stored plants also help in maintaining room temperature by providing insulation.

As we know, Ladakh and Lahaul-Spiti have very long winter period about 5-6 months where temperature goes below -20 ^{0}C to -30 ^{0}C every year. Survival during this long and harsh cold climatic conditions is very tough without heating facility. The cold & drought desert of Ladakh where plant population is sparsh and only few species are being used as fuel and fire wood. Due to the acute shortage of woody trees local people collect the small bushes and shrubs even herbs for fuel purposes. Though the quality of heat energy produced from the wild shrubs is not yet estimated even then most of the species produce good heat while used alongwith woody plants.

Acantholimon lycopodioides, Artemisia sp, Caragana versicolor, Hippophae species, Myricaria sp, Tamarix sp etc are some of useful fuel plants grown in Ladakh and Lahaul-Spiti. The tribal people collect these various fuel plants at the end of summer season and stored them for prolonged use during winter. The common fuel plants used by the local people have been described in alphabetical order alongwith families, English (E), hindi (H) / local names (L), distribution, habitat, brief botanical notes and uses in the enumeration. (Table 9).

Enumeration of Fuel Flora

1. ***Acantholimon lycopodioides*** (Girard) Boiss. *Statice lycopodioides* Girard.

See Sr. No. 1, Chapter VI.

Uses : It is one of the major and important fuel plant of this region which is being uprooted by the locals by the end of summer and stored on the roofs for winter uses.

2. *Androsace aizoon* Duby.

Family : Primulaceae

Name : Sga-tig (L)

Distribution : Throughout Himalayas.

Habitat : On drier areas, along road sides, Suru valley and Changthang valleys, around 2,980 m.

Plant Characters : A perennial tufted herb. Leaves densely rosulate sessile, spathulate, ciliolate. Inflorescence glandular-pubescent. Scape long. Bracts ovate, corolla-lobes bifid. Capsule with 2-fid valves.

Flowering & fruiting : July to mid September.

Uses : Used for fuel.

3. *Androsace villosa* Linn.

See Sr. No.18, Chapter VI.

Uses : This plant is used as fuel.

4. *Anthemis cotula* Linn.

Family : Asteraceae

Name : Stinking May weed (E)

Distribution : Himalayas from Garhwal to Kashmir.

Habitat : Rocky slopes, Zanskar and Spiti valleys between 3,940-4,190 m.

Plant Characters : A foetid annual herb. Stem much branched, reaching upto 50 cm tall. Leaves 2-3 pinnately cut into linear acute segments. Flower-heads solitary, long-stalked with spreading white ray-florets and yellow disc-florets. Involucral bracts hairy with narrow papery margins with swellings.

Flowering & fruiting : July to September.

Uses : Used for fuel.

5. *Arenaria festucoides* Benth.

Family : Caryophyllaceae

Name : Pianzo (L)

Distribution : Himalayas from Kashmir to Himachal Pradesh.

Habitat : On stony slopes, Suru and Spiti valleys around 4,230 m.

Plant Characters : A densely tufted perennial herb. Stem much branched, branches slender. Leaves bristle like, shinning, usually recurved. Flowers small, white. Petals obovate; sepals lanceolate pointed.

Flowering & fruiting : July to mid September.

Uses : The whole plant is used for fuel.

6. ***Arenaria kashmirica*** Edgew ex Hk. f. & T.

Family : Caryophyllaceae

Name : Khilimentok (L)

Distribution : Throughout Himalayas.

Habitat : Along rosdsides and dry places, Zanskar and Changthang valleys, between 3,980-4,180 m.

Plant Characters : A tufted perennial herb. Stem branched, pubescent, glandular. Leaves narrow, lineat, acute. Flowers small in cymes peduncled. Sepals ovate-lanceolate. Capsule ovoid.

Flowering & fruiting : July to mid September.

Uses : Used for fuel.

7. ***Artemisia brevifolia*** Wall. ex DC. *Artimisia maritima* Linn.

See Sr. No. 5, Chapter VII.

Uses : The whole plants are usually collected by the end of summer for fuel during winters.

8. ***Artemisia dracunculus*** Linn. Sp. Pl. 849. 1753; *A. glauca* Pallas ex Willd.

See Sr. No. 7, Chapter VII.

Uses : The whole plants are usually collected by the end of summer for fuel during winters.

9. ***Artemisia gmelinii*** Web. ex Stechm.

See Sr. No. 35, Chapter VI.

Uses : The whole plants are usually collected by the end of summer for fuel during winters.

10. ***Artemisia nilagirica*** (Clarke) Pamp.

See Sr. No. 37, Chapter VI.

Uses : The whole plants are usually uprooted by the end of summer for fuel.

11. ***Artemisia parviflora*** Roxb.

See Sr. No. 38, Chapter VI.

Uses : The whole plants are usually collected by the end of summer for winter uses.

12. *Artemisia persica* Boiss.

See Sr. No. 39, Chapter VI.

Uses : The whole plants are usually collected by the end of summer for fuel.

13. *Artemisia salsoloides* Willd.

See Sr. No. 40, Chapter VI.

Uses : Used as fuel.

14. *Artemisia sieversiana* Ehrh ex Willd.

See Sr. No. 42, Chapter VI.

Uses : Used as winter fuel.

15. *Artemisia tournefortiana* Reichb.

See Sr. No. 17, Chapter VII.

Uses : Dried stems are used as fuel.

16. *Astragalus cicerifolius* Royle ex Fisch.

See Sr. No. 11, Chapter IX.

Uses : The plant is used for fuel.

17. *Astragalus frigidus* (Linn.) Gary

Family : Fabaceae

Name : Skarmalo (L)

Distribution : From Kashmir and Kumaun.

Habitat : Open rocky areas, Zanskar & Changthang valleys between 4,230-4,450 m.

Plant Characters : An erect, herbaceous, perennial plant. Stem hairless upto 1 m. Leaves 10-15 cm, leaflets 9-13, large, oblong to ovate, becoming hairless above. Flowers large, yellow in axillary peduncled racemes. Standard longer than wings and keel. Pod oblong, stalked, densely hairy.

Flowering & fruiting : July to August.

Uses : Usually branches are cut for fuel purposes.

18. *Atriplex nitens* Schkukr.

Family : Chenopodiaceae

Name : Lichiang (L)

Distribution : Temperate Himalaya from Kashmir to Kumaun.

Habitat : Dry places & roadsides, Leh valley around 4,760 m.

Plant Characters : An annual herb. Stem branched, hairy. Leaves hastate-cordate sinuate toothed, acuminate. Inflorescence in loose panicle. Fruiting bracts free, thin, obtuse.

Flowering & fruiting : July to September.

Uses : The locals collect the same for winter fuel.

19. *Atriplex rosea* Linn.

Family : Chenopodiaceae

Name : Rakpa (L)

Distribution : From Assam to Meghalaya, J & K and Himachal Pradesh.

Habitat : Common on rocky places, Zanskar & Spiti valleys around 3,650 m.

Plant Characters : An annual herb. Stem branched from the base, branches diffuse, ascending. Leaves small, petioled rhombic ovate, obtuse. Male flowers in terminal clusters and female flowers in axillary solitary, rarely mixed with males.

Flowering & fruiting : July to September.

Uses : Used for fuel.

20. *Berberis ulicina* Hk. f. & T.

See Sr. No. 56, Chapter VI.

Uses : The plant is usually collected for winter fuel.

21. *Betula utilis* HD. Don.

See Sr. No. 1, Chapter XIII.

Uses : The branches are used as important fuel especially in winters.

22. *Caragana brevifolia* Komarov.

Family : Fabaceae

Name : Dama or Kama (L)

Distribution : Kashmir to Himachal Pradesh and Nepal.

Habitat : Sandy places & roadside, Zanskar valley around 3,590 m.

Plan Characters : A densely branched perennial undershrub. Stem branched. Leaves petioled; leaflets obovate, densely pubescent. Flowers yellow, streaked with red. Calyx haireless. Pod hairless.

Flowering & fruiting : July to August.

Uses : The whole plant is uprooted and collected by locals for fuel purposes.

23. *Caragana versicolor* Wall. (*Caragana pygmaea* DC.)

Family : Fabaceae

Name : Dama or Kama (L)

Distribution : Afganistan to Western Nepal, Tibet and China.

Habitat : Sandy places, Changthang & Nubra valleys between 3,019-3,574m.

Plant Characters : A low densely branched, very spiny shrub. Stem spiny at close nodes and axils with solitary yellow often orange flowers along the stem; stipule spiny, trifid. Flowers with broadly triangular stipules of 3 spines. Calyx hairless. Pod c. 3 mm, hairless.

Flowering & fruiting : July to September.

Uses : The plant is used as fuel. Usually the locals collect the plants at the end of growing season and stored on roofs for winter uses.

24. *Cedrus deodara* (Roxb. ex D. Don.) G. Don.

See Sr. No. 3, Chapter XIII.

Uses : The deodar is considered as one of the valuable fuel and firewood plant.

25. *Cirsium arvense* (Linn.) Scop. *Serratulla arvensis* Linn.

Family : Asteraceae

Name : Creeping thistle (E), Yangchher (L)

Distribution : From Pakistan to Central Nepal, Eurasia, Kashmir and Kumaun.

Habitat : Common on cultivated fields, Zanskar and Lahaul valleys around 3,650 m.

Plant Characters : A very variable, perennial herb. Stems many from the creeping roots, reaching upto 1-1.5 m tall. Leaves lanceolate-oblong, spiny, entire to pinnately lobed. Flower-heads pale purple, pedicelled, borne on branched edges; florets unisexual. Achenes smooth, pappus brown, white.

Flowering & fruiting : Mid June to September.

Uses : The whole plant is used for fuel.

26. *Colutea multiflora* Stepf. ex Ali.

See Sr. No. 4, Chapter XIII.

Uses : The branches are cut for winter fuel.

27. *Cotoneaster falconeri* Klotz. *C. integerrima* Medic

See Sr. No. 5, Chapter XIII.

Uses : The branches are cut for fuel purposes in large quantities.

28. *Cotyledon thyrsiflora* (Fish.) Maxim. *Orostachys thyrsiflora* Fish.

Family : Crassulaceae

Distribution : Pakistan to Nepal and Kashmir.

Habitat : Rocky slopes, Changthang valley around 4,790 m.

Plant Characters : An annual, herb. Stem 15-30 cm high, branched. Radical leaves rosulate, spathulate, terminated by a long spine, cauline oblong flat. Peduncle single flowered. Corolla white.

Flowering & fruiting : July to mid September.

Uses : Used for fuel.

29. *Crataegus songarica* C. Koch.

See Sr. No. 6, Chapter XIII.

Uses : The branches are used for fuel.

30. *Cynanthus lobatus* Wall. ex Benth.

Family : Campanulaceae

Distribution : South West China & Kashmir to Himachal Pradesh.

Habitat : On alpine meadows, Lahaul valley around 4,650 m.

Plant Characters : A dwarf, tufted, decumbent herb or shrub. Stems 10-30 cm long, glabrous below, hairy upwards. Leaves obovate-oblong, lobed, sparsely hairy. Flowers blue, in solitary, terminal heads. Calyx ovate with black hairs. Corolla funnel-shaped, lobes hairy in the throat.

Flowering & fruiting : Mid July to August.

Uses : Used as fuel.

31. *Cynoglossum petiolatum* (Hk. f.) DC. *Anchusa petiolata* Hk. f.

Family : Boraginaceae

Name : Ematso (L.)

Distribution : Afganistan to Bhutan, India, China, Japan and South East Asia.

Habitat : On rocky slopes, Zanskar and Lahaul valleys around 3,560 m.

Plant Characters : An erect, perennial, hairy, herb. Stem branched upto 90-100 cm tall, tomentose. Lower leaves oblong, petioled, middle and upper cauline leaves few, ovate, acute, and sessile. Flowers blue, in terminal bifurcate, cymes. Corolla longer than the calyx. Nutlets ovate, sparsely glochidiate.

Flowering & fruiting : Mid June to August.

Uses : Used for fuel.

32. *Cynoglossum zeylanicum* Thunb. ex Lehm.

Family : Boraginaceae

Name : Hound's tongue (E)

Distribution : From Afganistan to Bhutan, India, China & Japan.

Habitat : On wastelands areas, Nubra valley around 3,020 m.

Plant Characters : An erect, hairy, perennial herb. Stem stout, branched from the base, upto 60 cm long. Leaves oblong-elliptic, covered with soft adpressed hair. Flowers usually blue in branched clusters. Nutlets ovate with marginal glochidia.

Flowering & fruiting : Mid June to August.

Uses : The whole plant is used as fuel.

33. *Dianthus angulatus* Royle. ex Benth.

Family : Caryophyllaceae

Distribution : Kashmir to Pakistan & Zanskar.

Habitat : Dry areas, Leh and Nubra valleys between 3,300-3,600 m.

Plant Characters : A tufted, perennial herb. Stems many from the base, 15-25 cm long. Leaves linear, margin finely toothed. Flowers white, solitary terminal. Petals with a deep fringe of narrow teeth. Calyx cylindrical slightly enlarged at base. Petals fimbricate. Bracts often coloured.

Flowering & fruiting : Mid June to August.

Uses : Used for fuel.

34. *Dipsacus mitis* D. Don. *D. innermis* Wall.

Family : Dipsacaceae

Name : Sichimo (L)

Distribution : Kashmir to Bhutan and Tibet.

Habitat : Dry places, Zanskar and Spiti valleys around 3,580 m.

Plant Characters : A bristly hairy, perennial herb. Stem reaching upto 1 m tall. Lower leaves pinnatifid with distant segment, petioled, upper simple, dentate margins. Flower-heads 1-3 with linear and acute bracts, yellow. Achene ribbed.

Flowering & fruiting : July to September.

Uses : The plant is used as fuel.

35. *Dioscorea deltoidea* Wall. ex Kunth, Enum.

See Sr. No. 109, Chapter VI.

Uses : Occasionally collected for fuel.

36. *Ephedra gerardiana* Wall. ex Stapf.

See Sr. No. 116, Chapter VI.

Uses : The whole plants are used as fuel for winter period.

37. *Erigeron multicaulis* Wall. ex DC.

Family : Asteraceae

Distribution : Pakistan to Bhutan, Central Asia.

Habitat : Rocky areas, Leh & Nubra valleys between 3,540-4,201 m.

Plant Characters : A small, tufted, hairy or pubescent, erect, annual herb. Stem branched from the base. Branches slender. Leaves ovate-lanceolate, acute, radial leaves petioled; cauline sessile, toothed. Flower-heads pale-purple, solitary, in long peduncled. Fruits nearly hairless.

Flowering & fruiting : July to September.

Uses : Used for fuel.

38. *Eriocycla thomsonii* (Clarke.) Wolff. *Pituranthos thomsonii* Clarke. ex Hk. f. T.

Family : Apiaceae

Name : Temo (L)

Distribution : Pakistan to Turkistan, Himachal Pradesh, Central Asia and Kashmir.

Habitat : Along roadside, Zanskar valley 3,580 m.

Plant Characters : A perennial herb. Stem with several lateral branches, pubescent, 15-30 cm tall. Lower leaves 2-pinnate; ultimate segments ovate, deeply pinnatifid, toothed; upper reduced. Umbels long or short on peduncles. Bracts 4-8, linear, rays pubescent, subtomentose. Fruits laterally compressed.

Flowering & fruiting : July to September.

Uses : Used for fuel.

39. *Eritrichium fruticulosum* Klotz.

Family : Boraginaceae

Distribution : From Afganistan to Central Nepal, Tibet, and Central Asia.

Habitat : Alpine slopes & rock crevices, Leh and Spiti valleys between 3,890-4,120 m.

Plant Characters : A perennial herb. Stems many from the base. Leaves lanceolate, upper sessile, lower short petioled. Flowers light blue or sky blue with yellow eyes in terminal clusters. Nutlets 4, pubescent on both surfaces, stiff, forming a pyramid.

Flowering & fruiting : July to September.

Uses : It is used as fuel.

40. *Eritrichium nanum* (Vill) Schrad.

Family : Boraginaceae

Distribution : From Afganistan to Kashmir, Kumaun and Tibet.

Habitat : On alpine slopes, Zanskar and Lahaul valleys around 4,165 m.

Plant Characters : An erect, densely tufted, perennial herb. Stem branched above. Leaves sessile, acute, lower densely tufted, upper distant, broadly lanceolate. Flowers in dense raceme, bracteate. Nutlets basally attached, membranous, glabrous.

Flowering & fruiting : July to September.

Uses : Used for fuel.

41. *Erysimum repandum* Linn. ex Hk. f. & T.

Family : Brassicaceae

Name : Chharpa (L)

Distribution : Pakistan to Kashmir, Nepal.

Habitat : Rocky places, Nubra valley around 3,684 m.

Plant Characters : An erect, stout, very variable, perennial herb. Leaves variable, petioled, toothed margin, lanceolate or entire, variable, petioled. Flowers yellowish-brown in terminal cluster. Pods rigid or flexuous obtusely 4-angled glabrous.

Flowering & fruiting : July to September.

Uses : Used for fuel.

42. *Galium tricorne* With.

Family : Rubiaceae

Distribution : From Afganistan to Kashmir and temperate Himalayas.

Habitat : On dry stony slopes, Zanskar valley around 3,540 m.

Plant Characters : A stout, annual herb. Stem tall reaching upto 1m. Leaves sessile, 6-8 in whorls, linear or oblanceolate, cuspidate. Flowers 1-3 in axillary, white, corolla with ovate segments. Peduncles longer or shorter than the leaves, stout. Fruits glabrous.

Flowering & fruiting : July to September.

Uses : Used for fuel.

43. *Halogeton glomeratus* (M. Bieb.) C. A. Mey.; *H. tibeticus* Bunge.

Family : Chenopodiaceae

Name : Phyang (L)

Distribution : Afganistan to Kashmir and Central Asia.

Habitat : On sandy slopes, Leh & Suru valleys between 2,890-3,650 m.

Plant Characters : A much branched, glabrous, erect or spreading annual herb. Leaves spathulate, minute, fleshy, cylindrical tipped with a bristle. Flowers minute, solitary or clustered. Perianth 4-5, enlarging in fruit with rounded transparent apices. Fruit fan-shaped, white-pinkish veins.

Flowering & fruiting : July to September.

Uses : The whole plant is used for fuel purposes.

44. *Haloxylon thomsonii* Bunge.

Family : Chenopodiaceae

Distribution : Mediterranean region to Asia.

Habitat : Along sandy and wastelands, Changthang and Spiti valleys around 4,490 m.

Plant Characters : A dwarf, hairy, perennial shrub. Stem woody, branched reaching upto 10-20 cm long. Leaves linear, terete, obtuse floral shorter and small rounded upto 1 cm long. Flowers in dense rounded clusters. Ovary woolly. Fruits winged.

Flowering & fruiting : July to mid September.

Uses : It is used as fuel.

45. *Hippophae rhamnoides* Linn. spp. ***turkestanica*** Rousi.

See Sr. No. 158, Chapter VI.

Uses : A very important plant for the fuel. The woody stems / branches are cut and used as fuel. Sometimes the woody stems are burn for making charcoal having excellent caloric value.

46. *Hippophae salicifolia* D. Don.

See Sr. No. 159, Chapter VI.

Uses : Same as for *Hippophae rhamnoides*.

47. *Hippophae tibetana* Schlecht.

See Sr. No. 160, Chapter VI.

Uses : The whole plant is used as fuel.

48. *Indigofera heterantha* Wall. ex Brandis. (*I. gerardiana* Wallich).

Family : Fabaceae

Name : Kieche (L)

Distribution : Kashmir to Himachal Pradesh.

Habitat : On rocky drier places, Suru valleys around 2,870 m.

Plant Characters : A variable shrub covered with bristly white hairs. Leaves and leaflets very variable, leaflets elliptic to lanceolate with white hairs. Flowers purple or pale pink in erect often almost stalkless spike like clusters. Calyx bristly haired. Pod hairless.

Flowering & fruiting : July to September.

Uses : The bushy branches are used for fuel.

49. *Juniperus communis* Linn. var. *saxatilis* Pallas.

See Sr. No. 14, Chapter XIII.

Uses : It is suitable for fuel and firewood.

50. *Juniperus macropoda* Boiss.

See Sr. No. 15, Chapter XIII.

Uses : It is suitable for fuel and firewood.

51. *Leontopodium leontopodinum* (DC.) Hand. Mazz.; *Antennaria leontopodina* DC. Prodr. syn. L. alpinum Cass. DC.

See Sr. No. 53, Chapter X.

Uses : The whole plant is used for fuel.

52. *Lonicera asperifolia* (Decne.) Hk. f. & T.; *Xylosteon asperifolium* Decne.

Family : Caprifoliaceae

Name : Honey suckle (L)

Distribution : From Afganistan to Uttranchal and J&K.

Habitat : On stony slopes, Suru and Spiti valleys around 3,260 m.

Plant Characters : A branched, hispid, perennial shrub. Leaves 1-2 cm, subacute or obtuse, shining and glabrous above, petioled. Calyx-tube glabrous. Corolla-tube ciliate. Style glabrous. Drupe oblong.

Flowering & fruiting : July to September.

Uses : The bushy branches are used as fuel.

53. *Lonicera glauca* Hk. f. & T.

Family : Caprifoliaceae

Name : Lamora (L)

Distribution : From Pakistan to Kashmir.

Habitat : Rocky slopes, Zanskar valley around 4,280 m.

Plant Characters : A perennial under-shrub. Stem much branched from the base. Leaves 1-2.5 cm long, oblong, tomentose beneath, subsbtuse at both ends. Bracts linear-obovate, tomentose. Corolla tube pubescent with oblong lobes. Styles short, glabrous. Berries ellipsoid.

Flowering & fruiting : July to September.

Uses : The bushy plant is used for fuel.

54. *Lonicera microphylla* Willd. ex. Roem & Schult.

Family : Caprifoliaceae

Name : Honey suckle (L)

Distribution : Afganistan to Central Nepal and Kashmir.

Habitat : Drier slopes, Nubra valley around 2,890 m.

Plant Characters : A low glabrous, perennial shrub. Leaves often subsessile, 1-1.5 long obtuse. Flowers usually in stalked pairs. Peduncle pubescent. Corolla toothed. Berries ellipsoidal, connate.

Flowering & fruiting : June to September.

Uses : The plant is used for fuel.

55. *Lonicera rupicola* Hk. f. & T.

See Sr. No. 114, Chapter IX.

Uses : The whole plant is used for fuel.

56. *Lonicera spinosa* (Jacq. ex Decne.) Walp.

Family : Caprifoliaceae

Name : Lamora (L)

Distribution : From Pakistan to Central Nepal.

Habitat : Rocky slopes, Leh and Suru valleys between 3,160-3,280 m.

Plant Characters : A rigid spiny shrub forming dense mats. Stem branched, spinescent. Leaves narrow oblong, sessile, often glaucous. Flowers pale pink, several in stalkless cluster. Corolla tube slender, hairless, elliptic. Stamen and style longer. Berries ellipsoidal.

Flowering & fruiting : Mid July to September.

Uses : Used as fuel.

57. *Myricaria germanica* (Linn.) Desv.

See Sr. No. 19, Chapter XIII.

Uses : The branches are chooped by the locals for fuel purposes.

58. *Myricaria squamosa* Desv.; *M. hoffmeisteri* Klotz.

See Sr. No. 20, Chapter XIII.

Uses : The branches are chooped for fuel purposes.

59. *Oxytropis mollis* Royle. ex Benth.

Family : Fabaceae

Name : Dhambokoya (L)

Distribution : From Pakistan to Central Nepal.

Habitat : On stony and rocky slopes, Leh valley around 4,120 m.

Plant Characters : A low spreading, stemless, perennial herb. Leaves, imparipinnately compound, sessile; leaflets ovate, silky haired with acute tip. Flowers many, purple borne in terminal cluster on along stalked. Pods 2-2.5 cm, oblong, hairs white.

Flowering & fruiting : June to September.

Uses : The whole plant is used for fuel.

60. *Picea smithiana* (Wall.) Boiss.

See Sr. No. 21, Chapter XIII.

Uses : The branches are used for fuel.

61. *Pinus wallichiana* A.B. Jackson.

See Sr. No. 22, Chapter XIII.

Uses : The pine is one of the most valuable fuel and firewood tree.

62. *Populus caspica* Bornm.

See Sr. No. 23, Chapter XIII.

Uses : Used as fuel.

63. *Populus ciliata* Wall. ex Royle.

See Sr. No. 24, Chapter XIII.

Uses : The wood and branches are used as fuel in large quantities.

64. *Populus euphratica* Oliv.

See Sr. No. 26, Chapter XIII.

Uses : One of the most important tree of which wood is used as fuel.

65. *Potentilla bifurca* Linn.

See Sr. No. 149, Chapter IX.

Uses : The whole plant is used as a source of fuel.

66. *Potentilla cuneata* Wall.ex. Lehm.

Family : Rosaceae

Name : Chantsu (L)

Distribution : Pakistan to Kashmir & Sikkim.

Habitat : On alpine slopes, Nubra valley around 3,820 m.

Plant Characters : A small spreading perennial herb. Leaflets trifid at the apex, hairy or hairless. Flowers yellow, solitary borne on short stems. Petals broadly ovate, longer than the triangular acute calyx lobes and blunt elliptic epicalyx lobes. Achenes many with long silky hairs.

Flowering & fruiting : July to September.

Uses : Used for fuel.

67. *Ribes alpestre* Wall. ex Decne.

Family : Saxifragaceae

Name : Asian gooscberry (E), Zasot (L)

Distribution : From Afganistan to South West China.

Habitat : Along roadside & open slopes, Suru and Spiti valleys around 2,905 m.

Plant Characters : A perennial prickly shrub reaching upto 1.5 m. Leaves sub-orbicular, obtusely 3-5 lobed, crenate, villous beneath. Flowers greenish to pink borne in solitary axillary. Calyx bell-shaped with narrow lobes. Stamens protruding. Berries globular, dull red, glandular-hairy.

Flowering & fruiting : July to September.

Uses : The branches are chooped and largely collected for winter fuel.

68. *Ribes himalense* Royle ex. Decne (*R. emodense* Rehder)

Family : Saxifragaceae

Name : Asian gooseberry (E), Zasot (L)

Distribution : Afganistan to South West China.

Habitat : Dry, rocks & stony slopes, Suru and Spiti valleys around 3,700 m.

Plant Characters : A tall, perennial shrub with hairless branches. Leaves heart shaped, with acute or long pointed deeply toothed lobes. Flowers numerous, reddish to greenish yellow in long pendulous clusters. Berries deep red.

Flowering & fruiting : June to September.

Uses : The branches are cut for winter fuel in large quantities.

69. *Ribes orientale* Desf.

Family : Saxifragaceae

Name : Asian gooseberry (E), Zasot (L)

Distribution : Afganistan to South West China.

Habitat : Dry, rocks & stony slopes, Suru valley around 3,200 m.

Plant Characters : A tall, perennial shrub with glandular hairy twigs. Leaves broad, rounded or kidney-shaped with 3-5 blunt lobes covered with stiff glandular hairs. Flowers numerous, reddish-purple. Calyx lobes rounded, larger than the tiny petals and stamens. Beries globular, deep red.

Flowering & fruiting : June to September.

Uses : The branches are cut for winter fuel in large quantities.

70. *Rosa ecae* Aitchinson.

See Sr. No. 75, Chapter X.

Uses : The whole plant is used as fuel in Ladakh.

71. *Rosa webbiana* Wall. ex Royle.

See Sr. No. 78, Chapter X.

Uses : The whole plant is used as fuel in Ladakh.

72. *Salix alba* Linn.

See Sr. No. 35, Chapter XIII.

Uses : The stem wood / branches are used as fuel purposes.

73. *Salix angustifolia* Willd.

See Sr. No. 36, Chapter XIII.

Uses : The branches are useful as fuel.

74. *Salix caesia* Vill.

See Sr. No. 38, Chapter XIII.

Uses : The stem wood / branches are used as fuel.

75. *Salsola collina* Pall.

Family : Chenopodiaceae

Distribution : From Kashmir to Himachal Pradesh.

Habitat : Open rocky slopes, Leh and Lahaul valleys around 2,990 m.

Plant Characters : An erect, branched, perennial herb. Leaves ovate, rigid, half-amplexicaule finely hairy. Flowers 1-3 axillary, perianth membranous. Fruits small, hairy.

Flowering & fruiting : June to August.

Uses : Used for fuel.

76. *Sibbaldia cuneata* Kunze.

Family : Rosaceae

Distribution : Kumaun to Kashmir, Afganistan, South West China & Central Asia.

Habitat : Wastelands areas, Suru and Spiti valleys around 2,760 m.

Plant Characters : A spreading, mat-forming or tufted perennial plant. Leaves mostly basal. Leaflets 3-5 wedge-shaped. Flowers yellow, borne in clusters. Calyx lobes triangular, bristly haired. Achenes smooth.

Flowering & fruiting : July to September.

Uses : The plant is used for fuel.

77. *Sisymbrium loeselii* (Linn.) Cent.

See Sr. No. 163, Chapter IX.

Uses : Used as fuel in Ladakh.

78. *Sophora alopecuroides* Linn.

Family : Fabaceae

Name : Singtik (L)

Distribution : Himalayas from Ladakh to Central Nepal & Tibet.

Habitat : Stony slopes & along roadside, Zanskar valley around 3,850 m.

Plant Characters : A perennial undershrub. Stem branched, Pubescent. Leaves 15-20 cm. Leaflets 15-25 cm, pale green, oblong, pubescent. Flowers in dense terminal raceme, with short pedicell. Calyx densely pubescent. Corolla yellow. Pod broad, hairy.

Flowering & fruiting : July to September.

Uses : The plant is used as a source of fuel.

79. ***Sophora moorcroftiana*** (Benth.) Benth. ex Baker.; *Caragana moorcroftiana* Benth.

See Sr. No. 45, Chapter XIII.

Uses : The whole woody branches are usually chooped and dried for fuel.

80. ***Tamarix gallica*** Pallas. II. (Desv.) Thist. Dyer.

See Sr. No. 47, Chapter XIII.

Uses : The woody branches are used as fuel.

81. ***Thylacospermum caespitosum*** (Camb.) Schischk.; *Thylacospermum rupifragum* (Kar. & Kir.) Schrenk.

Family : Caryophyllaceae

Name : Tagaracan (L)

Distribution : Ladakh to Tibet.

Habitat : Stony slopes, Leh and Changthang valleys around 5,750 m.

Plant Characters : A large hard compact chusion forming herb. Leaves minute, triangular, overlapping. Flowers solitary, sessile, 2.5 mm across. Capsule spherical, coriaceous.

Flowering & fruiting : July to September.

Uses : It is an important fuel plant of Ladakh for higher regions. The whole plant is uprooted for fuel purposes.

82. ***Ulmus wallichiana*** Palchon.

See Sr. No. 48, Chapter XIII.

Uses : The woody branches are useful as fuel.

❑❑❑

13

Forestry

Forestry plays an important role in maintaining the ecological balance of an area. It is more true in case of trans Himalayan cold desert areas where only a limited forest cover is available. The forest of our country can be divided into 16 major groups comprising 221 types of forest (Table 10).

Table -10 : Forest Types – distribution

Forest Type Distribution

Tropical forests

i) Tropical wet evergreen-North, East, South, Andaman & Nicobar

ii) Tropical semi evergreen-South & East

iii) Tropical moist deciduous-Central & East

iv) Tropical littoral & swamp-Along the coast

v) Tropical dry deciduous-West & Central

vi) Tropical thorn-West & Central

vii) Tropical dry evergreen-Central & South

Subtropical forests

i) Subtropical broad leaves hill forest-South

ii) Subtropical pine-Sub-Himalayan tract

iii) Subtropical dry evergreen-North-East & South

Temperate forests

i) Mountane wet temperate-Himalaya & Nilgiris (in Western Ghats)

ii) Himalayan moist temperate-Temperate areas in Himalaya

iii) Himalayan dry temperate-Dry temperate areas of Himalaya

Sub-alpine and alpine forests

i) Sub-alpine-Himalaya

ii) Moist alpine-scrub Himalaya

iii) Dry alpine scrub-Himalaya

Forest performs important ecological functions such as maintaining fragile ecological balance, conserving soil from water and air erosion, controlling floods, drought and pollution. Besides it has got great economic significance as the forest provides timber which is one of the important raw materials for construction. Refusal parts of the timber are used as fuel. Forestry also helps in providing shelter to wild fauna and protects them from human interference. The forestry and forest produce can be differentiated as follows :

Forest produce (Timber, fodder & fuel)

Cedar, pine, willow, popular, walnut, juniper are the major woody trees which provide quality timbers, furnitures, building materials, bridge construction materials, handles for agricultural tools & implements and fuel etc to tribal communities of Ladakh and Lahaul-Spiti. Similarly Seabuckthorn, *Myricaria, Tamarix, Syringa, Cotoneaster, Colutea, Elaeagnus, Morus* etc. are the other minor woody plants which provide handles for agricultural implements, fodder, fruits, fuel materials etc.

Eco-environment Improvement

The climate of trans Himalayan cold desert is extremely harsh and hostile, marked by sharp temperature flactuations, low oxygen pressure, strong storm of wind and snow and heavy influx of infra and ultra violet radiations. If the forest cover is increased sufficiently to cover the desertic wastelands, it would very much helpful in making the world's the most hostile terrain, reasonable green with increased oxygen concentration, and in turn the climate will be healthier and salubrious.

Soil stability and fertility

About 150 million hectare of land in our country is subjected to varying degree of soil erosion and the total loss of top fertile soil is about 6,000 tonnes annually. Wind erosion affects an area of 32 million ha mainly covering hot desert of Rajasthan, Gujrat and cold desert of trans Himalaya i.e. Ladakh and Lahaul-Spiti. Various forms of water erosion such as sheet, rill and gully are caused by melting snow & snow avalanche while even little rain causes land slides in these areas because of unstable soils and lack of vegetation cover. Due to short growing period and heavy grazing pressure, rejuvenation is slow in cold arid regions. The planned and increased agro-forestry will be helpful in improving soil stability as well check soil erosion also. In addition, it will add organic matter to the soils, thereby improving physical, chemical and biological properties.

Home for fauna and herbaceous flora

Natural vegetation provides food, shelter and allieds to the wild life. Usually hervivorous live on natural vegetation, and in turn they become food for carnivores. Tree cover is a multistoreyed system where various herbs and shrubs are also accommodated and will be a home for fauna. There is almost negligible natural forest in trans Himalayn cold desert due to prevailing climatic factors. However the sub alpine and alpine forest of trans Himalaya may be divided into three major forest groups

i. Sub-alpine forest

ii. Moist alpine shrub

iii. Dry alpine shrub

The trans-Himalaya is situated at high altitude where very few species grow naturally. The main forest plants like Pine, Cedar, Betula, Walnut etc are found in the lower altitudes of Lahaul-Spiti which comes under temperate zone. The major forest trees of willow and popular seen near habitation and along river banks are usually planted. It is also a notable fact from the reports of forest department, Ladakh is the only place where forest cover is slowly increasing every year due to the continuous efforts made by the various departments working for the plantation in the region during past few years.

Field Research Laboratory has been working on various aspects viz. Identification & introduction of suitable forest species, standardization of propagation techniques, establishment of forest germplasm, provision of quality saplings and awareness programmes among military / paramilitary forces and civil polulation with a view to increase afforestation which will not only helping in greening of cold desert areas but also lead to eco-environment improvement of trans Himalayan Cold desert. This lab has its own forest germplasm of *Salix sp, Populus sp, Eleagnus sp* and *Ulmus sp*. Every year approximately 30,000 quality saplings of different species of forest plants are being distributed to Military / Paramilitary forces with technical knowhow for mass plantation. The significant efforts in the field of afforestation carried out by this lab was well recognized at National level and this lab was covetted with Indira Prydarshni Vrishmitra Award for the year 1993. Now-a-days this lab is working on mass multiplication of indigenous forest species such as *Hippophae species* (Seabuckthorn), *Myricaria sp*, *Tamarix sp* (Tamarix), *Eleagnus sp* (Oleaster), *Juniperus sp* (Juniper), *Sophora sp*, *Ulmus sp* (Elm) etc. The potential forest species of trans Himalayan cold desert are described in brief alongwith families, English (E), hindi (H) / local names (L), distribution, habitat, brief botanical notes and uses in the enumeration. (Table- 11).

Enumeration of Forest Species

1. *Betula utilis* HD. Don.

Family : Betulaceae

Name : Bhojpatra (L)

Distribution : From Kashmir to Himachal Pradesh & South West China.

Habitat : Forest areas, Lahaul valley around 3,980 m.

Plant Characters : A shrub or tree 2-12 m tall. Bark white, brownish or reddish-white. Young shoots pubescent, older glabrescent. Leaves ovate or rhomboid, base rounded, subcordate or cuneate, margins doubly serrate to subserrate. Male flowers in catkins. Bracts broadly ovate-obtuse. Fruiting scales woody. Nut elliptic-ovate.

Flowering & Fruiting : June to October.

Uses : The tree is one of the few hardwoods of the area and has a multitude of uses. It supplies beams and boards for buildings. A large number of articles are fabricated from it viz., book covers, spoons, saddle equipment, shutters, implements handles. The bark has been used for carrying food and writing material in earlier days.

2. *Caragana versicolor* Wall.

See Sr. No. 23, Chapter, XII.

Uses : The species can be used as forestry plant for higher regions.

3. *Cedrus deodara* (Roxb. ex D. Don.) G. Don.

Family : Pinaceae

Name : Deodar (H), Deyor (L)

Distribution : From Kashmir to Sikkim & Afganistan.

Habitat : Forest areas, Lahaul valley around 3,510 m.

Plant Characters : A large ever green trees. Branches brown-reddish. Leaves needle like, many in clusters. Cones solitary, erect, ovoid, dark brown.

Flowering & Fruiting : August to November.

Uses : The deodar is considered as one of the valuable timber of Lahaul-Spiti, especially in Pattan valley. The wood is used for making doors, windows and other household furnitures.

4. *Colutea multiflora* Stepf. ex Ali.

Family : Fabaceae

Name : Blader Sena (E)

Distribution : From Central & Eastern Nepal to Ladakh.

Habitat : Along roadside & on stony slopes, Nubra valley around 3,028 m.

Plant Characters : A spineless decidous shrub upto 2 m. stem branched, woody. Leaves 6-8 cm, with 6-7 pairs of widely spaced leaflets. Leaflets elliptic, blunt, hairless. Flowers yellow to reddish-yellow borne in lax clusters. Calyx hairless or finely hairy, with triangular teeth. Pod ellipsoidal, beaked.

Flowering & fruiting : July to September.

Uses : It is a potential forest shrub.

5. *Cotoneaster falconeri* Klotz. *C. integerrima* Medic

Family : Rosaceae

Name : Kumchung (L)

Distribution : Afganistan, Temperate Himalayas from Kashmir to Kumaun.

Habitat : Rocky places, Leh valley around 3,910 m.

Plant Characters : An erect, small, shrub with young twigs densely hairy but becoming hairless. Leaves ovate-elliptic, pubescent. Flowers white, solitary or few in axillary clusters. Fruits glabrous, red.

Flowering & fruiting : July to September.

Uses : A potential forest shrub.

6. *Crataegus songarica* C. Koch.

Family : Rosaceae

Name : Hawthorn (E), Ralchatnakpo (L)

Distribution : From Kashmir to Nepal & Himachal Pradesh.

Habitat : Open areas, Lahaul valley around 3,640 m.

Plant Characters : A shrub or small tree, young shoots with scattered long hairs, branches often armed with straight axillary and terminal thorns. Leaves broader than long, pinnately irregularly 3-7 lobed, toothed, base truncate but abruptly narrowed into the petiole or crenate. Flowers in many flowered terminal corymbose cymes. Calyx-lobes reflexed. Petals orbicular, white. Fruits globose-ovoid, scarlet.

Flowering & fruiting : May to September.

Uses : This is a potential forest tree and the wood is used as timber by the local people.

7. *Elaeagnus angustifolia* Linn. *E. hortensis*, M Bieb.

Family : Elaeagnaceae

Name : Oleaster (E), Shersing (L)

Distribution : Afganistan to South West China and Kashmir.

Habitat : Near human settlement, Leh & Suru valleys between 2,890-3,050 m.

Plant Characters : A tall, perennial spiny shrub. Stem with spinous, silvery shining branches. Leaves 3-5 cm, obtuse, alternate, petioled silvery scaly beneath. Flower few in axillary clusters, dull yellow, fragrant, perianth tube campanulate with 4-triangular acute lobes. Fruits ovoid, fleshy and red.

Flowering & fruiting : July to September.

Uses : An ornamental small tree suitable for forestry in trans –Himalayan cold desert.

8. *Fraxinus xanthoxyloides* (G. Don.) DC. Prodr.

See Sr. No. 130, Chapter VI.

Uses : Used as forest tree.

9. *Hippophae rhamnoides* Linn. spp. *turkestanica* Rousi.

See Sr. No. 158, Chapter VI.

Uses : It is a potential forest shrub for cold desert and it also helps in checking soil erosion and improves soil fertility by fixing atmospheric nitrogen because of having noduleted roots (*Frankia actinomycete*).

10. *Hippophae salicifolia* D. Don.

See Sr. No. 159, Chapter VI.

Uses : Same as for *Hippophae rhamnoides.*

11. *Hippophae tibetana* Schlecht.

See Sr. No. 160, Chapter VI.

Uses : The plant is suitable for high alpine forestry.

12. *Indigofera heterantha* Wall. ex Brandis

See S. No. 48, Chapter XII.

Uses : It can be used as forest plant.

13. *Juglans regia* Linn.

Family : Juglandaceae

Name : Walnut (E), Akhrot (H), sTarga (L)

Distribution : India to Bhutan & China.

Habitat : Planted around villages, Leh & Lahaul-Spiti valleys between 2,890-3,550 m.

Plant Characters : A tall, deciduous tree with aromatic bark and leaves. Imparipinnate, more or less tomentose when young; leaflets 5-9 (-13); the terminal largest, ovate-oblong, acute or acuminate. Flowers green, male and female on the same tree, appearing with the leaves. Male flowers numerous in pedunclous lateral catkins, tapels narrow, irregularly 5-lobed. Female flowers 1-3 clustered at the end of branches; tapels-lobes 4, linear-lanceolate. Drupe ovoid, green, thick, succulent; rind encloses a woody, wrinkled, 2-valved nut.

Flowering & fruiting : July to September.

Uses : The wood is used for the manufacture of furtinure and other articles.

14. *Juniperus communis* Linn. var. ***saxatilis*** Pallas, Fl. Ross. 1(2): 12. t. 4. f. A. 1788; *communis* non. Linn. Hk.

Family : Cupressaceae

Name : Common Juniper (E), Shukpa (L)

Distribution : Afganistan to Nepal, temperate Zones, Kumaun, Kashmir, temperate & sub-arctic Europe, Asia, North Africa & North America.

Habitat : Drier areas & cultivated, Nubra & Suru valleys around 2,950 m.

Plant Characters : A low growing shrub. Stem erect, branched and upto 1 m tall. Leaves crowded, needle like, acute apex. Male cones ovoid, recalling leaf-buds. Fruits globose, blue black, 3-seeded.

Flowering & fruiting : March to September.

Uses : It is suitable for ornamental shrub / small tree and a very important tree for the Buddhist religion.

15. *Juniperus macropoda* Boiss.

Family : Cupressaceae

Name : Himalayan Cedar (E), Shukpa (L)

Distribution : Afganistan to Nepal, Baluchistan, Kumaun & Kashmir.

Habitat : Rocky slopes, Suru valley around 3,150 m.

Plant Characters : A large gregarious shrub. Lower leaves linear; scale leaves small, closeluy pressed to stem and with a large resinous gland on the back. Fruits bluish-black, resinous with two or more seeds.

Flowering & fruiting : March to September.

Uses : Same as in case of *J. communis*.

16. *Juniperus recurva* Buch,-Ham

Family : Cupressaceae

Name : Weeping Blue Juniper (E), Shukpa or Shug-pa tsher-can (L)

Distribution : Afganistan to Nepal, Pakistan & Kashmir.

Habitat : Alpine slopes, Suru valley around 2,905 m.

Plant Characters : A prostrate or decumbent, perennial shrub or small tree. The tree attains the height upto 30 ft. straight, trunk conical crown. Leaves awl-shaped, overlapping in whorls. Flowers monoecious or dioecious. Fruits dark brown or purplish black. Seed solitary, ovoid.

Flowering & fruiting : March to September.

Uses : Same as in case of *J. communis*.

17. *Morus alba* Linn.

Family : Moraceae

Name : Sahtoot (H), Mulberry (E), Oshe (L)

Distribution : Indigenous to China. Now widely cultivated in plains and hills of India.

Habitat : Rarely cultivated in lower altitude upto 3,250 m Turtuk & Kargil of Ladakh.

Plant Characters : A monoecious, shrub or moderate sized tree. Woody, thick hard stem, branched. Leaves very variable, ovate. Flowers greenish; male spike lax-flowered, female catkins ovoid, pendulate; fruit a syncarp, purple or dark purple to black.

Flowering & fruiting : June to August.

Uses : Cultivated in lower altitudes.

18. *Myricaria elegans* Linn.

Family : Tamaricaceae

Distribution : Himachal Pradesh to Kashmir.

Habitat : Near stream & river sides on sandy patches Lahaul-Spiti valleys between 3,500-3,800 m.

Plant Characters : An erect, perennial tall shrub. Stem woody, branched upto 2-5 m. Leaves oblong-obovate to ovate, pointed at the tip,. Flower purplish-red in terminal clusters. Bracts small, sepals free lanceolate to oblong, almost equal to petals.

Flowering & fruiting : July to September.

Uses : It is potential forest shrub for afforestation.

19. *Myricaria germanica* (Linn.) Desv.

Family : Tamaricaceae

Name : Umbo (L)

Distribution : Pakistan to Kashmir, Wesstern Asia, Europe.

Habitat : Along river sides on sandy areas, Nubra and Suru valleys between 3,170-3,680 m.

Plant Characters : An erect, perennial shrub. Stem woody, branched upto 1-2 m. Leaves oblong-ovate, with dotted glands, densely clustered. Flower reddish borne in terminal spikes. Bracts short, with membranous margins. Sepals free lanceolate, equaling to petals.

Flowering & fruiting : July to September.

Uses : It is potential forest shrub for afforestation.

20. *Myricaria squamosa* Desv.; *M. hoffmeisteri* Klotz.

Family : Tamaricaceae

Name : Umbo (L)

Distribution : Pakistan to Kashmir, Western Asia, Europe.

Habitat : Along river sides on sandy areas, Nubra & Changthang valleys between 3,250-3,850 m.

Plant Characters : An erect, perennial shrub upto 2 m. Stem woody, branched light brown. Leaves glaucous, green, 2-5 mm long, elliptic. Flower-spike 5-1 cm, terminal and lateral with pink flowers.

Flowering & fruiting : Mid June to September.

Uses : The plant is suitable for forestry plantation in Ladakh.

21. *Picea smithiana* (Wall.) Boiss.

Family : Pinaceae

Name : Himalayan Spruce (E), Tosh (L)

Distribution : From Kashmir to Himachal Pradesh and Central Asia.

Habitat : Frequent and found growing in association with *Pinus wallichiana* in Lahaul valley, 3,900 m.

Plant Characters : A large sized evergreen tree. Branches whorled, horizontal, slightly curved. Leaves long, spirally arranged all around the branches, needle like, stiff, pungent. Flowers monoecious. Male catkins long, solitary, erect, nearly sessile, in the axils of the upper leaves. Female flowers in cones; cones in solitary, terminal, pendulous, dark brown when ripe. Seeds dark grey or blackish.

Flowering & fruiting : June to August.

Uses : The tree is used as important forestry species and one of the valuable timber tree of trans Himalayan belts.

22. *Pinus wallichiana* A.B. Jackson.

Family : Pinaceae

Name : Blue pine (E), Chir / Kail (L)

Distribution : From Kashmir to temperate Himalayas & Afganistan.

Habitat : On natural forests, Lahaul-valley, 3,560 m.

Plant Characters : A tall tree with grey barks. Branches whorled, dimorphic. Leaves 10-20 cm long, in fascicles of 5, bluish to grayish-green, drooping. Male cones in clusters, 1.0 cm long. Female cones usually 2-3 at the tips of ranches. Mature cones 15-30 cm long. Seeds winged.

Flowering & fruiting : May to October.

Uses : The pine is one of the most valuable timber tree in the economy of Lahaulis. The wood furnishes the best boards which are made into doors, floors, windows, boxes etc. The needle leaves are used to give consistency to mud used in plastering. A very useful forest tree of Lahaul areas.

23. *Populus caspica* Bornm. *P. alba* Linn.

Family : Salicaceae

Name : Safeda (H), White poplar (E), Yarpa (L)

Distribution : From Afganistan to Himachal Pradesh, Western & Central Asia.

Habitat : Along Indus river, on slopes & planted near habitation Leh valley 3,500 m.

Plant Characters : A medium to large sized perennial tree. Bark whitish or grayish, rough. Leaves oblong ovate or broadly ovate, sinuate lobes or toothed. Catkins pendulous, hairy, slender; the male 5-10 cm long with 6-10 stamens, the female shorter with 4 stigmas. Capsule shortly stalked, 2 valved. Seeds minute, hairy.

Flowering & fruiting : June to August.

Uses : One of the most important forest plant of Ladakh and Lahaul areas.

24. *Populus ciliata* Wall. ex Royle.

Family : Salicaceae

Name : Pak burta (H), Himalayan Poplar (E), Yarpa (L)

Distribution : From Pakistan to South West China.

Habitat : Along Indus river on slopes & oftenly planted near habitation Leh valley around 3,500 m.

Plant Characters : A tall perennial tree with smooth greenish-grey bark which becomes brown and deep vertical fissured on old trunks. Stem woody long 60-70 ft. Leaves petioled, ovate, cordate, acuminate, unequally toothed, glabrous. Male catkins short, female upto 15 cm long. Capsules stipitate, 3-4 valved. Seeds minute, hairy.

Flowering & fruiting : June to August.

Uses : The most important woody tree of trans Himalayan cold desert.

25. *Populus deltoides* Marsh. ex Arbustrum.

Family : Salicaceae

Name : Caroline Poplar (E), Yarpa (L)

Distribution : Pakistan to China and Kashmir.

Habitat : Along Indus river on slopes & oftenly planted near habitation Leh valley around 3,650 m.

Plant Characters : A tall tree upto 20-25 m. Branches upright, spreading. Leaves subcordate, glandular, crenate. Catkins 7-10 cm long. Capsule shortly stalked, smooth.

Flowering & fruiting : June to August.

Uses : The plant is economical as forest tree.

26. *Populus euphratica* Oliv.

Family : Salicaceae

Name : Indian poplar (E), Yulat (L)

Distribution : Pakistan to Punjab, Kashnir, Indus valley, Western Tibet.

Habitat : Indus river & oftenly planted near habitation, Nubra valley around 3,500 m.

Plant Characters : A tall tree. Leaves polymorphic, shortly petioled, rhombic or orbicular, glabrous. Catkins small, lax flowered, bracteate. Capsule large.

Flowering & fruiting : June to August.

Uses : The wood is used for making furniture and a valuable forest tree.

27. *Populus nigra* Linn.

Family : Salicaceae

Name : Black Poplar (E), Yulat (L)

Distribution : Commonly cultivated from Kashmir to Ladakh.

Habitat : Cultivated throughout Ladakh upto 3,700 m in the valleys.

Plant Characters : A tall tree with columnar form upto 30 m in height and 3 m girth. Branches upright. Leaves petioled, sub-cordate, smooth. Catkins lax flowered. Stamens many. Capsule shortly stalked.

Flowering & fruiting : June to August.

Uses : One of the most important and commonly planted woody tree of Ladakh and Lahaul-Spiti areas.

28. *Populus sylvestris*

Family : Salicaceae

Name : Poplar (E), Yulat (L)

Distribution : Kashmir to Tibet & Ladakh.

Habitat : Along river on sandy areas & oftenly planted near habitation Leh valley around 3,550 m.

Plant Characters : A medium sized tree. Trunk hard, woody, long, branched. Leaves large, petioled, upper surface green, lower silvery. Catkins lax flowered. Capsule short stalked.

Flowering & fruiting : July to September.

Uses : The plant is a valuable forest tree.

29. *Rhododendron anthopogon* D. Don. spp. ***hypenanthum*** (Balf. F.) Cullen.

Family : Ericaceae

Name : Rhododendron (E), Balu (L)

Distribution : From Kashmir to Western Nepal & South West Tibet.

Habitat : Alpine slopes, Lahaul valley around 3,670 m.

Plant Characters : A dwarf, fragrant shrub. Leaves petiolate, elliptic to elliptic-ovate, subobtuse, glabrous above, rusty-glandular-scaly beneath; leaf-bud scales persistent. Flowers cremish-white, in terminal clusters. Calyx-teeth obtuse with hairy margins. Corolla-lobes elliptic-obovate, obtuse entire, throat hairy.

Flowering & fruiting : June to September.

Uses : It is useful alpine forest shrub.

30. *Rhododendron campanulatum* D. Don.

Family : Ericaceae

Name : Rhododendron (E), Radukpa (L)

Distribution : From Pakistan to Kashmir, Himachal Pradesh & Nepal.

Habitat : On slopes, Lahaul valley between 3,435-3,780 m.

Plant Characters : A dwarf shrub with ascending or erect branches. Leaves 9.5-14.0 cm long, subcoriaceous, elliptic to ovate, mucronate, glabrous above, rusty-tomentose beneath, with glabrous petioles. Flowers whitish to pale-rose or lilac, ofteh apotted. Calyx broadly triangular. Corolla campanulate, 5-lobed. Capsule 2.0-2.5 cm long, cylindric, more or less curved, dark-brown, streaked with white spots.

Flowering & fruiting : June to August.

Uses : A very useful alpine forest shrub.

31. *Ribes alpestre* Wall. ex Decne.

See Sr. No. 67, Chapter XII.

Uses : It is used as forest shrub.

32. *Ribes orientale* Desf.

See Sr. No. 69, Chapter XII

Uses : The plant is useful as forest shrub.

33. *Robinia pseudacacia* Linn.

Family : Fabaceae

Name : Pseudo Acacia (E), Kikar (H)

Distribution : Himalayas from Kashmir to Sikkim & temperate world.

Habitat : Planted on limited scale Nubra & Suru valley around 2,950 m.

Plant Characters : A medium sized, thorny, perennial tree. Leaves imparipinnate; compound, leaflets 9-19, elliptic or oval, entire, glabrescent above, puberulous at the nerves below; stipules spinescent, upto 2.0 cm long. Flowers white, very fragrant, in axillary, racemes; pods flat, brown; seeds reniform, dark brown.

Flowering & fruiting : June to August.

Uses : A very potential forest tree of trans Himalayan cold desert.

34. *Rosa webbiana* Wall. ex Royle.

See S. No. 78, Chapter X.

Uses : The rose shrubs are useful forest plant along river belts.

35. *Salix alba* Linn.

Family : Salicaceae

Name : European or White Willow (E) & Mulchang (L)

Distribution : From Afganistan to Central Nepal.

Habitat : Planted near habitations & moist places; along river belts upto 3,550 m Ladakh.

Plant Characters : A tall tree with spreading branches. Leaves lanceolate, 4-9 cm, acuminate, serrulate, silky beneath, petioled. Stipule small. Catkins 6-10 cm, dense. Female yellow, ciliate.

Flowering & fruiting : May to August.

Uses : One of the most important and commonly planted woody trees of Ladakh and Lahaul-Spiti areas.

36. *Salix angustifolia* Willd.

Family : Salicaceae

Name : Willow (E), Mulchang (L)

Distribution : From Afganistan to C. Nepal.

Habitat : Planted along river belts upto 3,500 m Ladakh.

Plant Characters : A shrub, upto 2-4 m tall. Branches smooth. Leaves shortly stalked, 3-7 cm, denticulate. Catkins small, male 1-2 cm, female 3-6 cm. Capsule longer than bracts.

Flowering & fruiting : May to July.

Uses : The plant is useful as forest tree.

37. *Salix babylonica* Linn.

Family : Salicaceae

Name : Weeping Willow (E)

Distribution : From North India; plains to hill & from Afganistan to Nepal.

Habitat : Planted and along river belts upto 3,160 m Leh valley

Plant Characters : A medium size tree. Branches drooping. Leaves lanceolate, 4-8 cm long acuminate, glaucous, petioled. Catkins small.

Flowering & fruiting : June to September.

Uses : The plant is used for beautification having drooping branches.

38. *Salix caesia* Vill.

Family : Salicaceae

Name : Willow (E)

Distribution : From Afganistan to Nepal & Kashmir.

Habitat : Planted along river belts 3,460 m Leh valley.

Plant Characters : A small shrub upto 1.5 m tall. Branches divericate, stout. Leaves petioled, 1-2.5 cm long, apiculate. Female catkin longer than male catkin. Capsule small.

Flowering & fruiting : June to September.

Uses : The plant is useful as forest tree.

39. *Salix daphnoides* Vill.

Family : Salicaceae

Name : Violet willow (E),

Distribution : Himalayas from Kumaun to Kashmir.

Habitat : Open slopes & planted near habitations, between 2,850-3,160 m Ladakh.

Plant Characters : A small tree upto 6 m. Leaves oblong-lanceolate 5-7 cm, acuminate, serrate, pubescent, petioled. Catkins 3-5 cm long, silky. Ovary short stalked, ovoid.

Flowering & fruiting : May to August.

Uses : It is the best for forest plantation.

40. *Salix denticulata* Ander.

Family : Salicaceae

Name : Willow, Tookchang (L)

Distribution : Kashmir to Nepal & Tibet.

Habitat : On open places, planted near habitations; between 2,850-3,465 m, Leh & Nubra valleys.

Plant Characters : A small tree upto 4 m tall. Branches dark brown. Leaves narrow elliptic to elliptic, toothed margin, glaucous. Catkins after leafing. Male catkin on short leafy shoots, female catkin green, 2-5 cm long. Capsule 3-4 mm.

Flowering & fruiting : May to August.

Uses : The plant is useful as forest tree.

41. *Salix excelsa* Gmelin. (*Salix fragilis* Linn.)

Family : Salicaceae

Name : Crack willow or red wood willow (E), Mulchang / Brokchang (L)

Distribution : Kashmir to Lahaul, Nepal, Asia & Tibet.

Habitat : Dry areas, planted near habitations, between 2,850-3,500 m Leh valley.

Plant Characters : A tall, perennial tree. Branches fragile. Leaves lanceolate, 6-10 cm long, silky when young, serrate margin. Stipule deciduous. Catkins on leafy peduncles. Capsule 6-8 mm long with bifid style.

Flowering & fruiting : May to August.

Uses : The plant is useful as forest tree.

42. *Salix hastata* Linn.

Family : Salicaceae

Name : Selchang (L)

Distribution : Himalayas from Kashmir to Himachal Pradesh.

Habitat : On open slopes & along river belts, between 2,800-3,500 m Ladakh.

Plant Characters : A small shrub. Branches dark brown or black glabrous. Leaves 5-7 cm long, petioled, ovate. Male catkins sessile, 1-2 cm long, female catkins 5-10 cm long, curved. Capsule sessile, beaked.

Flowering & fruiting : July to September.

Uses : The plant is useful as forest tree.

43. *Salix sclerophylla* Ander.

Family : Salicaceae

Name : Solchang (L)

Distribution : Himalayas from Kumaun to Kashmir.

Habitat : Cultivated near habitations and along streams/ river belts between 2,850-3,500 m Ladakh.

Plant Characters : A stout, much branched, perennial shrub. Branches glaucous. Leaves 3-4 cm rigid, nerved, ovate-lanceolate, silky. Male catkins sessile, 1-2 cm long. Female catkins curved. Capsule sessile, conic, beaked, hoary.

Flowering & fruiting : June to August.

Uses : The plant is useful as forest shrub.

44. *Salix tetrasperma* Roxb.

Family : Salicaceae

Name : Solchang (L)

Distribution : Kumaun to Kashmir & Nilgiri hills.

Habitat : Planted near habitations and wet swampy places, between 2,850-3,600 m Ladakh.

Plant Characters : A small tree with stout trunk. Branches sub-erect. Leaves petioled, ovate-lanceolate, acuminate, serrulate. Flowering usually after leafing. Catkins slender. Flowers scattered in clusters. Stamens 5-10. bracts spathulate. Capsule long.

Flowering & fruiting : May to August.

Uses : The plant is useful as forest tree.

45. *Sophora moorcroftiana* (Benth.) Benth. ex Baker.; *Caragana moorcroftiana* Benth.

Family : Fabaceae

Name : Singtik (L)

Distribution : Kashmir to Ladakh, Central Nepal & Tibet.

Habitat : Sandy places & wastelands, Nubra valley around 3,040 m.

Plant Characters : A much branched, spiny, shrub upto 1-2 m. . Stem branched from the base. Lateral stems spine-tipped. Leaves pinnate. Leaflets several, ovate, silky-haired. Flowers blue borne in axillary clusters. Calyx grey, silky-haired. Pod 8-12 cm long, 5-6 seeded.

Flowering & fruiting : July to September.

Uses : The whole woody and bushy habit of plant make a very potential forest shrub.

46. *Syringa emodi* Linn.

Family : Oleaceae

Name : Himalayan Lilac

Distribution : Himalayas from Kashmir to Nepal & Afganistan.

Habitat : Cultivated near habitation, Leh & Suru valleys between 2,980-3,510 m.

Plant Characters : A large deciduous perennial shrub or small tree. Leaves opposite, petioled, elliptic or ovate, entire, acute, dark green and glabrous above, whitish beneath, slightly pubescent when young. Flowers white, in terminal panicles usually arising from a terminal bud, with basal leaves. Corolla salver shaped, tube longer than calyx, lobes 4, tips inflexed. Capsule cylindrical, acute, slightly curved, 1.5 cm, opening by valves.

Flowering & fruiting : June to August.

Uses : The plant is used for ornamental purposes and suitable for afforestation.

47. *Tamarix gallica* Pallas. II. (Desv.) Thist. Dyer.

Family : Tamaricaceae

Name : Tamarisk (E), Agrelta (L)

Distribution : Himalayas mainly in Northern India.

Habitat : Along river belts, Nubra valley around 3,060 m,

Plant Characters : An erect and branched shrub upto 1-3 m. Leaves small, ovate-lanceolate. Flowers small, pink or white, in terminal spikes.

Flowering & fruiting : Mid June to September.

Uses : The plant is useful forest shrub or small tree.

48. *Ulmus wallichiana* Palchon.

Family : Ulmaceae

Name : Elm (E), Umbuk (L)

Distribution : Alpine Himalayas from Kashmir to Nepal & Afganistan.

Habitat : Rare on moist places, Nubra valley around 3,060 m.

Plant Characters : A large tree with rough grey bark upto 3-4 m tall. Leaves elliptic-obovate, long pointed, double-toothed. Mature leaves rough above. Young shoots finely hairy. Flowers green, stalked, borne in globular clusters before leafing. Fruits nutlets surrounded by a papery wing.

Flowering & fruiting : May to August.

Uses : The whole tree is useful for forestry development in desert conditions of Ladakh and Lahaul-Spiti.

❑❑❑

14

Endangered Plant Species

Value of any ecosystem depends on the value of its constituents - biotic and abiotic units. The rich diversity in flora and fauna, economically useful species, rare and endangered species, the genetic diversity, high degree of endemism, presence of alien elements in the ecosystems, usefulness of one of world' highest fresh water and brackish water lakes, high altitude passes and unique cultural heritage etc. make the trans Himalayan cold desert ecosystem unique and the most fragile in the world.

Rarity is a two-fold concept associated with biology of the species and ecology of the area. Threat is more difficult to characterize since it may be a natural consequence of biological or geological processes or be the result of past or present human activities directly or indirectly influencing the plant populations or their environment. The plant populations keep changing size and density over a period of time and such changes may make plant species rare, endangered and threatened, eventually leading in the extinction.

During the ethno-botanical survey, it has been observed that the frequency of some of important plants especially medicinal plants of Indian cold desert Himalaya with which it was earlier abound, have considerably declined due to their unscientific exploitation, natural calamities, road construction, uprooting for fuel, overgrazing and other activities. The unscientific collection of plants by the local people and traders and habitat destruction have rendered many species endangered and threatened. Day by day the population size and number are decreasing and the habitats are receding towards higher and higher elevations as also observed by earlier workers. According to one study some 40,000 to 80,000 sq. meter of debris is removed from the mountain slopes to construct just one kilometer of road. Many plant species like *Acantholimon, Saussurea, Rheum, Artemisia, Thylacospermum* etc. can be seen hanging along roadside near Khardungla 5,520 m, Tanglangla 5,255 m. and other high motorable passes.

The field observations on rare, endangered and threatened (RET) plant species have been made and tabulated on table 12 including botanical name, habitat and status/category.

Table-12 : List of endangered plant species

S. No.	Botanical Name	Habitat	Category
1.	*Aconitum heterophyllum*	Alpine slopes	EN
2.	*Aconitum violaceum*	-do-	VU
3.	*Allium stracheyi*	Alpine slopes	VU
4.	*Arnebia benthamii*	Rocky slopes	CR
5.	*Arnebia euchroma*	Rocky slopes	EN
6.	*Artemisia brevifolia*	Open slopes	EN
7.	*Bergenia stracheyi*	Moist alpine	VU
8.	*Betula utilis*	-do-	EN
9.	*Bieberstenia odora*	Alpine slopes	VU
10.	*Bunium persicum*	-do-	EN
11.	*Colchicum luteum*	-do-	EN
12.	*Cremanthodium ellisii*	High alpine	VU
13.	*Dactylorhiza hatagirea*	Alpine meadows	CR
14.	*Ephedra gerardiana*	Rocky areas	EN
15.	*Epipactis helliborne*	Damp places	EN
16.	*Ferula jaeschkeana*	Sub-alpine	VU
17.	*Fritillaria roylei*	Moist alpine	CR
18.	*Gentiana algida*	Meadows	EN
19.	*Gentiana kurroo*	Meadows	CR
20.	*Herminium monorchis*	Forest sheds	VU
21.	*Heracleum lanatum*	Dry slopes	VU
22.	*Hyoscyamus niger*	Waste lands	LR-NT
23.	*Jurinella macrocephala*	Meadows	VU
24.	*Malaxis muscifera*	Moist places	VU
25.	*Meconopsis aculeata*	Rocky crevices	EN
26.	*Nepeta longibracteata*	High alpine	VU
27.	*Onosma hispidum*	High alpine	EN
28.	*Picrorhiza kurrooa*	Alpine rocky	EN
29.	*Physalis alkekengi*	Dry areas	VU
30.	*Physochlaina praealta*	Waste lands	VU
31.	*Podophyllum hexandrum*	Sub-alpine	EN
32.	*Polygonatum multiflorum*	Moist places	VU
33.	*Polygonatum verticillatum*	Moist places	VU
34.	*Rheum moorcroftiana*	Alpine slopes	VU
35.	*Rheum spiciforme*	Alpine slopes	VU
36.	*Rheum webbianum*	Alpine rocks	VU

37.	*Rhododandron anthopogon*	High alpine	VU
38.	*Rhododandron campanulatum*	High alpine	VU
39.	*Rubia cordifolia*	Moist places	VU
40.	*Saussurea gossypiphora*	Meadows	VU
41.	*Saussurea lappa*	Forest areas	CR
42.	*Saussurea obvallata*	Glaciated moraine	VU
43.	*Thymus serpyllum*	Alpine meadows	VU

Abbreviation used : CR - Critically Endangered; EN - Endangered; VU - Vulnerable; LR-NT - Lower Risk - Near Threatened.

15

Conservation and Sustainable Utilization

Eco-environment system is a life-support system consisting of biotic and abiotic components - man, animals, plants, micro-organisms, soils, minerals, water and atmosphere all interacting through a trophic structure, biotic diversity and natural cycles so that the system itself is kept viable. The concept of conservation is not new but the rediscovery of a truth well known to our ancient people. The nature is beautifully balanced and every little thing has its own place and importance. Any disturbance creates a chain reaction which may not be visible for some time, but certainly has adverse impact in long term. Many noted conservationists and conservation organizations have suggested that nation should seek to protect some 5% of their country' land surface in order to conserve the full range of biological resources. Extensive and healthy ecosystems could manage with less, but less extensive and fragile ecosystems with a greater diversity of values would require more. The trans Himalayan cold desert belong to this group.

The importance of species diversity conservation is usually justified by the direct use of species by man for various purposes like food, medicine, aromatics, fibre, fodder, aesthetic, fuel etc. Another aspect is the historical adjustment of species diversity to human needs. Desired selection of species, dispersal of species, habitat modification and domestication are the main interactions between man and plant and faunal diversity.

It is a known fact for the indigenous people of trans Himalayan region that ethnobotanical diversity is closely associated with the day to day life of the local inhabitants and they have been using surrounding plants for various uses for survival since ages. The inaccessible, remoteness and harsh climatic conditions have provided natural safety for these plants. The recent development activities and increased population have exposed trans Himalayan floristic diversity especially herbal wealth and brought new commercial interest. Therefore, it is very important to develop a mechanism for sustainable and coordinated exploitation of these wild plants which will provide ecologically sustainable, socially acceptable and economically equitable production and utilization system. This will required an appropriate participation and action from major stake holders like traditional medical practitioners, foresters, local community, Government and non Government organization, Researchers and Industries etc. The following steps should be adopted for conservation and sustainable utilization of trans Himalayan ethnobotanical wealth for overall development of cold desert areas.

a) In situ conservation

This is one of the most effective measures for conservation of fragile and precious ethnobotanical diversity. The concept of preserving biodiversity resulted in the Biosphere Reserve in 1974 as UNESCO MAB (Man & Biosphere) Project for the conservation of the natural areas and of the genetic materials they contain. The *in situ conservation* is nothing but the protection of an area having rich diversity in flora and fauna without disturbing natural process. The Hemis National Park of Ladakh and Pin valley of Lahaul Spiti serve the purpose and play an important role in conservation of flora and fauna. Recently the matter of *in situ conservation* has been taken up with local concerned authorities and emphasis has been given on creation of some more Bio Reserves / Parks in the cold desert fragile ecosystems. The *in situ* method of conservation will not only serve protection and preservation of flora and faunal diversity, but also play a major role in improvement of ecological balance of the most fragile trans Himalayan cold desert regions.

b) Ex situ conservation

The ***ex situ*** conservation is one of the most important aspects for conservation and scientific exploitation of economically important plants. National Bureau of Plant Genetic Resources, N Delhi is playing a major role in *ex situ conservation* of plant genetic resources and their sustainable utilization involving various other govt organizations as well as NGO's. Emphasis was given on collection, maintenance and evaluation of germplasm in the form of seeds, rhizomes, bulbs etc. At present, approximately100 germplasm of ethnobotanically important plants are being maintained at Field Research Laboratory. Various studies – seed germination through various treatment, viability test, seed test weight, etc. are being carried out. An alpine herbal garden has been established within the premises of FRL with the same objective. Micro propagation of selected endangered and commercially viable medicinal and other important plants are being started.

c) Conservation through Gene Banks

Besides *in situ* and *ex situ* conservation, long storage of seeds can also be undertaken at Gene Banks. The National Plant Genetic Resources, N Delhi has already developed such facility having short term storage, medium storage and long storage of seeds and playing a vital role in this field. Therefore seed materials of ethnobotanically important plants should be sent to NBPGR for conservation of precious plant wealth. The area of conserving rare, endangered and threatened plants of economic value through biotechnological tools is vital to the conservational needs of the country. Clonal propagation is now gaining recognisation as a quicker alternative in conservation programme. *In vitro* techniques are being increasingly applied to supplement conventional method of propagation.

d) Education and awareness programmes

Till mid 1990's, the issues of conservation and cultivation of wild plants were not so popular. Now-a-days biodiversity and bio-conservation are becoming very popular throughout the world. This laboratory has been conducting regular training, workshop, seminar, radio talks, print media etc. to students, farmers, Amchis, army and others, regarding fragile ecology, precious floristic and faunal wealth of trans Himalayan regions and their judious and sustainable utilization. Due to which, today conservation is a familiar topic to every Amchi and local population and cultivation of few plant species on small scale is also started by government and non government organization and individual Amchis/ farmers. This lab has made a half an hour documentary film of medicinal plants of Ladakh Himalaya, the CD's of which are being shown or distributed to local population with the same objectives.

e) Recent Developments

From recent times, few important and positive developments are taking place for conservation of trans Himalayan plants, out of which some important happenings are :-

- NBPGR's programme for conservation of agro-biodiversity of India with special emphasis on Himalayan plants under the mission programme of National Agriculture Technology Project (NATP).
- National Medicinal Plant Board, India giving emphasis on cultivation, propagation, conservation and sustainable utilization of medicinal and aromatic plants through Govt Organisations, Non Govt. Organisations, Amchis, farmers etc.
- Forest departments of J&K and HP have initiated cultivation and conservation programmes.
- WWF-India has established its branch at Leh for conservation of High Altitudes Wetlands and allied activities.
- Development of projects under Global Environmental Facility (GEF) programme by UNDP, India and Ministry of Forest and Environment, Govt. of India (planning phase).

In nutshell, the effective conservation measures will play vital role in conserving the biodiversity of ethnobotanical wealth. The sustainable utilization of these resources will provide a powerful incentive to conserve species and their supporting ecosystems, provided the people most likely to have an impact on the species and ecosystem concerned receive an adequate share of these benefits.

❑❑❑

16

Propagation & Cultivation

Propagation and cultivation studies are one of the important aspects for conservation and scientific based sustainable utilization of high altitude ethnobotanical wealth. An alpine herbal garden has been established within the premises of the laboratory at an altitude of 3,454 m. with the objective to know the propagation techniques and to standardize the cultivation practices of medicinal as well as other commercially viable plants growing at different altitudes. Germplasm in the form of seeds, roots, rootstocks, rhizomes etc. of approximately 100 medicinal, aromatic and other important plants were collected from natural localities and are being maintained at the laboratory.

Over the period 31 important plants with emphasis on medicinal and aromatic plants of Ladakh and Lahaul-Spiti viz. *Aconitum sp, Dactylorhiza hatagirea, Podophyllum hexandrum, Rheum sp, Rhodiola sp.* etc have been tried for propagation and cultivation study. Briefs of propagation studies of individual plant are as under :

1.***Ach illea millefolium*** Linn. (Asteraceae)

Common Name : Yarrow / Milfoil

Cultivation & Propagation : Easily propagated through seeds as well as root cuttings.

a)S eeds : 87 % germination under controlled conditions and 52 % in field conditions. Nursery should be raised in green house and be transplanted in open field by the end of April. 85 % survival was achieved under the trial

b)Division of rootstock : It is also propagated through root cuttings with 72 % success rate.

Harvesting : Leaves are ready for harvest within three months from the date of transplantation and should be harvested before flowering.

2.***Aconitum heterophyllum*** Wall ex Royle (Ranunculaceae)

Common Name : Aconite / Atees / Bonakarpo

Cultivation & Propagation : The plant is propagated through seeds as well as division of roots.

a)Seeds : 85 % germination under control conditions with the treatment of gibbrelic acid (GA3) and 60 % in field conditions. Nursery should be raised in green house and be transplanted in April.

b)Division of roots : Easily propagated through division of roots. The daughter roots may be uprooted either in the month of October or by the end of April. 82 % success has been achieved.

Harvesting : Mature roots are harvested in every second or third years in the month of October.

3.***Allium carollinianum*** DC. (Amaryllidiceae)

Common Name : Wild garlic / Skotche

Cultivation & Propagation : The plant is easily propagated through seeds as well as bulbs.

a)Seeds : 84 % germination under controlled conditions and 67 % germination in field conditions. Survival rate is excellent (72%).

b)Division of bulbs : Easily propagated through division of bulbs.

Harvesting : The leaves are ready for harvest in the month of August-September.

4.***Allium prezewalskianum*** Regel (Amaryllidiceae)

Common Name : Wild garlic / Skotche

Cultivation & Propagation : The plant is easily propagated through seeds as well as bulbs.

a)Seeds : 89 % germination under controlled conditions and 74 % germination in field conditions. Survival rate is excellent (83%).

b)Division of bulbs : Easily propagated through division of bulbs.

Harvesting : The leaves are ready for harvest in the month of August-September.

5.***Arnebia euchroma*** Johnston (Boraginaceae)

Common Name : Rattanjot / Demok

Cultivation & Propagation : The plant is propagated through seed as well as root cuttings.

a)S eeds : Under controlled conditions, seeds have shown 85% germination while, 70% germination was observed under field conditions. Approx 60% survival rate was achieved through nursery transplantation.

b)Di vision of rootstock : Studies were undertaken for multiplication through division of rootstocks and very limited success (35 %) was achieved.

Harvesting : Mature roots may be harvested in the end of Sep-Oct from 3-5 years old plants.

6.***Artemisia annua*** (Asteraceae)

Common Name : Sweet wormwood

Cultivation & Propagation : Easily propagated through seeds.

a)Seeds : 85 % germination under field conditions. Nursery should be raised in green house and be transplanted in open field by the end of April. 80 % survival was achieved under the study.

Harvesting : The entire plant can be harvested by the end of August - September every year as raw material.

7.***Artemisia dracunculus*** Linn. (Asteraceae)

Common Name : Tarragon

Cultivation & Propagation : Easily propagated through seeds as well as root cuttings

a)S eeds : 72 % germination under controlled conditions and 64 % in field conditions. Nursery should be raised in green house and be transplanted in open field by the end of April. 84 % survival was achieved under the trial

b)Division of rootstock : It is also propagated through root cuttings with 76 % success rate.

Harvesting : The entire plant can be harvested by the end of August - September every year as raw material.

8.***Bidens pilosa*** Linn. (Asteraceae)

Common Name : Local tea

Cultivation & Propagation : Easily propagated through seeds.

Seeds : 86 % germination under controlled condition and 75 % in field conditions. Nursery should be raised in green house and be transplanted in open field by the end of April. 82 % survival was achieved under the trial.

Harvesting : The leaves / tender shoots can be harvested by the end of July and the second harvesting by the mid of September.

9.*Carum carvi* Linn. var. ***gracile*** (Apiaceae)

Common Name : Caraway / Konyot

Cultivation & Propagation : The plant is propagated through seeds.

a)Seeds : Seeds were sown directly in open conditions in the month of March and start germination with 74 % success.

Harvesting : Mature fruits are ready for harvest in the month of August - September in the second year.

10. *Cicer microphyllum* Benth. (Fabaceae)

Common Name : Wild gram / Seri

Cultivation & Propagation : Propagation trial was carried out through both seeds as well as rootstock multiplication.

a)Seeds : Seeds have got mechanical dormancy (seed coat) which can be overcome by simple scarification / concentrated sulphuric acid (H_2SO_4) treatment. Scarified seeds, after soaking in water for 12 hours, have shown 82.2 % germination under lab condition. This plant can be easily multiplied through nursery transplantation or direct sowing. 67 % success rate was achieved.

b)Vegetative : 3-5 years old plant should be uprooted in April and at least 4-8 rooted plants can be obtained from this plant. These rooted plants should be immediately planted for maximum survival (65 %).

Harvesting : Seeds are ready for harvest in the month of September. Because of shuttering property, seeds should be collected before maturity.

11. *Dactylorhiza hatagirea* (D. Don) Soo (Orchidaceae)

Common Name : Spotted heart orchid / Salam panja / Ambolakpa

Cultivation & Propagation : The plant is propagated through seeds as well as division of roots.

a)Seeds : Seeds are very minute and require micorrhizal association for germination. Studies are underway.

b)Division of roots : Easily propagated through division of roots. The daughter roots may be uprooted either by the end of growing season (October) or onset of summer (April). 85 % survival was observed by this method.

Harvesting : The mature roots may be harvested in second or third years of growth. The roots are generally harvested in the month of September - October.

12. *Dracocephalum heterophyllum* Benth. (Lamiaceae)

Common Name : Zinkzer / Zypsi

Cultivation & Propagation : The plant is propagated through seeds and division of roots/suckers.

a)S eeds : 80 % germination under control conditions and 60 % in field conditions. Nursery should be raised in green house and be transplanted in April.

b)D ivision of roots/suckers : Easily propagated through division of roots/ suckers. The roots/ suckers may be uprooted in the month of April and should be planted by the end of April or beginning of May. 75 % success has been achieved.

Harvesting : The aerial parts should be harvested just before maturity for good essential oil percentage.

13. ***Ephedra gererdiana*** Wall ex Stapf. (Ephedraceae)

Common Name : Somlata

Cultivation & Propagation : The plant is propagated through seeds as well as suckers.

a)Seeds : Seeds were sown in green house for chilling treatment in November and 80 % germination was achieved under field conditions. Nursery was ready for transplantation in the mid of April. Survival % is also good (65 %).

b)By suckers : It may also propagated through suckers. Suckers may uprooted in the beginning of summer (April) for maximum sprouting. 55 % success rate was achieved through this method.

Harvesting : 3-4 years plants are ready for harvest. Harvesting of twigs should be done by the end of summer (October).

14. ***Hippophae rhamnoides*** Linn. var. ***turkestanica*** (Elaeagnaceae)

Common Name : Seabuckthorn / Tsermang or Sastalulu

Cultivation & Propagation : The plant is propagated through seeds as well as vegetative means.

a)Seeds : Seeds require pre-treatment (Hot water & others) for germination. 71 % germination was achieved under field conditions.

b)Vegetative means : Easily propagated through division of rootstocks, suckers and stem cuttings.

Harvesting : All the parts of the plant are of medicinal value. Fruits are rich source of Vit 'C' which are ready for harvest between August- October.

15. ***Hippophae salicifolia*** (Elaeagnaceae)

Common Name : Seabuckthorn / Tsermang or Sastalulu

Cultivation & Propagation : The plant is propagated through seeds as well as vegetative means.

a)Seeds : Seeds require pre-treatment (Hot water & others) for germination. 60 % germination was achieved under field conditions.

b)Vegetative means : Easily propagated through division of rootstocks, suckers and stem cuttings.

Harvesting : All the parts of the plant are of medicinal value. Fruits are rich source of Vit 'C' which are ready for harvest between August- October.

16. *Hippophae tibetana* (Elaeagnaceae)

Common Name : Seabuckthorn / Tsermang or Sastalulu

Cultivation & Propagation : The plant is propagated through seeds as well as vegetative means.

a)Seeds : Seeds require pre-treatment (Hot water & others) for germination. 52 % germination was achieved under field conditions.

b)Vegetative means : Easily propagated through division of rootstocks, suckers and stem cuttings.

Harvesting : All the parts of the plant are of medicinal value. Fruits are rich source of Vit 'C' which are ready for harvest between August-September.

17. *Inula racemosa* HK. f. (Asteraceae)

Common Name : Puskarmool /Manu

Cultivation & Propagation : This plant is easily propagated through cutting of roots as well as seeds.

a)Seeds : Seeds have got dormancy. To overcome dormancy, chilling treatment is essential. Seeds have been sown in polyhouse in the month of Nov and they started germination in March. 80% germination was achieved. Nursery was ready for transplantation in the month of April. 72 % survival rate was achieved.

b)By division of rootstock : It is easily propagated through division of rootstock. The stout rootstock was uprooted in April – May and made into small pieces. These pieces were planted horizontally 5-8 cm deep in soil. 80% survival rate was achieved under trial.

Harvesting : Roots are ready for harvest from 2-3 years old plants and they are usually harvested during October or march.

18. *Mentha longifolia* (Linn.) Hudson (Lamiaceae)

Common Name : Horse mint / Wild mint / Pholoing

Cultivation & Propagation : The plant is propagated through seeds, cuttings, and rootstock divisions.

a)Seeds : 72 % germination was achieved under field conditions. Survival rate is good (80 %).

b)Division of rootstock : Easily propagated through division of rootstocks. They are generally uprooted in April, made into pieces and then are planted in fields. 88 % survival rate has been achieved through this method.

Harvesting : Leaves are ready for harvest within three months from the date of transplantation and should be harvested before flowering.

19. *Origanum vulgare* Linn. (Lamiaceae)

Common Name : Wild oregano / Marjoram

Cultivation & Propagation : The plant is propagated through seeds, cuttings, and rootstock divisions.

a)Seeds : 86 % germination was achieved under field conditions. Survival rate is good (76 %).

b)Division of rootstock : Easily propagated through division of rootstocks. They are generally uprooted in April, made into pieces and then are planted in fields. 88 % survival rate has been achieved through this method.

Harvesting : Leaves are ready for harvest within three months from the date of transplantation and should be harvested before flowering.

20. *Perovskia abrotanoides* Karelin (Lamiaceae)

Common Name : Wild lavender

Cultivation & Propagation : The plant is propagated through seeds, cuttings, layers and rootstock divisions.

a)Seeds : 76 % germination was achieved under field conditions. Survival rate is good (72 %).

b)Division of rootstock : Easily propagated through division of rootstocks. They are generally uprooted in April, made into pieces and then are planted in fields. 74 % survival rate has been achieved through this method.

Harvesting : Leaves are ready for harvest within three months from the date of transplantation and should be harvested before flowering.

21. *Physochlaina praealta* (Decne.) Miers. (Solanaceae)

Common Name : Wild dathura / Langthang

Cultivation & Propagation : The plant is propagated through seeds, and rootstock divisions.

a)Seeds : 80 % germination was achieved under field conditions. Survival rate is good (76 %).

b)Division of rootstock : Easily propagated through division of rootstocks. They are generally uprooted in April, made into pieces and then are planted in fields. 69 % survival rate has been achieved through this method.

Harvesting : Leaves are ready for harvest within three months from the date of transplantation and should be harvested before flowering.

22. *Podophyllum hexandrum* Royle. (Berberidaceae)

Common Name : Himalayan Mayapple

Cultivation & Propagation : The taxa can not be propagated easily because of its hard seed coat and poor germination rate.

a)Seeds : Under controlled condition only 2-4 % germination was observed. Sulphuric acid and hot water treatment were found effective to enhance the germination rate upto 65 %. Survival rate was 65 %.

b) Vegetative : 3-5 years old rhizomatus disc develops several vegetative buds around the aerial shoot during March – April. The disc was divided in such a way that each piece possesses at least two buds alongwith tuberous roots. 70 % survival rate was observed under the trial.

Harvesting : 3-5 years old plant is considered ideal for harvesting rhizome. The rhizomes are generally harvested in October-November to get quality drug.

23. *Rheum spiciforme* Royle (Polygonaceae)

Common Name : Himalayan Rhubarb/Rewandchini /Lachhu

Cultivation & Propagation : Through seeds as well as vegetative parts.

a)Seeds : Under controlled condition 64 % germination was observed. 78 % survival rate was achieved.

b)By rootstocks : Under vegetative propagation, 3-4 years old vigorous rootstocks were made into pieces. These pieces were planted 5-8 cm. deep in soil during April – May. 82 % survival rate was observed under the trial.

Harvesting : The roots are generally dug up in October-November from 3-4 years old plants.

24. *Rheum webbianum* Royle (Polygonaceae)

Common Name : Himalayan Rhubarb / Rewandchini / Lachhu

Cultivation & Propagation : Through seeds as well as vegetative parts.

a)Seeds : Under controlled condition 59 % germination was observed. GA_3 treatment had enhanced the germination rate. 73 % germination was noted in 100 ppm followed by 68 % and 61 % in 50 ppm and 10 ppm of GA_3 respectively. 78 % survival rate was achieved.

b)By rootstocks : Under vegetative propagation, 3-5 years old vigorous rootstocks were made into pieces. These pieces were planted 5-8 cm. deep in soil during April – May. 80 % survival rate was observed under the trial.

Harvesting : The roots are generally dug up in October-November from 3-4 years old plants.

25. *Rhodiola imbricata* Edgew. (Crassulaceae)

Common Name : Rose root /Himalayan stone crop / Shrolo

Cultivation & Propagation : The plant is propagated through seeds as well as rootstock cuttings.

a)Seeds : 65 % germination was achieved under field conditions.

b)Division of rootstock : Easily propagated through division of rootstocks. 3-4 years plants are ideal for suitable rootstocks for plantation. 86 % survival rate has been achieved.

Harvesting : 3-5 years plants are considered ideal for good size roots. They are advised to uprooted in the month of September-October.

26. *Rubia cordifolia* Roxb. ex Fleming (Rubiaceae)

Common Name : Indian madder / Manjistha

Cultivation & Propagation : The plant is propagated through seeds as well as rootstock cuttings.

a)Seeds : 81 % germination was achieved under field conditions. Nursery should be transplanted in the month of April. Transplantation % is good (74 %).

b)Division of rootstock : Easily propagated through division of rootstocks.

Harvesting : 1-2 years old plants are considered ideal for harvesting the rootstocks. The roots should be harvested in the month of September-October.

27. *Salvia sclarea* Linn. (Lamiaceae)

Common Name : Clary sage

Cultivation & Propagation : Easily propagated through seeds.

a)Seeds : 90 % germination under field conditions. Nursery should be raised in green house and be transplanted in open field by the end of April. 80 % survival was achieved under the trial

Harvesting : The flowering spike can be harvested by the end of mid August – mid September within the year.

28. ***Saussurea lappa*** Clarke (Asteraceae)

Common Name : Costus/Kuth/Rustha

Cultivation & Propagation : The plant is propagated through seeds as well as roots.

a)Seeds : 87 % germination was achieved under field conditions.

b)Division of rootstock : It is also propagated through division of rootstocks.

Harvesting : Roots are ready for harvest within 2-3 years. They should be harvested in October or March.

29. ***Urtica hyperborea*** Jacq. ex Wedd. (Urticaceae)

Common Name : Stinging nettle / Dzatsutt

Cultivation & Propagation : The plant is propagated through seeds, and rootstock divisions.

a)Seeds : 82 % germination was achieved under field conditions. Survival rate is good (73 %).

b)Division of rootstock : Easily propagated through division of rootstocks. They are generally uprooted in April, made into pieces and then are planted in fields. 81 % survival rate has been achieved through this method.

Harvesting : Leaves are ready for harvest within three months from the date of transplantation and should be harvested before flowering.

30. ***Valeriana jatamansi*** Jones. (Valerianaceae)

Common Name : Jatamansi

Cultivation & Propagation : The plant is propagated through seeds as well as vegetative means.

a)Seeds : Studies through seeds are going on under Ladakh conditions.

b)Division of rootstock : The plant is easily propagated by division of old roots either in the beginning (April) or end of summer (October). 88 % survival rate has been achieved.

Harvesting : 2-3 years plants are ready for root harvesting. Harvesting should be done in the month of October.

31. ***Verbascum thapsus*** Linn. (Scrophulariaceae)

Common Name : Flannel Mullein / Dandashal

Cultivation & Propagation : The plant is propagated through seeds.

a)Seeds : 63 % germination was achieved under field conditions. Survival rate is good (70 %).

Harvesting : Leaves are ready for harvest within three months from the date of transplantation and should be harvested before flowering.

❑❑❑

List of Tables

Table 3. List of Medicinal plants of Ladakh and Lahaul Spiti

S. No	BOTANICAL NAME WITH FAMILY	COMMON NAME	PARTS USED	MEDICINAL USES
1.	*Acantholimon lycopodioides* (Plumbaginaceae)	Lonze	Whole plant	Cardiac disorders
2.	*Achillea millefolium* (Asteraceae)	Milfoil/Yarrow/Chuang	Leaves	Tonic, cough, cold & toothache
3.	*Aconitum heterophyllum* (Ranunculaceae)	Aconite, Attees, Bona-karpo	Roots	Antipyretic, astringent, stomach complaint
4.	*Aconitum laeve* (Ranunculaceae)	Aconite	Roots	Antipyretic, stomach complaints
5.	*Aconitum rotundifolium* (Ranunculaceae)	Aconite	Roots	Fever & cold
6.	*Aconitum violaceum* (Ranunculaceae)	Aconite, Pattees, Bona-nagpo	Roots	Antipyretic, fever
7.	*Aconogonum tortuosum* (Polygonaceae)	Nyalo	Whole plant	Abdominal pain
8.	*Actinocarya tibetica* (Boraginaceae)	—	Roots	Fever
9.	*Ajuga bracteosa* (Lamiaceae)	—	Leaves	Tonic, stimulant
10.	*Allium carolinianum* (Amaryllidaceae)	Skotche	Leaves	Indigestion
11.	*Anaphalis busua* (Asteraceae)	Telgang	Flower-heads	Cold & cough
12.	*Anaphalis cuneifolia* (Asteraceae)	Simula	Whole plant	Skin diseases

Contd..

Table 3. Contd...

13.	*Anaphalis triplinervis* var. *monocephala* (Asteraceae)	sPra-rgod	Flower-heads	Chronic diseases
14.	*Anaphalis triplinervis* var. *intermedia* (Asteraceae)	Yaktso	Flower-heads	Skin diseases, genital problems
15.	*Anaphalis virgata* (Asteraceae)	Simula	Flower-heads	Fever
16.	*Androsace mucronifolia* (Primulaceae)	Ga-tig-smug-po	Whole herb	Fever, abdominal pain
17.	*Androsace rotundifolia* (Primulaceae)	—	Whole herb	Tonic
18.	*Androsace villosa* (Primulaceae)	Fujung	Leaves	Tonic
19.	*Anemone rivularis* (Ranunculaceae)	Srub-ka	Fruits	Indigestion
20.	*Anemone rupicola* (Ranunculaceae)	Simaso	Whole plant	Stomach complaints
21.	*Aquilegia fragrans* (Ranunculaceae)	Fragrans columbine	Whole herb	Antiscarbutic, diuretic & diaphoretic
22.	*Arabidopsis multiflorum* (Brassicaceae)	Imatso	Whole plant	Diarrhoea
23.	*Arabis glandulosa* (Brassicaceae)	Umakno	Leaves	Abdominal pain
24.	*Arctium lappa* (Asteraceae)	Burdock, Jangli kuth, Pizums	Roots	Diuretic, depurgative, lung diseases
25.	*Arenaria bryophylla* (Caryophyllaceae)	Lekhum	Whole plant	Kidney problems
26.	*Arenaria griffithii* (Caryophyllaceae)	Sokhtam	Whole plant	Menstrual problems
27.	*Arnebia benthamii* (Boraginaceae)	Ratanjot, Balchar	Roots	Hair tonic, blood purification

Contd..

Table 3. Contd...

28.	*Arnebia euchroma* (Boraginaceae)	Ratanjot, demok	Roots	Hair tonic & blood purificatrion
29.	*Arnebia guttata* (Boraginaceae)	Ratanjot, demok	Roots	Hair tonic & blood purification
30.	*Artemisia annua* (Asteraceae)	Sweet wormwood	Whole plant	Anti-malarial
31.	*Artemisia biennis* (Asteraceae)	Margrass	Seeds	Obesity, stomach complaints
32.	*Artemisia brevifolia* (Asteraceae)	Wormseed	Leaves	Antiworm, stomach complaints
33.	*Artemisia desertorum* (Asteraceae)	Khamlol	Leaves	Intestinal complaints
34.	*Artemisia dracunculus* (Asteraceae)	Tarragan / Sanche	Whole plant	Diuretic, toothache
35.	*Artemisia gmelinii* (Asteraceae)	Russian wormwood, Khampa shridi	Leaves	Cold & cough
36.	*Artemisia laciniata* (Asteraceae)	Bintso	Leaves	Toothache
37.	*Artemisia nilagirica* (Asteraceae)	Chirmaro	Leaves	Antimicrobial, insecticidal
38.	*Artemisia parviflora* (Asteraceae)	Khamang	Whole plant	Antiviral, diuretic
39.	*Artemisia persica* (Asteraceae)	Burnak	Roots	Febrifuge, stomach complaints
40.	*Artemisia salsoloides* (Asteraceae)	Amango	Whole plant	Intestinal complaints
41.	*Artemisia scoparia* (Asteraceae)	Khamptspa	Whole plant	Intestianal complaints
42.	*Artemisia sieversiana* (Asteraceae)	Khamchu	Leaves & flowers	Antimicrobial, intestinal worms

Contd..

Table 3. Contd...

43.	*Artemisia stracheyi* (Asteraceae)	Rumonlo	Whole plant	Toothache
44.	*Artemisia tournefortiana* (Asteraceae)	Khamchu	Leaves	Intestinal worms
45.	*Asperugo procumbens* (Boraginaceae)	Zingsha	Roots	Stomach complaints
46.	*Asperula oppositifolia* (Rubiaceae)	Zingsha	Whole herb	Tonic
47.	*Aster flaccidus* (Asteraceae)	Wild aster	Flowers	Cold & cough
48.	*Astragalus multiceps* (Fabaceae)	—	Seeds	Colic problems
49.	*Astragalus rhizanthus* (Fabaceae)	Sarma	Whole plant	Skin diseases
50.	*Astragalus strictus* (Fabaceae)	Serpang	Whole plant	Diuretic
51.	*Astragalus tribulifolius* (Fabaceae)	Yanglo	Whole plant	Diuretic
52.	*Atelanthera perpusilla* (Brassicaceae)	—	Leaves	Fever
53.	*Berberis jaeschkeana* (Berberidaceae)	—	Roots	Eye trouble
54.	*Berberis kunawurensis* (Berberidaceae)	Berberis	Roots	Tonic
55.	*Berberis pachyacantha* (Berberidaceae)	Karpa	Fruits	Stomach problems
56.	*Berberis ulicina* (Berberidaceae)	Berberry, Khizar	Roots & bark	Tonic
57.	*Berberis zabeliana* (Berberidaceae)	Berberry	Roots	Fever

Contd..

Table 3. Contd...

58.	*Bergenia ligulata* (Saxifragaceae)	Pashanbhed	Roots	Astringent, tonic, kidney problems
59.	*Bergenia stracheyi* (Saxifragaceae)	Pashanbhed, Tiang	Roots	Astringent, tonic, kidney problems
60.	*Biebersteinia odora* (Geraniaceae)	Khardung / Drakpose	Leaves	Antiseptic, skin sores
61.	*Blumea bifoliata* (Asteraceae)	Leuman	Roots	Fever
62.	*Caltha palustris* (Ranunculaceae)	Marsh marigold	Leaves & roots	Gonnorrhoea
63.	*Capparis spinosa* (Capparidaceae)	Caper bush, kabra	Whole plant	Astringent, liver complaints, rheumatism
64.	*Capsella bursa-pastoris* (Brassicaceae)	Shepherds' purse	Whole plant	Lung complaints, stomach problems
65.	*Carum carvi* var. *gracile* (Apiaceae)	Caraway, Koniyot	Seeds	Carminative, digestive, febrifuge
66.	*Cassiope fastigiata* (Ericaceae)	Sulu	Whole plant	Tonic
67.	*Catabrosa aquitaca* (Poaceae)	Leema	Whole plant	Fever
68.	*Centaurea depressa* (Asteraceae)	Vashakha	Whole plant	Fever
69.	*Cerastium cerastioides* (Caryophyllaceae)	Tyndong	Whole plant	Headache
70.	*Chaerophyllum reflaxum* (Apiaceae)	Lcha-wa	Roots	Stomach complaints
71.	*Chenopodium botrys* (Chenopodiaceae	Sagni	Whole plant	Anthelmintic, laxative, stomach complaints
72.	*Chrysanthemum griffithii* (Asteraceae)	Serpan	Flower-heads	Menses regulation

Contd..

Table 3. Contd...

73.	*Chrysanthemum pyrethroides* (Asteraceae)	Serpan	Whole plant	Fever, rheumatism
74.	*Chrysanthemum tibeticum* (Asteraceae)	Phemantso	Whole plant	Antiseptic
75.	*Cicer microphyllum* (Fabaceae)	Wild gram, seri	Seeds	Adaptogenic
76.	*Cirsium wallichii* (Asteraceae)	Kakar	Leaves	Stomach complaints
77.	*Clematis tibetana* (Ranunculaceae)	Clematis, Zakgic	Roots	Indigestion
78.	*Codonopsis clematidea* (Campanulaceae)	Burrkutang	Roots	Rheumatism, stimulant
79.	*Codonopsis ovata* (Campanulaceae)	Ludut	Roots	Rheumatism, stimulant
80.	*Codonopsis rotundifolia* (Campanulaceae)	Ludut	Roots	Rheumatism, stimulant
81.	*Colchicum luteum* (Liliaceae)	Hirantotiya, Tukapa	Corm & seeds	Colchicine & gout
82.	*Conringia planisiliqua* (Brassicaeae)	Khaskar	Roots	Diuretic
83.	*Corispermm hyssopifolium* (Chenopodiaceae)	Seimso	roots	Kidney problems
84.	*Corydalis cashmeriana* (Papaveraceae)	Ralchatnakpo	Whole herb	Antiperiodic, diuretic
85.	*Corydalis crassissima* (Papaveraceae)	sKyes-pa dong-btags	Leaves	Lung complaints
86.	*Corydalis flabellata* (Papaveraceae)	—	Roots	Skin diseases
87.	*Corydalis govaniana* (Papaveraceae)	Stong-zil	Leaves, roots	Antipyretic, diuretic, febrifuge

Contd..

Table 3. Contd...

88.	*Corydalis meifolia* (Papaveraceae)	Tonzil	Leaves	Diuretic, diuretic
89.	*Corydalis thyrsiflora* (Papaveraceae)	Nyetapu	Whole plant	Skin diseases
90.	*Cousinia falconeri* (Asteraceae)	Kreatising	Whole plant	Diuretic
91.	*Cousinia thomsoni* (Asteraceae)	Megtham	Whole plant	Diuretic
92.	*Cremanthodium ellisii* (Asteraceae)	Ming-chen-ser-po/ Lukumentok	Whole plant	Contagious disease, fever
93.	*Cremanthodium reniforme* (Asteraceae)	Ming-chen-ser-po/ Lukumentok	Whole plant	Contagious disease, fever
94.	*Crepis flexuosa* (Asteraceae)	Hawksbread, Remang	Leaves	Muscle pain
95.	*Cuscuta approximata* (Cuscutaceae)	Large dodder, Amarlata	Whole palnt	Skin diseases
96.	*Cuscuta capitata* (Cuscutaceae)	Large dodder, Amarlata	Whole palnt	Kidney problems
97.	*Cuscuta europaea* (Cuscutaceae)	Large dodder, Amarlata	Whole palnt	Kidney diseases
98.	*Cuscuta gigantea* (Cuscutaceae)	Hande-thapa	Whole palnt	Antiseptic
99.	*Cuscuta reflexa* (Cuscutaceae)	Large dodder, Amarlata	Whole palnt	Purgative, warts
100.	*Cynanthus lobatus* (Fumariaceae)	Ralchatnagpo	Whole plant	Skin diseases
101.	*Cynoglossum wallichii* (Boraginaceae)	—	Whole plant	Vomiting
102.	*Dactylorhiza hatagirea* (Orchidaceae)	Spotted heart orchid, Salam panja, Ambolakpa	Roots Whole plant	Energetic, health toic, nervine Tonic Colic, insecticidal

Contd..

Table 3. Contd...

103.	*Delphinium brunonianum* (Ranunculaceae)	Himalayan larkspur, Lunde-kaown		
104.	*Delphinium cashmerianum* (Ranunculaceae)	Himalayan larkspur, Lunde-kaown	Whole plant	Colic, insecticidal
105.	*Delphinium vestitum* (Ranunculaceae)	—	Whole plant	Colic, insecticidal
106.	*Dianthus anatolicus* (Caryophyllaceae)	—	Leaves	Stomach problems
107.	*Dianthus deltoides* (Caryophyllaceae)	Khenchpa	Leaves	Stomach problems
108.	*Dianthus jacquemontii* (Caryophyllaceae)	Khenchpa	Leaves	Stomach problems
109.	*Dioscorea deltoidea* (Dioscoreaceae)	Dioscorea, Singli-mingli	Tubers	Oral contraceptive pills, source of disogenin
110.	*Draba tibetica* (Brassicaceae)	—	Leaves	Tonic
111.	*Dracocephalum heterophyllum* (Lamiaceae)	Zinkzer	Whole plant	Cold & cough
112.	*Dracocephalum moldavicum* (Lamiaceae)	Zinkzer	Whole plant	Intestinal problems
113.	*Dracocephalum stamineum* (Lamiaceae)	Zinkzer	Whole plant	Cold & cough
114.	*Echinops cornigerus* (Asteraceae)	Globe thistle, Aczema	Leaves	Septic wounds
115.	*Echinops niveus* (Asteraceae)	Ackzema	Leaves	Septic wounds
116.	*Ephedra gerardiana* (Ephedraceae)	Asmania, Som-lata, Chhapat	Whole plant	Asthma, hepatic disease, source of ephedrine
117.	*Ephedra intermedia* (Ephedraceae)	Asmania, Som-lata, Chhapat	Whole plant	Asthma

Contd..

Table 3. Contd...

118.	*Epipactis helleborine* (Orchidaceae)	Pengilo	Roots	Blood purification
119.	*Eragrostis poaeoides* (Poaceae)	—	Roots	Stomach complaints
120.	*Erigeron alpinus* (Asteraceae)	Repan	Whole plant	Cold & cough
121.	*Ermania albiflora* (Brassicaceae)	—	Whole plant	Fever
122.	*Ermania lanuginosa* (Brassicaceae)	Measlo	Leaves	Diuretic, purgative
123.	*Euphorbia hispida* (Euphorbiaceae)	Therno	Latex	Skin eruption
124.	*Euphorbia kanorica* (Euphorbiaceae)	Kianglo	Latex	Skin eruption
125.	*Euphorbia stracheyi* (Euphorbiaceae)	Khron-bu	Latex	Boils
126.	*Euphrasia laxa* (Scrophulariaceae)	Eyebright, Kangchuk	Whole palnt	Eye complaints
127.	*Euphrasia officinalis* (Scrophulariaceae)	Eyebright, Kangchuk	Whole palnt	Eye complaints
128.	*Euphrasia vulgaris* (Scrophulariaceae)	Bepamatso	Whole plant	Eye complaints
129.	*Ferula jaeschkeana* (Apiaceae)	Wild heeng, Sampharu	Roots	Rheumatism, septic wounds
130.	*Fraxinus xanthoxyloides* (Oleaceae)	Ash	Leaves	Astringent, diuretic, tonic
131.	*Fritillaria roylei* (Liliaceae)	A-bhi-sa	Bulbs	Lung complaints, tonic
132.	*Galium aparine* (Rubiaceae)	Ranche	Leaves	Urinary complaints

Contd..

Table 3. Contd...

133.	*Galium pauciflorum* (Rubiaceae)	Phomongo	Whole plant	Intestinal problems
134.	*Galium serpylloides* (Rubiaceae)	Pimantso	Leaves	Diuretic, purgative
135.	*Gaultheria trichophylla* (Ericaceae)	Winter green	Leaves	Antiseptic, source of gautherine
136.	*Gentiana algida* (Gentianaceae)	Titkas	Whole plant	Analgesic, bronchitis
137.	*Gentiana aquatica* (Gentianaceae)	Titkas	Roots	Health tonic
138.	*Gentiana carinata* (Gentianaceae)	Pang-gyan-khar-bo	Leaves	Stomach complaints
139.	*Gentiana humilis* (Gentianaceae)	Pang-gyan-khar-bo	Leaves	Fever
140.	*Gentiana squarrosa* (Gentianaceae)	Ziang	Roots	Diuretic
141.	*Gentiana stracheyi* (Gentianaceae)	Ziang	Roots	Tonic
142.	*Gentianella moorcroftiana* (Gentianaceae)	Span-gain-karpo	Whole plant	Febrifuge
143.	*Gentianella paludosa* (Gentianaceae)	1Kags-ti-nagpo	Roots	Febrifuge, tonic
144.	*Geranium collinum* (Geraniaceae)	Legkatin	Leaves & roots	Stomach complaints
145.	*Geranium pratense* (Geraniaceae)	Meadow cranesbill, Analgesic, febrifuge	Gad-ur	Leaves
146.	*Geranium sibiricum* (Geraniaceae)	Eyamlomentok	Leaves	Dysentry
147.	*Geranium tuberaria* (Geraniaceae)	Yusiang	Whole plnat	Kidney problems

Contd..

Table 3. Contd...

148.	*Geum eletum* (Rosaceae)	—	Whole plant	Health tonic
149.	*Gnaphalium stewartii* (Rosaceae)	Peo	Whole plant	Skin diseases
150.	*Goodyera repens* (Orchidaceae)	—	Whole plant	Health tonic
151.	*Gypsophyla cerastioides* (Caryophyllaceae)	Chicktkpa	Whole plant	Cold & cough
152.	*Hackelia uncinata* (Boraginaceae)	—	Whole plant	Cold & cough
153.	*Halenia elliptica* (Gentianaceae)	lCags-tig-ra-mgo-ma	Whole herb	Stomach complaints
154.	*Halerpestis tricuspis* (Ranunculaceae)	Chu-rag-sBl-lg	Whole plant	Stomachache
155.	*Heracleum lanatum* (Apiaceae)	Rasal	Roots	Leucoderma, menses regulation, source of xanthotoxin
156.	*Heracleum pinnatum* (Apiaceae)	sPru-ma	Roots	Leucoderma, menses regulation, source of xanthotoxin
157.	*Herminium monorchis* (Orchidaceae)	Musk orchid	Bulb	Health tonic
158.	*Hippophae rhamnoides* ssp. *turkestaica* (Elaeagnaceae)	Seabuckthorn, Sastalulu	Fruits, seeds, leaves	Source of multi-vitamins, carotenoids, anti-oxidants
159.	*Hippophae salicifolia* (Elaeagnaceae)	Seabuckthorn, Sastalulu	Fruits, seeds, leaves	Source of multi-vitamins, carotenoids, anti-oxidants
160.	*Hippophae tibetana* (Elaeagnaceae)	Seabuckthorn, Sastalulu	Fruits, seeds, leaves	Source of multi-vitamins, carotenoids, anti-oxidants
161.	*Hippuris vulgaris* (Hippuridaceae)	Dam-bu-kara-machong	Leaves & stems	Antiseptic, febrifuge
162.	*Humulus lupulus* (Urticaceae)	Hops	Flowers & fruits	Nervine tonic, sedative effect

Contd..

Table 3. Contd...

163.	*Hyoscyamus niger* (Solanaceae)	Henbane, Khurshani Ajawain, Gay-langthang	Whole plant	Asthma, Toothache, source of hyoscyamine drug
164.	*Hyoscyamus pusillus* (Solanaceae)	Henbane, Khurshani Ajawain, Gay-langthang	Leaves	Asthma & Toothache
165.	*Hypecoum leptocarpum* (Papaveraceae)	Meerang / Parpata	Leaves	Septic wounds
166.	*Hypericum perforatum* (Hypericaceae)	St Jonh's Wort	Whole herb	Neuralgia, sciatica
167.	*Hyssopus officinalis* (Lamiaceae)	Hyssop (E), Juff / Tengu	Whole herb	Sprains, stimulant
168.	*Inula falconeri* (Asteraceae)	—	Roots	Sprains
169.	*Inula obtusifolia* (Asteraceae)	—	Leaves	Internal wounds
170.	*Inula racemosa* (Asteraceae)	Puskarmool, Manu	Roots	Anthelmintic, bronchial asthma, expectorant
171.	*Inula rhizocephala* var. *rhizophaloides* (Asteraceae)	Riamko	Roots	Cough & chest pains
172.	*Inula royleana* (Asteraceae)	Rupmak	Shoots	Dermatitis
173.	*Juniperus communis* (Juniperaceae)	Juniper, Sukpa	Fruits	Purgative, anti-inflammatory
174.	*Juniperus macropoda* (Juniperaceae)	Juniper, Sukpa	Fruits	Purgative, anti-inflammatory
175.	*Juniperus recurva* (Juniperaceae)	Juniper, Sukpa	Fruits	Purgative, anti-inflammatory
176.	*Jurinea ceratocarpa* (Asteraceae)	Ispangchot	Leaves	Bronchitis
177.	*Jurinea macrocephala* (Asteraceae)	Guggul	Roots	Stimulant, fever

Contd..

Table 3. Contd...

178.	*Lactuca lessertiana* (Asteraceae)	Wild salad	Whole herb	Skin diseases
179.	*Lagotis cashmeriana* (Scrophulariaceae)	Hong-len	Whole plant	Febrifuge, anti-inflammatory
180.	*Lagotis kunawuresis* (Scrophulariaceae)	Tsermo	Leaves	Cold & cough
181.	*Lancea tibetica* (Scrophulariaceae)	Spa-yang-rtsa-ba	Whole plant	Tonic, wound healing
182.	*Lappula barbata* (Boraginaceae)	—	Roots	Fever
183.	*Leontopodium leontopodium* (Asteraceae)	Palloo	Whole plant	Septic wounds
184.	*Leontopodium nanum* (Asteraceae)	Palloo	Whole plant	Septic wounds
185.	*Lespedeza juncea* (Fabaceae)	—	Whole plant	Cold & cough
186.	*Leucopoa albida* (Poaceae)	Fukche	Whole plant	Intestinal pains
187.	*Linum perenne* (Linaceae)	—	Whole plant	Emollient, expectorant
188.	*Lloydia serotina* (Liliaceae)	rTsa-awa	Bulbs	Blood purification
189.	*Lomatogonium rotatum* (Gentianaceae)	Tsemrang	Roots	Fever
190.	*Lonicera spinosa* (Caprifoliaceae)	—	Flowers	Asthma & headache
191.	*Lychnis himalayensis* (Caryophyllaceae)	Gayangarlo	Whole plant	Cold & cough
192.	*Lychnis nutans* (Caryophyllaceae)	Lappu	Roots	Dysentery

Contd..

Table 3. Contd...

193.	*Lycium ruthenicum* (Solanaceae)	Umila	Leaves	Diuretic
194.	*Lycopsis arvensis* (Boraginaceae)	—	Whole plant	Tonic
195.	*Malaxis musifera* (Orchidaceae)	—	Roots	Tonic
196.	*Malcolmia africana* (Brassicaceae)	Chungpo	Whole plant	Fever
197.	*Malva verticillata* (Malvaceae)	Sucheli	Leaves	Liver tonic
198.	*Mattiastrum thomsonii* (Boraginaceae)	—	Leaves	Diuretic
199.	*Meconopsis aculeata* (Papaveraceae)	Himalayan Blue Popy, Akshatsermum	Whole plant	Ulcer, liver & lung problems
200.	*Meconopsis bikramii* (Papaveraceae)	Himalayan Blue Popy, Akshatsermum	Whole plant	Ulcer
201.	*Medicago lupulina* (Fabaceae)	Bu-su-hang-pho	Whole herb	Lung disorders & cough
202.	*Mentha longifolia* (Lamiaceae)	Horse mint, Phuloling	Leaves	Headache, stomachache
203.	*Microula tibetica* (Boraginaceae)	—	Whole plant	Pulmonary disorders
204.	*Minuartia kashmirica* (Caryophyllaceae)	—	Whole plant	Liver tonic
205.	*Morina longifolia* (Dipsacaceae)	Kim	Roots	Worm infestation
206.	*Morina persica* (Dipsacaceae)	—	Roots	Worm infestation
207.	*Myosotis silvastica* (Boraginaceae)	—	Roots	Diuretic

Contd..

Table 3. Contd...

208.	*Myricaria squamosa* (Tamaricaceae)	Umbo	Whole plant	Febrifuge, poison
209.	*Nepeta clarkei* (Lamiaceae)	Khamyu	Whole plant	Septic wounds
210.	*Nepeta coerulescens* (Lamiaceae)	Kharu	Whole plant	Dysentery
211.	*Nepeta discolor* (Lamiaceae)	Shamlolo	Leaves	Colds & cough
212.	*Nepeta erecta* (Lamiaceae)	Eripantso	Whole plant	Dysentery
213.	*Nepeta eriostachya* (Lamiaceae)	Zimathikle	Whole plant	Eye complaints
214.	*Nepeta floccosa* (Lamiaceae)	Shangukaram	Leaves	Cold & cough
215.	*Nepeta glutinosa* (Lamiaceae)	Jatukpa	Leaves	Diarrhoea
216.	*Nepeta laevigata* (Lamiaceae)	Jatukpa	Whole plant	Pneumonia
217.	*Nepeta leucolaena* (Lamiaceae)	—	Leaves	Skin disease
218.	*Nepeta longibracteata* (Lamiaceae)	Prainku / Chhagnamgo	Whole plant	Stomachache
219.	*Onosma hispidum* (Boraginaceae)	Ratanjot, Gajowan	Roots	Stimulant, cardiac tonic
220.	*Origanum vulgare* (Lamiaceae)	Wild oregano, Marjorum	Whole plant	Carminative, stimulant, mild tonic
221.	*Orobanchae hansii* (Orobanchaceae)	sGro-shangprtse	Whole plant	Boils
222.	*Oxyria digyna* (Polygonaceae)	Mountain sorrel	Leaves	Digestive, purgative

Contd..

Table 3. Contd...

223.	*Oxytropis lapponica* (Fabaceae)	Rushkum	Leaves	Antiseptic
224.	*Oxytropis macrophylla* (Fabaceae)	—	Leaves	Antiseptic, fever
225.	*Papaver nudicaule* (Papaveraceae)	Orange popy	Fruits	Pains
226.	*Parnassia laxamanni* (Saxifragaceae)	—	Whole plant	Diuretic
227.	*Pedicularis brevifolia* (Scrophulariaceae)	Luramenthag	Inflorescence	Sedative
228.	*Pedicularis cheilanthifolia* (Scrophulariaceae)	Lug-ru-karpo	Whole herb	Stomach complaint
229.	*Pedicularis longiflora ssp. tubiformis* (Scrophulariaceae)	Lug-ru-serpo	Whole plant	Diuretic
230.	*Pedicularis mollis* (Scrophulariaceae)	Lug-ru-marpo	Flowers	Antiseptic, febrifuge
231.	*Peganum harmala* (Zygophyllaceae)	Wild rue, Sepan	Roots, seeds	Asthma, rheumatism, diuretic
232.	*Perovskia abrotenoides* (Lamiaceae)	Wild lavender, Skiling	Leaves	Laxative
233.	*Phlomis bracteosa* (Lamiaceae)	—	Whole plant	Cold & cough
234.	*Physalis alkekengi* (Solanaceae)	Winter cherry, Shoklo	Fruits	Gout, rheumatism, vermifuge
235.	*Physochlaina praealta* (Solanaceae)	Langthang	Leaves & seeds	Narcotic, vermifuge, source of atropine drug
236.	*Picrorhiza kurrooa* (Scrophulariaceae)	Kutki, Kaor	Roots	Appetizer, bitter tonic, liver protactant
237.	*Pimpinella tenera* (Apiaceae)	—	Whole plant	Carminative, diuretic

Contd..

Table 3. Contd...

238.	*Plantago depressa* (Plantaginaceae)	Tha-ram	Whole plant	Anti-diarrhoeal, anthelmintic, stimulant
239.	*Plantago major* (Plantaginaceae)	Ripple grass	Whole plant	Anti-diarrhoeal, anthelmintic, stimulant
240.	*Podophyllum hexandrum* (Berberidaceae)	Mayapple, Ban kakri, Ol-mose	Roots & fruits	Anti-cancer, radioprotection & Gynaecological diseases
241.	*Polygonatum multiflorium* (Liliaceae)	Soloman's seal (E), Ramnge	Rhizome	Kidney problems & health tonic
242.	*Polygonatum verticillatum* (Liliaceae)	—	Rhizome	Kidney problems & health tonic
243.	*Potentilla ambigua* (Rosaceae)	—	Whole herb	Tonic
244.	*Potentilla anserina* (Rosaceae)	Silver weed, troma	Leaves	Astringent, diar rhoea, leucoderma
245.	*Potentilla atrosanguinea* (Rosaceae)	Chisheing	Whole plant	Fever
246.	*Potentilla fruticosa* (Rosaceae)	Cinquefoil, Khiangar	Leaves	Fever
247.	*Potentilla multifida* (Rosaceae)	Thakto	Seeds	Sleeplessness
248.	*Prangos pabularia* (Apiaceae)	Prangos, plans	Roots, seeds & leaves	Carminative, diuretic, stomach complaints, source of coumarin
249.	*Primula microphylla* (Primulaceae)	Prime rose	Root-stocks	Cough
250.	*Primula rosea* (Primulaceae)	Prime rose	Root-stocks	Muscular pains
251.	*Ranunculus brotherusii* (Ranunculaceae)	Maoking	Flowers	Digestive, ulcer
252.	*Ranunculus laetus* (Ranunculaceae)	—	Leaves	Stimulant

Contd..

Table 3. Contd...

253.	*Ranunculus lobatus* (Ranunculaceae)	—	Whole plant	Gum inflammation, toothache
254.	*Ranunculus trichophyllus* (Ranunculaceae)	Rengo	Whole plant	Diarrhoea
255.	*Ranunculus tricuspis* (Ranunculaceae)	Chu-rug-sbal-lag	Leaves, roots, flowers, fruits	Eye inflammation
256.	*Rheum moorcroftianum* (Polygonaceae)	Rhubarb, Revalchini	Roots	Laxative, purgative, source of anthraquinone compounds
257.	*Rheum spiciforme* (Polygonaceae)	Rhubarb, Revalchini	Roots	Laxative, purgative, source of anthraquinone compounds
258.	*Rheum webbianum* (Polygonaceae)	Rhubarb, Revalchini	Roots	Laxative, purgative, source of anthraquinone compounds
259.	*Rhodiola crenulata* (Crassulaceae)	Rose-root, golden arctic root, shrolo	Root-stocks	Anti-stress, restores memory, health tonic
260.	*Rhodiola heterodonta* (Crassulaceae)	Rose-root, golden arctic root, shrolo	Root-stocks	Anti-stress, restores memory, health tonic
261.	*Rhodiola imbricata* (Crassulaceae)	Rose-root, golden arctic root, shrolo	Root-stocks	Anti-stress, restores memory, health tonic
262.	*Rhodiola tibetica* (Crassulaceae)	Rose-root, golden arctic root, shrolo	Root-stocks	Anti-stress, restores memory, health tonic
263.	*Rhododendron anthopogon* (Ericaceae)	Rhododandron, Balu	Flowers, leaves	Gnorrhoea, stomach ailments
264.	*Rhododendron campanulatum* (Ericaceae)	Rhododandron, Balu	Flowers, leaves	Rheumatism
265.	*Rosa macrophylla* (Rosaceae)	Wild rose	Rose hips	Vitamin 'C'
266.	*Rosa webbiana* (Rosaceae)	Wild rose, Siah	Rose hips	Vitamin 'C', hepatitis
267.	*Rubia cordifolia* (Rubiaceae)	Maddar, Manjith, b-Tsbd	Roots & stem	Anthelmintic, blood purification, leucoderma

Contd..

Table 3. Contd...

268.	*Rubia tibetica* (Rubiaceae)	Maddar, Manjith, b-Tsod	Roots & stem	Anthelmintic, blood purification, leucoderma
269.	*Rumex acetosa* (Polygonaceae)	—	Leaves	Laxative
270.	*Rumex pateintia ssp tibeticus* (Polygonaceae)	Shoma	Whole herb	Purgative, vermifuge
271.	*Sagina saginoides* (Caryophyllaceae)	—	Whole herb	Tonic
272.	*Saussurea bracteata* (Asteraceae)	Pseudo Brahma Kamal, Span-rtsa-bo		Flower-bud Boils
273.	*Sassurea glaciales* (Asteraceae)	Pansipa	Flower-buds	Intestinal complaints
274.	*Saussurea gnaphaloides* (Asteraceae)	Yuliang	Whole plant	Kidney problems
275.	S *Saussurea jacea* (Asteraceae)	Pashuk	Aerial parts	Diuretic
276.	*Saussurea lappa* (Asteraceae)	Costus, Kuth, Rusta	Roots	Asthma, anthelmintic, carminative
277.	*Saussurea obvallata* (Asteraceae)	Brahma Kamal, Span-rtsa-bo	Flowers-buds	Cold & cough
278.	*Saxifraga cernua* (Saxifragaceae)	Bri-rta sa-dzin dman-pa	Whole plant	Tonic
279.	*Saxifraga jacquemontiana* (Saxifragaceae)	Sasomantso	Whole plant	Liver tonic
280.	*Saxifraga pulvinaria* (Saxifragaceae)	Sum-cu-tig	Whole plant	Liver tonic
281.	*Saxifraga stenophylla* (Saxifragaceae)	Man-pa	Whole plant	Blood purification
282.	*Scorzonera virgata* (Asteraceae)	Rtsa-mkhris	Whole plant	Intestinal problems

Contd..

Table 3. Contd...

283.	*Scrophularia dentata* (Scrophulariaceae)	Hamchi	Leaves	Appetizer
284.	*Scrophularia koelzii* (Scrophulariaceae)	Fogwart, Yarma	Seeds	Joints pain
285.	*Scutellaria prostata* (Lamiaceae)	Haunching	Whole plant	Jaundice
286.	*Sedum ewersii* (Crassulaceae)	Stone crop	Whole plant	Dysentry
287.	*Selinum elatum* (Apiaceae)	—	Roots	Tonic
288.	*Senecio chrysanthemoides* (Asteraceae)	rGu-drus	Leaves	Rheumatism
289.	*Senecio tibeticus* (Asteraceae)	Niyangar	Leaves	Diuretic
290.	*Silene vulgaris* (Caryophyllaceae)	Bladder campaion	Whole plant	Skin diseases
291.	*Solanum nigrum* (Solanaceae)	Makoi	Fruits & leaves	Appetizer, diuretic, laxative
292.	*Solidago virg-aurea* (Asteraceae)	Goldenrod	Leaves	Antiseptic
293.	*Sophora moorcroftiana* (Fabaceae)	Singtik	Seeds	Blood purifier
294.	*Statice macrorrhabdos* (Plumbaginaceae)	—	Whole plant	Stomach complaints
295.	*Swertia cordata* (Gentianaceae)	Wild Chirayta, Tikta	Leaves	Fever
296.	*Swertia petiolata* (Gentianaceae)	Wild Chirayta, Tikta	Leaves	Fever, tonic

Contd..

Table 3. Contd...

297.	*Swertia thomsonii* (Gentianaceae)	Wild Chirayta, Tikta	Leaves	Fever, febrifuge
298.	*Tanacetum dolichophyllum* (Asteraceae)	Khampa serpo	Leaves & flowers	Intestinal worms
299.	*Tanacetum fruticulosum* (Asteraceae)	Khamchu	Whole plant	Stomachache
300.	*Tanacetum gracile* (Asteraceae)	Khanchu	Leaves & flowers	Intestinal worms
301.	*Tanacetum nanum* (Asteraceae)	Khanchu	Leaves & flowers	Intestinal worms
302.	*Tanacetum tibeticum* (Asteraceae)	Khamchu	Leaves & flowers	Febrifuge
303.	*Taraxacum officinale* (Asteraceae)	Dandalion, Khur-mang	Roots	Analgesic, diuretic, liver disorders, tonic
304.	*Taraxacum sikkimensis* (Asteraceae)	Khur-mang-karpo	Roots	Febrifuge, analgesic
305.	*Thalictrum alpinum* (Ranunculaceae)	Alpine rue	Roots, leaves	Fever
306.	*Thalictrum foetidum* (Ranunculaceae)	Lesser meadow rue, sNgo-sprin	Roots & leaves	Diuretic & fever
307.	*Thermopsis barbata* (Fabaceae)	—	Whole plant	Pains
308.	*Thlaspi alpestre* (Brassicaceae)	Bre-ga	Whole herb	Digestive, fever
309.	*Thlaspi arvense* (Brassicaceae)	Breg-pa	Whole herb	Digestive, fever
310.	*Thymus serpyllum* (Lamiaceae)	Wild thyme	Whole herb	Carminative, carminative, expectorant, stimu lant

Contd..

Table 3. Contd...

311.	*Tragopogon dubius* (Asteraceae)	Goats bread, Tharnoo	Latex	Wound healing
312.	*Tragopogon gracillis* (Asteraceae)	Goats bread / Tharno	Latex	Wound healing
313.	*Tragopogon pretense* (Asteraceae)	Goats bread / Tharno	Flower-heads	Wound healing
314.	*Tribulus terrestris* (Zygophyllaceae)	Puncture vine, Gokhroo, gZe-ma	Fruits	Diuretic, sexual weakness, tonic
315.	*Ulmus wallichiana* (Ulmaceae)	Elm, Umbok	Bark	Wounds, hair-tonic
316.	*Urtica hyperborea* (Urticaceae)	Stinging hairs, Zazot	Roots	Rheumatism
317.	*Valeriana jatamansii* (Valerianaceae)	Jatamansi	Roots	Epilepsy, liver diseases
318.	*Verbascum thapsus* (Scrophulariaceae)	Flannel Mullein, Dandashal	Leaves & seeds	Asthma, frost-bite
319.	*Veronica koelzii* (Scrophulariaceae)	Bod-skyes ba-sha-ka	Leaves	Cold & cough
320.	*Veronica macrostemon* (Scrophulariaceae)	Shimiongar	Leaves	Skin diseases
321.	*Vincetoxicum hirundinaria* (Asclepiadaceae)	sNgo-dug-mo-nyunġ	Fruits	Dysentery
322.	*Viola biflora* (Violaceae)	Banafasha	Whole plant	Tonic
323.	*Viola kunawarensis* (Violaceae)	Banafasha	Whole plant	Tonic
324.	*Viscum album* (Loranthaceae)	Kahbung	Whole plant	Antiseptic, diuretic, purgative

Contd..

Table 3. Contd...

325.	*Waldheimia glabra* (Asteraceae)	Palu	Whole plant	Septic wounds
326.	*Waldheimia stoliczkai* (Asteraceae)	Palu	Whole plant	Septic wounds
327.	*Waldheimia tomentosa* (Asteraceae)	Palu	Whole plant	Septic wounds
328.	*Youngia glauca* (Asteraceae)	—	Whole herb	Health tonic
329.	*Youngia tenuifolia* (Asteraceae)	—	Whole herb	Health tonic

Table 4. List of Aromatic plants of Ladakh and Lahaul Spiti

S. No	BOTANICAL NAME WITH FAMILY	COMMON NAME	PARTS USED	REMARKS
1.	*Achillea millefolium* (Asteraceae)	Milfoil/ Yarrow	Whole plant	Source of azulenes, achillene
2.	*Aquilegia fragrans* (Ranunculaceae)	Fragrans columbine	Flowers	Having fragrant flowers
3.	*Artemisia annua* (Asteraceae)	Sweet wormwood	Whole plant	Source of artemesin
4.	*Artemisia biennis* (Asteraceae)	Margrass	Whole plant	Aromatic oil
5.	*Artemisia brevifolia* (Asteraceae)	Worm seed, Burtse	Whole plant	Source of santonin
6.	*Artemisia desertorum* (Asteraceae)	Khamlol	Whole plant	Aromatic oil
7.	*Artemisia dracunculus* (Asteraceae)	Tarragon, Sanche	Leaves	Tarragon oil for flavour, source of estragole
8.	*Artemisia gmelinii* (Asteraceae)	Khampa shirdi	Whole plant	Aromatic oil
9.	*Artemisia laciniata* (Asteraceae)		Whole plant	Aromatic oil
10.	*Artemisia nilagirica* (Asteraceae)	Chirmaro	Whole plant	Aromatic oil having antiseptic properties
11.	*Artemisia parviflora* (Asteraceae)	Khamang	Whole plant	Aromatic oil
12.	*Artemisia persica* (Asteraceae)	Burnak	Whole plant	Aromatic oil
13.	*Artemisia salsoloides* (Asteraceae)	Amango	Whole plant	Aromatic oil
14.	*Artemisia scoparia* (Asteraceae)	Khamptspo	Whole plant	Aromatic oil

Contd..

Table 4. Contd...

15.	*Artemisia sieversiana* (Asteraceae)	Khamchu	Whole plant	Aromatic oil
16.	*Artemisia stracheyi* (Asteraceae)	Rumalo	Whole plant	Aromatic oil
17.	*Artemisia tournefortiana* (Asteraceae)	Khamchu	Whole plant	Aromatic oil
18.	*Biebersteinia odora* (Geraniaceae)	Khardung / Drakspose	Leaves	Aromatic oil
19.	*Bunium persicum* (Apiaceae)	Kala Zeera	Seeds	Condiment, flavour
20.	*Carum carvi* var *gracile* (Apiaceae)	Caraway, Koniyot	Seeds	Condiment, flavour, source of carvone
21.	*Chenopodium botrys* (Chenopodiaceae	Sagani	Whole plant	Aromatic oil
22.	*Chrysanthemum pyrethroides* (Asteraceae)	A-byag-gzer-joms	Flowers	Incense
23.	*Dracocephalum heterophyllum* (Lamiaceae)	Zinkzer / Ziptsi	Whole plant	Having potential essential oil, source of citronellol
24.	*Dracocephalum moldavicum* (Lamiaceae)	Zinkzer	Whole plant	Aromatic oil
25.	*Dracocephalum stamineum* (Lamiaceae)	Zinkzer	Whole plant	Aromatic oil
26.	*Elaeagnus angustifolium* (Elaeagnaceae)	Oleaster, Sersing	Flowers	Fragrant flowers
27.	*Elsholtzia densa* (Lamiaceae)	Bye-rug-nagpo	Leaves	Flavouring agent
28.	*Elsholtzia eriostachya* (Lamiaceae)	Bye-rug-serpo	Leaves	Flavouring agent
29.	*Ferula jaeschkeana* (Apiaceae)	Chuklam	Whole plant	Fragrant roots

Contd..

Table 4. Contd...

30.	*Humulus lupulus* (Urticaceae)	Hops	Flowers, fruits	High value aromatic oil used in beer as flavour
31.	*Hyssopus officinalis* (Lamiaceae)	Hyssop, Juffa / Tengu	Whole plant	Fragrant oil, source of camphor
32.	*Inula racemosa* (Asteraceae)	Puskarmool, Manu	Roots	High value aromatic oil
33.	*Juniperus communis* (Juniperaceae)	Juniper, Sukpa	Leaves & bark	Incense, dhoop
34.	*Juniperus indica* (Juniperaceae)	Juniper, Sukpa	Leaves & bark	Incense, dhoop
35.	*Juniperus macropoda* (Juniperaceae)	Juniper, Sukpa	Leaves & bark	Incense, dhoop
36.	*Juniperus recurva* (Juniperaceae)	Juniper, Sukpa	Leaves & bark	Incense, dhoop
37.	*Mentha longifolia* (Lamiaceae)	Horse mint, Phuloling	Whole plant	Aromatic oil of medicinal properties, source of menthol
38.	*Nepeta clarkei* (Lamiaceae)	Khamyu	Whole plant	Aromatic oil
39.	*Nepeta coerulescens* (Lamiaceae)	Khora	Whole plant	Aromatic oil
40.	*Nepeta discolor* (Lamiaceae)	Shamalolo	Whole plant	Aromatic oil
41.	*Nepeta erecta* (Lamiaceae)	Eripantso	Whole plant	Aromatic oil
42.	*Nepeta eriostachya* (Lamiaceae)	Zim-thik-le	Whole plant	Aromatic oil
43.	*Nepeta floccosa* (Lamiaceae)	Shangukaram	Leaves & shoots	Flavouring agent
44.	*Nepeta glutinosa* (Lamiaceae)	Jatukpa	Whole plant	Aromatic oil

Contd..

Table 4. Contd...

45.	*Nepeta laevigata* (Lamiaceae)	Jatukpa	Whole plant	Aromatic oil
46.	*Nepeta leucolaena* (Lamiaceae)	—	Aerial parts	Aromatic oil
47.	*Nepeta longibracteata* (Lamiaceae)	Prainku	Whole palnt	Aromatic oil of medicinal prcperties
48.	*Nepeta podostachys* (Lamiaceae)	Shangukaram	Aerial parts	Fragrance
49.	*Origanum vulgare* (Lamiaceae)	Wild oregano, marjorum	Whole plant	Aromatic oil, source of thymol
50.	*Oxytropis macryphylla* (Fabaceae)	sTag-sha-marpo	Whole plant	Having strong fragrance
51.	*Perovskia abrotenoides* (Lamiaceae)	Wild lavender, Iskiling	Whole plant	Potential aromatic oil of lavender type
52.	*Prangos pabularia* (Apiaceae)	Prangos, Plans	Whole plant	Strong fragrant roots
53.	*Primula macrophylla* (Primulaceae)	Prime rose	Flowers	Fragrant flowers
54.	*Primula rosea* (Primulaceae)	Prime rose	Flowers	Fragrant flowers
55.	*Rhododendron anthopogon* spp. *hypenanthum* (Areceae)	Rhododrans, Balu	Leaves	Incense
56.	*Rosa macrophylla* (Rosaceae)	Wild rose	Flowers	Aromatic oil
57.	*Rosa webbiana* (Rosaceae)	Wild rose, Siah	Flowers	Fragrant flowers
58.	*Salvia sclarea* (Lamiaceae)	Clary sage	Flowers	Aromatic oil of high value source of linalyl acetate, linalool
59.	*Saussurea lappa* (Asteraceae)	Costus, Kuth	Roots	Aromatic oil of high value

Contd..

Table 4. Contd...

60.	*Stachys tibetica* (Lamiaceae)	Yakzask	Whole plant	Fragrant
61.	*Tanacetum dolichophyllum* (Asteraceae)	Khampa-serpo	Leaves & flowers	Aromatic oil
62.	*Tanacetum fruticulosum* (Asteraceae)	Khamchu	Whole plant	Aromatic oil
63.	*Tanacetum gracile* (Asteraceae)	Khamchu	Leaves & flowers	Aromatic oil
64.	*Tanacetum nanum* (Asteraceae)	Khamchu	Leaves & aerial parts	Aromatic oil
65.	*Tanacetum tibeticum* (Asteraceae)	Khamchu	Leaves & Flowers	Aromatic oil
66.	*Thymus serpyllum* (Lamiaceae)	Wild Thyme, Jangli Ajwain, Tumbark	Whole plant	Flavouring agent, source of thymol, carvacrol
67.	*Valeriana jatamansii* (Valerianaceae)	Jatamansi	Roots	Incense
68.	*Waldheimia glabra* (Asteraceae)	Palu	Whole plant	Incense, dhoop
69.	*Waldheimia stoliczkai* (Asteraceae)	Palu	Whole plant	Incense, dhoop
70.	*Waldheimia tomentosa* (Asteraceae)	Palu	Whole plant	Incense, dhoop

Table 5. List of Edible wild plants of Ladakh and Lahaul Spiti

S. No.	Botanical Name	Common Name	Parts used	Remarks
1.	*Adonis chrysocyanthus* (Ranunculaceae)	Semshi	Leaves	Vegetable
2.	*Allium carolianum* (Amaryllidaceae)	Koshuk, Skotche	Fresh or dry leaves	Flavouring agent
3.	*Allium humile* (Amaryllidaceae)	Jangli pyaz, Skotche	Leaves	Flavouring agent
4.	*Allium jacquemontii* (Amaryllidaceae)	Jangli pyaz	Leaves	Flavouring agent
5.	*Allium loratum* (Amaryllidaceae)	Jangli pyaz	Leaves	Flavouring agent
6.	*Allium oreoprasum* (Amaryllidaceae)	Skotche	Leaves	Flavouring agent
7.	*Allium przewalskianum* (Amaryllidaceae)	Wild garlic, skotche	Fresh or dry leaves	Condiment, vegetable
8.	*Allium ramosum* (Amaryllidaceae)	Skotche	Leaves	Condiment, flavouring agent
9.	*Allium schoenoprasum* (Amaryllidaceae)	Skotche	Leaves	Condiment
10.	*Allium semenovii* (Amaryllidaceae)	Wild garlic	Leaves	Flavouring agent
11.	*Allium stracheyi* (Amaryllidaceae)	Skotche	Leaves	Flavoruing agent
12.	*Allium thomsonii* (Amaryllidaceae)	Skothce	Leaves	Condiment, flavouring agent
13.	*Allium victorialis* (Amaryllidaceae)	Gokpa	Leaves, scapes	Vegetable
14.	*Amaranthus spinosus* (Amaranthaceae)	Amaranth, Chaulai	Fresh or dry leaves	Vegetable

Contd..

Table 5. Contd...

15.	*Anaphalis contorta* (Asteraceae)	Telgang	Young shoots	Vegetable	
16.	*Anaphalis triplenervis var intermedia* (Asteraceae)		Yaktso	Flower-buds	Salad
17.	*Arabidopsis himaliaca* (Brassicaceae)	Byi'ulaphung	Tender leaves	Vegetable	
18.	*Arabidopsis mollissima* (Brassicaceae)	Sikiang	Leaves	Vegetable	
19.	*Arabidopsis thaliana* (Brassicaceae)	Byi'ulaphung	Leaves	Vegetable	
20.	*Arabidopsis wallichii* (Brassicaceae)	Byi'ulaphung	Leaves & shoots	Pot herb	
21.	*Arabis alpine* (Brassicaceae)	—	Leaves & shoots	Vegetable	
22.	*Arabis tenuirostris* (Brassicaceae)	Chiki	Leaves	Vegetable	
23.	*Arabis tibetica* (Brassicaceae)	Tiktso	Leaves	Pot herb	
24.	*Arenaria holosteoides* (Caryophyllaceae)	Chiki (Berger)	Leaves	Vegetable	
25.	*Arenaria serpyllifolia* (Caryophyllaceae)	Rtswa-a-krong	Leaves	Vegetable	
26.	*Arnebia benthamii* (Boraginaceae)	Ratanjot, Demok	Roots	Edible red dye	
27.	*Arnebia euchroma* (Boraginaceae)	Ratanjot, Demok	Roots	Edible red dy	
28.	*Arnebia guttata* (Boraginaceae)	-do-	Roots	-do-	
29.	*Artemisia dracunculus* (Asteraceae)	Tarragan	Leaves	Flavouring agent	

Contd..

Table 5. Contd...

30.	*Atriplex crassifolia* (Chenopodiaceae)	Sniu	Leaves	Vegetable
31.	*Atriplex hortensis* (Chenopodiaceae)	Giangsar	Leaves	Vegetable
32.	*Barbarea vulgaris* (Brassicaceae)	Yellolw rocket	Tender shoots	Vegetable
33.	*Berberis jaeschkeana* (Berberidaceae)	Berbery	Fruits	Edible
34.	*Berberis ulicina* (Berberidaceae)	Khizer	Fruits	Edible
35.	*Bidens pilosa* (Asteraceae)	Gurgur chai	Leaves	Local tea
36.	*Bunium persicum* (Apiaceae)	Kala Zeera	Seeds, bulbs	Seeds as flavouring agent.
37.	*Capparis spinosa* (Capparidaceae)	Caper bush, Ber, Kabra	Leaves & fruits	Vegetable, edible
38.	*Capsella bursa pastoris* (Chenopodiaceae)	Shepherd's purse, Shamso	Whole herb	Vegetable
39.	*Capsella elliptica* (Chenopodiaceae)	Shamso	Leaves	Vegetable
40.	*Capsella thomsonii* (Chenopodiaceae)	Shamso	Leaves	Vegetable
41.	*Cardamine flexuaosa* (Brassicaceae)	—	Shoots, leaves	Vegetable
42.	*Carum carvi* (Apiaceae)	Local zeera, Kosnyot	Seeds	Flavouring agent.
43.	*Chaerophyllum acuminatum* (Apiaceae)	Chikmo	Whole plant, leaves	Pot-herb, Falvouring agent
44.	*Chaerophyllunm villosum* (Apiaceae)	Mesoso	Roots	Vegetable

Contd..

Table 5. Contd...

45.	*Chenopodium album* (Chenopodiaceae)	Wild spinach, Bathua, Em	Whole herb	Salad, vegetable
46.	*Chenopodium botrys* (Chenopodiaceae)	Sagani	Leaves	Vegetable
47.	*Chenopodium foliosum* (Chenopodiaceae)	Sangsti	Leaves	Vegetable
48.	*Chenopodium glaucum* (Chenopodiaceae)	Kuplad	Shoots	Vegetable
49.	*Chorispora sabulosa* (Brassicaceae)	Imbuk	Leaves & shoots	Vegetable
50.	*Christolea crassifolia* (Brassicaceae)	Sanak	Shoots	Vegetable
51.	*Cichorium intybus* (Asteraceae)	Cichory / Kasni	Whole herb	Edible
52.	*Cicer microphyllum* (Fabaceae)	Wild gram, seri	Seeds	Edible, vegetable
53.	*Codonopsis clematidea* (Campanulaceae)	Brukutang	Roots	Vegetable
54.	*Codonopsis ovata* (Campanulaceae)	Ludut	Roots	Vegetable
55.	*Crambe kotschyana* (Brassicaceae)	Brnma	Leaves	Pot-herb
56.	*Draba alpina* (Brassicaceae)	Piang	Whole herb	Pot-herb
57.	*Draba cachemirica* (Brassicaceae)	Mukungla	Whole herb	Pot-herb
58.	*Draba stenocarpa* (Brassicaceae)	Lepcha	Leaves	Vegetable
59.	*Draba tibetica* (Brassicaceae)	Mukungla	Shoots	Vegetable

Contd..

Table 5. Contd...

60.	*Elsholtzia densa* (Lamiaceae)	Betso	Leaves	Flavouring agent
61.	*Elsholtzia eriostachya* (Lamiaceae)	Betso	leaves	As a flavour, condiment
62.	*Ephedra gerardiana* (Ephedraceae)	Chhapat	Fruits	Eaten raw
63.	*Eremurus himalaicus* (Liliaceae)	Desert candle, Nangtso	Flower-buds	Edible
64.	*Erigeron multiradiatus* (Asteraceae)	During-phang	Leaves	Pot-herb
65.	*Ermania himalayensis* (Brassicaceae)	Fiam	Leaves	Vegetable
66.	*Ermania stewartii* (Brassicaceae)	Tschalu	Leaves	Vegetable
67.	*Erodium stephanianum* (Geraniaceae)	Tikziang	Leaves	Vegetable
68.	*Erodium tibetanum* (Geraniaceae)	Tikziang	Leaves	Vegetable
69.	*Eruca stativa* (Brassicaceae)	—	Whole herb	Vegetable
70.	*Erysimum hieraciifolium* (Brassicaeae)	Gong-thog-pa	Leaves	Vegetable
71.	*Fagopyrum esculentum* (Polygonaceae)	Buckwheat, Dayat	Leaves, seeds	Vegetable, flour
72.	*Fagopyrum tataricum* (Polygonaceae)	Duckwheat, Dayat	Leaves, seeds	Vegetavle, flour
73.	*Fragaria vesca* (Rosaceae)	Gooseberry	Fruits	Edible
74.	*Gentiana tianschanica* (Gentianaceae)	Wanglo	Whole herb	Vegetable

Contd..

Table 5. Contd...

75.	*Hippophae rhamnoides* ssp. *turkestanica* (Elaeagnaceae)	Seabuckthorn, Shermang	Fruits	Fruits : source of multivitamins
76.	*Hippophae salicifolia* (Elaeagnaceae)	Seabuckthorn	Fruits	Fruits : source of multivitamins
77.	*Hippophae tibetana* (Elaeagnaceae)	Seabuckthorn	Fruits	Fruits : source of multivitamins
78.	*Juglans regia* (Jugalanceae)	Walnut, sTarga	Fruits	Edible as a dry fruits
79.	*Lactuca decipiens* (Asteraceae)	Khala	Leaves	Vegetable
80.	*Lactuca dolichophylla* (Asteraceae)	Khala	Leaves	Vegetable
81.	*Lactuca lahulensis* (Asteraceae)	Khala	Leaves	Vegetable
82.	*Lactuca lessertiana* (Asteraceae)	Chumaseo	Leaves	Vegetable
83.	*Lactuca orientalis* (Asteraceae)	Khala	Leaves	Vegetable
84.	*Lactuca scariola* (Asteraceae)	Khala	Leaves	Vegetable
85.	*Lactuca tatarica* (Asteraceae)	Khalask	Leaves	Vegetable
86.	*Lamium amplexicaule* (Lamiaceae)	Supila	Whole plant	Vegetable
87.	*Lathyrus sativus* (Fabaceae)	Local pea, sanma	Seeds	Vegetable
88.	*Lavetera kashmiriana* (Malvaceae)	Wild Hollyhock, Sazamool, Sazposh		Shoots Vegetable
89.	*Lepidium capitatum* (Brassicaceae)	Sangtso	Leaves	Vegetable

Contd..

Table 5. Contd...

90.	*Lepidium latifolium* (Brassicaceae)	Shangsho	Leaves, shoots	Vegetable
91.	*Lepidium ruderale* (Brassicaceae)	Khrag-khrog	Leaves, shoots	Vegetable
92.	*Malva verticillata* (Malvaceae)	Chinese Mallow, Sucheli	Leaves, shoots	Vegetable.
93.	*Mentha longifolia* (Lamiaceae)	Horse mint, Phololing	Leaves	Flavouring agent
94.	*Microula tibetica* (Asteraceae)	—	Leaves	Vegetable
95.	*Morus alba* (Moraceae)	Mulberry	Fruits	Edible
96.	*Nepeta floccosa* (Lamiaceae)	Shangukaram	Leaves	Flavouring agent
97.	*Nepeta longibracteata* (Lamiaceae)	Shangukaram	Leaves	Flavouring agent
98.	*Orobanche hansii* (Orobanchaceae)	Gro-shang-rtse	Tender shoots	For making soup
99.	*Oxyria digyna* (Polygonaceae)	Mountain sorrel, Lamanchu	Leaves	Vegetable
100.	*Pedicularis bicornuta* (Scrophulariaceae)	Piasing	Leaves	Pot-herb
101.	*Pedicularis mollis* (Scrophulariaceae)	Lungrumarpo	Leaves	Edible
102.	*Pedicularis punctata* (Scrophulariaceae)	Memeang	Leaves	Pot-herb
103.	*Pedicularis pyramidata* (Scrophulariaceae)	Shakku	Leaves	Vegetable
104.	*Perovskia abrotanoides* (Lamiaceae)	Iskiling	Leaves	Flavouring agent

Contd..

Table 5. Contd...

105.	*Picris hieracioides* (Asteraceae)	—	Whole plant	Pot-herb
106.	*Podophyllum hexandrum* (Podophyllaceae)	Demobkusu	Fruits	Edible
107.	*Polygonum affine* (Polygonaceae)	Samtsong	Leaves	Vegetable
108.	*Polygonum aviculare* (Polygonaceae)	Chiankat	Leaves	Pot-herb
109.	*Polygonum sibiricum* (Polygonaceae)	—	Leaves	Salad
110.	*Polygonum viviparum* (Polygonaceae)	—	Leaves	Salad
111.	*Potentilla anserina* (Rosaceae)	Silver weed, troma	Leaves, roots	Edible
112.	*Potentilla atrosanguinea* (Rosaceae)	Skialdeepo	Leaves	Edible
113.	*Prunus armeniaca* (Rosaceae)	Apricot, Khubani, Chuli	Fruits, Kernel	Fresh fruits & kernel edible
114.	*Psychrogeton andryaloides* (Asteraceae)	Luchchugba	Leaves	Vegetable
115.	*Rheum spiciforme* (Polygonaceae)	Rhubarb, Lachu	Leaf-stalks	Edible
116.	*Rheum tibeticum* (Polygonaceae)	Rhubarb, Lachu	Leaf-stalks	Edible
117.	*Rheum webbianum* (Polygonaceae)	Rhubarb, Lachu	Leaf-stalks	Edible
118.	*Rhodiola heteodonta* (Crassulaceae)	Rose root, Shrolo	Leaves & tender shoots	Vegetable
119.	*Rhodiola imbricata* (Crassulaceae)	Rose root, Shrolo	Leaves & tender shoots	Vegetable

Contd..

Table 5. Contd...

120.	*Rhodiola tibetica* (Crassulaceae)	Rose root, Shrolo	Leaves & tender shoots	Vegetable
121.	*Ribes alpestre* (Rubiaceae)	Asian gooseberry	Fruits	Edible
122.	*Ribes orientale* (Rubiaceae)	Asian gooseberry	Fruits	Edible
123.	*Rosa webbiana* (Rosaceae)	Wild rose, Siah	Fruits	Edible
124.	*Rumex acetosa* (Polygonaceae)	Shoma	Leaves	Vegetable
125.	*Rumex angulatus* (Polygonaceae)	Shoma	Leaves	Vegetable
126.	*Rumex hastatus* (Polygonaceae)	Gungantse	Shoots	Vegetable
127.	*Rumex nepalensis* (Polygonaceae)	Palak	Leaves	Vegetable
128.	*Rumex patientia* (Polygonaceiaae)	Shoma	Leaves	Vegetable
129.	*Sedum ewersii* (Crassulaceae)	Stone crop, Gomni	Shoots	Sauce, salad
130.	*Silene conoidea* (Caryophyllaceae)	—	Shoots	Edible
131.	*Silene setisperma* (Caryophyllaceae)	—	Shoots	Edible
132.	*Solanum nigrum* (Solanaceae)	Makoi, Tsigma	Fruits	Edible
133.	*Sonchus oleraceus* (Asteraceae)	Milk thistle, Khala	Leaves, shoots	Vegetable
134.	*Stellaria palustris* (Caryophyllaceae)	Khalta	Shoots	Pot-herb

Contd..

Table 5. Contd...

135.	*Taraxacum officinale* (Asteraceae)	Dandelion, Han	Fresh or dry leaves	Vegetable
136.	*Taraxacum sikkimensis* (Asteraceae)	Han	Leaves	Vegetable
137.	*Thlaspi alpestre* (Lamiaceae)	Bre-ga	Leaves	Vegetable
138.	*Thlaspi arvense* (Lamiaceae)	Bre-ga	Leaves	Vegetable
139.	*Thymus serpyllum* (Lamiaceae)	Wild thyme, Tumbrak	Leaves & flowers	Flavouring agent
140.	*Tragopogon pratense* (Asteraceae)	Kujgarigali	Shoots, flower-heads	Edible
141.	*Tulipa stellata* (Liliaceae)	Wild tulip	Bulbs	Edible
142.	*Tribulus terrestris* (Zygophyllaceae)	Gokharoo, Zema	Seeds	Cooking oil
143.	*Urtica dioca* (Urticaceae)	Stinging neetle, Bichhu buti	Dried leaves	Vegetable
144.	*Urtica hyperborea* (Urticaceae)	Stinging neetle, Dzatsutt	Dried leaves	Vegetable

Table 5(a). Nutritional profile of edible wild plants

Sr No.	Botanical Name	Edible parts	Moisture	Ash	Fat	Starch	Protein Crude	Fibre Crude	Min eralsK	(mg/ Fe	P	100) Na
1.	*Arneria holosteoides holosyteoides*	Aerial	4.6	14.00	3.62	32.52	6.75	20.20	3280	133	400	1280
2.	*Capparis Spinosa*	Leaf	9.2	19.5	4.47	16.7	17.87	6.83	2200	—	—	85
3.	*Chenopodium botrys*	Leaf	7.1	—	4.2	25.7	26.4	7.1	—	109	525	—
4.	*Cicer microphyllum*	Seed	9.6	3.40	7.20	45.35	19.30	5.25	1640	266	250	840
5.	*Fagopyrum esculentum*	Seed	5.7	—	3.3	36.9	25.1	7.7	—	111	574	—
6.	*Fagopyrum tartaricum*	Leaf	3.22	2.0	0.06	52.50	20.12	6.8	320	25	150	—
7.	*Lactuca dolichophylla*	Leaf	10.0	—	5.3	33.3	27.1	6.8	—	29	390	—
8.	*Lactuca tartarica*	Leaf	8.0	19.6	9.6	13.7	10.3	13.6	3000	7	125	—
9.	*Lepidium latifolium*	Leaf	8.26	15.51	4.15	11.88	24.62	11.2	—	290	550	31
10.	*Oxyria digyna*	Leaf	7.9	—	3.4	29.8	18.3	10.8	—	120	274	—
11.	*Rheum webbianum*	Leaf	8.7	16.8	2.24	—	12.80	14.95	4100	120	—	280
12.	*Rumex patientia*	Leaf	7.4	—	6.4	35.1	29.4	9.4	—	26	509	—
13.	*Tribulus terrestris*	Leaf	7.2	25.1	3.1	15.7	19.0	7.0	1900	5.3	120	—
14.	*Urtica hyperborea*	Leaf	10.4	—	3.8	30.2	28.3	7.8	—	81	107	—

Table 6. List of fodder plants of Ladakh and Lahaul Spiti

S. No	BOTANICAL NAME WITH FAMILY	COMMON NAME	PARTS USED	REMARKS
1.	*Aconogonum alpinum* (Polygonaceae)	Serpalolo, Nyalo	Whole plant	Green / Dry
2.	*Aconogonum tortuosum* (Polygonaceae)	Serpalolo, Nyalo	Whole plant	Green / Dry
3.	*Acroglochin persicarioides* (Chenopodiaceae)	Thimangchu	Whole plant	Green
4.	*Agropyron jacquemontii* (Poaceae)	Zamak	Whole grass	Green / Dry
5.	*Agropyron repens* (Poaceae)	Zamak	Whole grass	Green / Dry
6.	*Ajuga bracteosa* (Lamiaceae)	—	Whole plant	Green
7.	*Alopecurus himalaicus* (Poaceae)	Zian	Whole plant	Green / Dry
8.	*Androsace primuloides* (Primulaceae)	Pemug	Whole plant	Green fodder
9.	*Arctium lappa* (Asteraceae)	Burdock, Pezums	Whole plant	Green fodder
10.	*Astragalus adesmiaefolius* (Fabaceae)	—	Leaves	Green fodder
11.	*Astragalus cicerifolius* (Fabaceae)	Yakpo	Leaves	Green fodder
12.	*Astragalus ciliolotus* (Fabaceae)	—	Leaves	Green fodder
13.	*Astragalus confertus* (Fabaceae)	Singyalo	Whole plant	Green fodder
14.	*Astragalus frigidus* (Fabaceae)	Skarmalo	Branches	Green fodder

Contd..

Table 6. Contd...

15.	*Astragalus gracilipes* (Fabaceae)	Flangu	Whole plant	Green fodder
16.	*Astragalus heydei* (Fabaceae)	Yanglong	Whole plant	Green fodder
17.	*Astragalus himalayanus* (Fabaceae)	—	Leaves	Green fodder
18.	*Astragalus munroi* (Fabaceae)	Serma	Whole plant	Green fodder
19.	*Astragalus multiceps* (Fabaceae)	—	Whole plant	Green fodder
20.	*Astragalus oxydon* (Fabaceae)	Sharma	Whole plant	Green fodder
21.	*Astragalus peduncularis* (Fabaceae)	Gyalo	Whole plant	Green fodder
22.	*Astragalus rhizanthus* (Fabaceae)	Rungentso	Leaves	Green fodder
23.	*Astragalus subuliformis* (Fabaceae)	Rungentso	Leaves	Green fodder
24.	*Astragalus webbianus* (Fabaceae)	Khangsker	Leaves	Green fodder
25.	*Astragalus zanskarensis* (Fabaceae)	Nanchu	Leaves	Green fodder
26.	*Avena fatua* (Poaceae)	Wild oat, Jangali jai, Yupo	Whole plant	Green / Dry
27.	*Avena sativa* (Poaceae)	Oat, Jai, Yupo	Whole plant	Green / Dry
28.	*Barbarea vulgaris* (Boraginaceae)	Yellow rocket	Whole plant	Fodder
29.	*Bothriochloa ischaemum* (Poaceae)	Jamak	Whole plant	Green / Dry

Contd..

Table 6. Contd...

30.	*Bothriochloa pertusa* (Poaceae)	Jamak	Whole plant	Green / Dry
31.	*Brachyactis pubescens* (Asteraceae)	—	Whole plant	Green / Dry
32.	*Bromus gracillimus* (Poaceae)	Piantso	Whole plant	Green / Dry
33.	*Bromus inermis* (Poaceae)	Piantso	Whole plant	Green / Dry
34.	*Bromus japonicus* (Poaceae)	Piantso	Whole plant	Green / Dry
35.	*Bromus oxydon* (Poaceae)	Piantso	Whole plant	Green / Dry
36.	*Bromus scoparius* (Poaceae)	Piantso	Whole plant	Green / Dry
37.	*Bromus tectorum* (Poaceae)	Zamak	Whole plant	Green / Dry
38.	*Bupleurum falcatum* (Apiaceae)	Sickle leaved Hae's ear, Zeerakarpo	Whole plant	Green fodder
39.	*Bupleurum longicaule* (Apiaceae)	Zeerakarpo	Whole plant	Green fodder
40.	*Bupleurum thomsonii* (Apiaceae)	Zeerakarpo	Whole plant	Green fodder
41.	*Calamogrostis emodensis* (Poaceae)	Yanglo	Whole plant	Green fodder
42.	*Calamogrostis garhwalensis* (Poaceae)	Yanglo	Whole plant	Green fodder
43.	*Calamogrostis stolizkai* (Poaceae)	Singshi	Whole plant	Green / Dry
44.	*Carex infuscata* (Cyperaceae)	—	Whole plant	Green fodder

Contd..

Table 6. Contd...

45.	*Carex obscura* (Cyperaceae)	—	Whole plant	Green fodder
46.	*Chaerophyllum acuminatum* (Chenopodiaceae)	Chikmo	Whole plant	Green fodder
47.	*Chesneya cuneata* (Fabaceae)	Leemopa	Whole plant	Green fodder
48.	*Chloris virgata* (Poaceae)	—	Whole plant	Green fodder
49.	*Cicer microphyllum* (Fabaceae)	Wild gram, Seri	Whole plant	Green/ Dry
50.	*Clematis orientalis* (Ranunculaceae)	Clematis, Tiktikma	Whole plant	Green fodder
51.	*Clematis tibetana* (Ranunculaceae)	Clematis	Whole plant	Green fodder
52.	*Convolvulus arvensis* (Convolvulaceae)	Tiktikma	Whole plant	Green fodder
53.	*Crepis stolickzai* (Asteraceae)	Richkut	Whole plant	Green fodder
54.	*Crepis tunifolia* (Asteraceae)	Remang	Whole plant	Green fodder
55.	*Cymbopogon jwarancusa* (Poaceae)	—	Whole plant	Green / Dry
56.	*Cynoglossum petiolatum* (Boraginaceae)	Hound's tounge, Ematso	Whole plant	Green fodder
57.	*Cynoglossum wallichii* (Boraginaceae)	—	Whole plant	Green fodder
58.	*Dactylus glomerata* (Poaceae)	Tahaung	Whole plant	Green / Dry
59.	*Deschampsia caespitosa* (Poaceae)	—	Whole plant	Green / Dry

Contd..

Table 6. Contd...

60.	*Deschampsia koelerioides* (Poaceae)	—	Whole plant	Green / Dry
61.	*Deyeuxia holciformis* (Poaceae)	—	Whole plant	Green / Dry
62.	*Digitaria ischaemum* (Poaceae)	—	Whole plant	Green fodder
63.	*Dilophia salsa* (Brassicaceae)	—	Whole plant	Green fodder
64.	*Echinochloa colonun* (Poaceae)	Yesoshin	Whole plant	Green / Dry
65.	*Echinospermum redowskii* (Boraginaceae)	Bukma	Whole plant	Greeen fodder
66.	*Ephedra gerardiana* (Ephedraceae)	Ephedra, Chhapat	Whole plant	In dry form
67.	*Eragrostis persica* (Poaceae)	—	Whole plant	Green / Dry
68.	*Eragrostis pilosa* (Poaceae)	—	Whole plant	Green / Dry
69.	*Eragrostis poaeoides* (Poaceae)	—	Whole plant	Green / Dry
70.	*Eremopoa persica* (Poaceae)	—	Whole plant	Green / Dry
71.	*Fagopyrum esculentum* (Polygonaceae)	Buckwheat, Dayat	Whole plant	Fodder
72.	*Ferula jaeschkeana* (Apiaceae)	Jangli Heeng	Whole plant	Dry form
73.	*Festuca altica* (Poaceae)	Kongola	Whole plant	Fodder grass
74.	*Festuca cummunsiii* (Poaceae)	—	Whole plant	Fodder grass

Contd..

Table 6. Contd...

75.	*Festuca kashmiriana* (Poaceae)	—	Whole plant	Fodder grass
76.	*Festuca lucida* (Poaceae)	Kongola	Whole plant	Fodder grass
77.	*Festuca nitidula* (Poaceae)	—	Whole plant	Fodder grass
78.	*Festuca ovina* (Poaceae)	Sheep's fescue, Niupang	Whole plant	Fodder grass
79.	*Filago arvensis* (Asteraceae)	—	Whole plant	Green fodder
80.	*Galinsoga parviflora* (Asteraceae)	—	Whole plant	Green fodder
81.	*Geranium meeboldii* (Geraniaceae)	—	Whole plant	Fodder
82.	*Hackelia uncinata* (Boraginaceae)	—	Whole plant	Fodder
83.	*Halocharis violacea* (Chenopodiaceae)	Sheing	Whole plant	Fodder
84.	*Halogeton glomeratus* (Chenopodiaceae)	Phyang	Whole plant	Fodder
85.	*Haloxylon thomsonii* (Chenopodiaceae)	—	Whole plant	Fodder
86.	*Heracleum candicans* (Apiaceae)	sPru-dkar	Whole plant	Green & dry form
87.	*Heracleum pinnatum* (Apiaceae)	Spru-ma	Whole plant	Green & dry form
88.	*Heracleum thomsonii* (Apiaceae)	—	Whole plant	Green & dry form
89.	*Hippophae rhamnoides* ssp. *turkestanica* (Elaeagnaceae)	Seabuckthorn, Tsermang	Leaves & twigs	Green & dry form

Contd..

Table 6. Contd...

90.	*Hippophae salicifolia* (Elaeagnaceae)	Seabuckthorn, Tsermang	Leaves & twigs	Green & dry form
91.	*Hippophae tibetana* (Elaeagnaceae)	Seabuckthorn, Tsermang	Leaves & twigs	Green & dry form
92.	*Indigofera heterantha* (Fabaceae)	Kieche	Branches	Green fodder
93.	*Iris kumaonensis* (Iridaceae)	Wild iris, Tesmamentok	Whole plant	Green & dry form
94.	*Iris lactea* (Iridaceae)	Wild iris, Tesmamentok	Whole plant	Green & dry form
95.	*Isopyrum grandiflorum* (Ranunculaceae)	—	Whole plant	Fodder
96.	*Junchus himalayensis* (Juncaceae)	Gimnalo	Whole plant	Fodder grass
97.	*Junchus membraneceous* (Juncaceae)	Gimnalo	Whole plant	Fodder grass
98.	*Junchus sphacelatus* (Juncaceae)	Gimnalo	Whole plant	Fodder grass
99.	*Junchus triglomis* (Juncaceae)	Gimnalo	Whole plant	Fodder grass
100.	*Kochia idica* (Chenopodiaceae)	—	Whole plant	Green & dry form
101.	*Kochia prostrata* (Chenopodiaceae)	—	Whole plant	Green & dry form
102.	*Kochia stellaris* (Chenopodiaceae)	—	Whole plant	Green & dry form
103.	*Koeleria argentea* (Poaceae)	—	Whole plant	Fodder grass
104.	*Krascheninnikovia ceratoides* (Chenopodiaceae)	—	Whole plant	Green & dry form

Contd..

Table 6. Contd...

105.	*Lathyrus humilis* (Fabaceae)	Kaown	Whole plant	Green & dry form
106.	*Lathyrus sativus* (Fabaceae)	Sarma	Whole plant	Green & dry form
107.	*Lepidium latifolium* (Brassicaceae)	Dittander, sangso	Whole plant	Green & dry form
108.	*Ligusticum thomsonii* (Apeaceae)	—	Whole plant	Fodder
109.	*Lindelofia anchusoides* (Boraginaceae)	Makpen	Whole plant	Green & dry form
110.	*Lindelofia longiflora* (Boraginaceae)	Makpen	Whole plant	Green & dry form
111.	*Lindelofia stylosa* (Boraginaceae)	Makpen	Whole plant	Green & dry form
112.	*Lonicera asperifolia* (Caprifoliaceae)	Honey suckle, Piralu	Leaves	Fodder
113.	*Lonicera heterophylla* (Caprifoliaceae)	Honey suckle, Piralu	Leaves	Fodder
114.	*Lonicera rupicola* (Caprifoliaceae)	Lamora	Leaves	Fodder
115.	*Lotus corniculatus* (Fabaceae)	Birds foot tree foil	Whole plant	Green & dry form
116.	*Mathiola flavida* (Caprifoliaceae)	—	Leaves	Fodder
117.	*Medicago falcata* (Fabaceae)	Alfalfa, Lucerne, Ole	Whole plant	Green & dry form & Cultivated fodder
118.	*Medicago lupina* (Fabaceae)	Alfalfa, Lucerne, Ole	Whole plant	Green & dry form
119.	*Medicago media* (Fabaceae)	Alfalfa, Lucerne, Ole	Whole plant	Green & dry form & cultivated fodder

Contd..

Table 6. Contd...

120.	*Medicago sativa* (Fabaceae)	Alfalfa, Lucerne, Ole	Whole plant	Green & dry form & Cultivated fodder
121.	*Melilotus alba* (Fabaceae)	White sweet clover, Ole	Whole plant	Green & dry form & Cultivated fodder
122.	*Melilotus officinalis* (Fabaceae)	Yellow sweet clover, Ole	Whole plant	Green & dry form & Cultivated fodder
123.	*Myricaria squamosa* (Tamaricaceae)	—	Branches	Fodder
124.	*Oryzopsis brachyclada* (Poaceae)	Pianiutan	Whole plant	Fodder grass
125.	*Oryzopsis gracillis* (Poaceae)	Fulungkan	Whole plant	Fodder grass
126.	*Oxytropis cachemiriana* (Fabaceae)	Yamchu	Whole plant	Fodder
127.	*Oxytropis lapponica* (Fabaceae)	Rushkum	Whole plant	Fodder
128.	*Oxytropis tatarica* (Fabaceae)	Zingula	Whole plant	Fodder
129.	*Parrya exscapa* (Brassicaceae)	Sumu	Whole plant	Fodder
130.	*Phragmitis australis* (Poaceae)	Piachnig	Whole plant	Fodder grass
131.	*Phragmitis karka* (Poaceae)	Piachnig	Whole plant	Fodder grass
132.	*Pleurospermum candollii* (Apiaceae)	rTsad-rgod	Whole plant	Fodder
133.	*Pleurospermum stellatum* (Apiaceae)	rTsad	Whole plant	Fodder
134.	*Poa alpina* (Poaceae)	Humulu	Whole plant	Green & dry form

Contd..

Table 6. Contd...

135.	*Poa annua* (Poaceae)	Wukung	Whole plant	Green & dry form
136.	*Poa bulbosa* (Poaceae)	Wukung	Whole plant	Green & dry form
137.	*Poa falconeri* (Poaceae)	Zamak	Whole plant	Green & dry form
138.	*Poa polycolea* (Poaceae)	Zamak	Whole plant	Green & dry form
139.	*Poa Stapfiana* (Poaceae)	—	Whole plant	Green & dry form
140.	*Poa stwartiana* (Poaceae)	—	Whole plant	Green & dry form
141.	*Polygonum cognatum* (Polygonaceae)	—	Whole plant	Fodder
142.	*Polygonum glaciale* (Polygonaceae)	—	Whole plant	Fodder
143.	*Polygonum nepalense* (Polygonaceae)	—	Whole plant	Fodder
144.	*Polygonum paronychioides* (Polygonaceae)	Sewagawan	Whole plant	Fodder
145.	*Polygonum vacciniifolium* (Polygonaceae)	—	Whole plant	Fodder
146.	*Populus ciliata* (Salicaceae)	Himalayan papal, Yarpa	Leaves	Fodder
147.	*Populus euphratica* (Salicaceae)	Poplar, Yulat	Leaves	Fodder
148.	*Populus nigra* (Salicaceae)	Poplar, Yulat	Leaves	Fodder
149.	*Potentilla bifurca* (Rosaceae)	—	Whole plant	Fodder

Contd..

Table 6. Contd...

150.	*Potentilla microphyllum* (Rosaceae)	Thakto	Whole plant	Fodder
151.	*Potentilla sericea* (Rosaceae)	Pemma	Whole plant	Fodder
152.	*Prangos pabularia* (Apiaceae)	Plans	Whoel palnt	Dry fodder, a major and nutritious wild fodder
153.	*Pseudomertensia echioides* (Boraginaceae)	—	Whoel palnt	Fodder
154.	*Ribes alpestre* (Saxifragaceae)	Asian gooseberry, Zasot	Branches	Fodder
155.	*Ribes orientale* (Saxifragaceae)	Asian gooseberry, Zasot	Branches	Fodder
156.	*Robinia pseudacacia* (Fabaceae)	Pseudo-acacia	Leaves	Fodder
157.	*Rosa webbiana* (Rosaceae)	Wild rose, Siah	Leaves	Fodder
158.	*Salix alba* (Salicaceae)	Willow, Mulchang	Leaves	Fodder
159.	*Salix daphnoides* (Salicaceae)	Willow, Mulchang	Leaves	Fodder
160.	*Setaria glauca* (Poaceae)	Yangjuk	Whole plant	Fodder
161.	*Setaria viridis* (Poaceae)	Cheehe	Whole plant	Fodder
162.	*Sisymbrium brassiciforme* (Brassicaceae)	Pelaung	Whole plant	Green fodder
163.	*Sisymbrium loeselii* (Brassicaceae)	—	Whole plant	Green fodder
164.	*Sisymbrium sophia* (Brassicaceae)	—	Whole plant	Green fodder

Contd..

Table 6. Contd...

165.	*Stachys tibetica* (Lamiaceae)	Yakjak	Whole plant	Green fodder
166.	*Stellaria media* (Caryophyllaceae)	Chichweed, Khalta	Whole plant	Green fodder
167.	*Stipa orientalis* (Poaceae)	Pilli	Whole plant	Green & dry form
168.	*Stipa sibirica* (Poaceae)	Pilli	Whole plant	Green & dry form
169.	*Tamarix gallica* (Tamaricaceae)	Agrelta	Branches	Fodder
170.	*Trifolium pratense* (Fabaceae)	Purple or red clover	Whole plant	Green fodder
171.	*Trifolium repens* (Fabaceae)	White clover	Whole plant	Green fodder
172.	*Triglochin palustris* (Juncaginaceae)	Na-ram	Whole plant	Green fodder
173.	*Trigonella corniculata* (Fabaceae)	Watpo	Whole plant	Legume fodder
174.	*Trigonella emodi* (Fabaceae)	Watpo	Whole plant	Legume fodder
175.	*Tripogon filiformis* (Asteraceae)	—	Whole plant	Fodder
176.	*Ulmus wallichiana* (Ulmaceae)	Elm, Umbuk	Branches	Green fodder

Table 6a. Nutritional Profile of Potential Wild Fodder Plants

Sr No.	Bot. Name	Part Used	Moist.	DM %	CP %	NDF %	ADF %	L %	HC %	C %
1.	*Aconogonum tortuosum*	Whole plant	-	88.87	8.63	51.04	38.22	12.63	12.82	25.59
2.	*Avena fatua*	Whole plant	09.09	90.91	6.79	38.35	28.96	04.60	09.39	24.34
3.	*Capparis spinosa*	Leaves	09.20	90.80	17.87	06.83	-	-	-	-
4.	*Chenopodium botrys*	Leaves	07.10	92.90	26.40	07.10	-	-	-	-
5.	*Cicer microphyllum*	Whole plant	08.60	91.40	15.30	36.38	32.34	05.90	04.04	26.44
6.	*Ephedra gerardiana*	Leaves & Fruits	08.19	91.81	09.27	45.23	42.48	21.28	02.75	21.20
7.	*Fagopyrum esculentum*	Whole plant	05.70	94.30	25.19	07.70	37.28	7.15	13.26	30.13
8.	*Fagopyrum tartaricum*	Whole plant	03.22	96.78	20.12	06.80	-	-	-	-
9.	*Hippophae rhamnoides*	Leaves	08.99	91.09	23.93	22.72	20.21	08.47	02.51	11.74
10.	*Iris lactea*	Leaves	09.63	90.37	10.62	47.28	42.35	04.01	-	
11.	*Lepidium latifolium*	Whole plant	08.26	91.74	24.62	11.20	-	-	-	-
12.	*Medicago falcata*	Whole plant	-	-	15.85	20.86	-	-	-	25.80
13.	*Medicago media*	Whole plant	-	-	17.50	20.58	-	-	-	25.70
14.	*Medicago sativa*	Whole plant	-	-	18.22	20.65	-	-	-	25.90
15.	*Melilotus officinalis*	Whole plant	10.05	89.95	24.01	20.94	18.53	04.71	02.41	13.82
16.	*Prangos pabularia*	Whole plant	10.57	89.43	12.66	35.78	32.69	05.12	03.09	27.57
17.	*Rheum webbianum*	Leaves	08.70	91.30	12.80	14.95	-	-	-	-
18.	*Rumex spp*	Whole plant		88.38	23.57	30.89	21.00	7.29	9.89	13.71
19.	*Salix babylonica*	Leaves	08.97	91.03	18.12	36.35	24.68	09.28	11.67	15.40
20.	*Trifolium pratense*	Whole plant	10.32	89.68	17.56	39.14	32.56	06.23	06.58	26.35
21.	*Trifolium repens*	Whole plant	10.53	89.47	12.60	41.80	32.75	06.87	09.05	25.88
22.	*Ulmus wallichiana*	Leaves	10.30	89.70	19.02	28.98	22.74	6.05	6.24	16.69

DM = Dry matter **L** = Legnin **NDF** = Neutral detergent fibre **HC** = Hemi cellulose

ADF = Acid detergent fibre **C** = Cellulose **CP** = Crude protein

Table 7. List of ornamental wild plants of Ladakh and Lahaul Spiti

S. No	BOTANICAL NAME WITH FAMILY	COMMON NAME	PARTS USED	REMARKS
1.	*Acantholimon lycopodioides* (Plumbaginaceae)	Lonze	Whole plant	As dry decoration
2.	*Androsace aizoon* (Primulaceae)	—	Whole plant	Ornamental
3.	*Allium przewalskianum* (Amaryllidaceae)	Wild garlic, Skotche	Whole plant	As border flower
4.	*Aquilegia fragrans* (Ranunculaceae)	Fragrans Columbine,	Flower spike	Fragrant and attractive flowers
5.	*Aquilegia pubiflora* (Ranunculaceae)	Fragrans Columbine,	Flower spike	Fragrant and attractive flowers
6.	*Aster diplostephioides* (Asteraceae)	Wild aster, Kiamentok	Flowers	Ornamental
7.	*Aster flaccidus* (Asteraceae)	Wild aster, Niamentok	Flowers	Ornamental
8.	*Aster tibeticus* (Asteraceae)	Wild aster	Flowers	Ornamental
9.	*Braya alpina* (Brassicaceae)	Lingantso	Flowers	Ornamental
10.	*Braya tibetica* (Brassicaceae)	Lingantso	Flowers	Ornamental
11.	*Bupleurum tenue* (Apiaceae)	Kimantso	Flower spike	Flowers of aesthetic value
12.	*Caltha palustris* (Ranunculaceae)	Marsh marigold	Flower spike	Attractive flowers
13.	*Campanula argyrotricha* (Campanulaceae)	—	Flowers	Ornamental

Contd..

Table 7. Contd...

14.	*Campanula aristata* (Campanulaceae)	Piseng	Flowers	Ornamental
15.	*Campanula cashmeriana* (Campanulaceae)	Nipeng	Flowers	Ornamental
16.	*Campanula latifolia* (Campanulaceae)	Large bell flower, Nipeng	Flowers	Ornamental
17.	*Campanula pallida* (Campanulaceae)	Large bell flower, Nipeng	Flowers	Ornamental
18.	*Cardus edelbergii* (Asteraceae)	Scotch thistle	Flower-head	Ornamental
19.	*Choriospora sabulosa* (Brassicaceae)	—	Flowers	Flowers of mild fragrance
20.	*Clematis orientalis* (Ranunculaceae)	Zakgic	Whole plant	As climber
21.	*Clematis tibetana* (Ranunculaceae)	Zakgic	Whole plant	As climber
22.	*Codonopsis clematidea* (Campanulaceae)	Burkutang	Flower spike	Having attractive flowers
23.	*Codonopsis ovata* (Campanulaceae)	Ludut	Flower spike	Having attractive flowers
24.	*Cremanthodium ellisii* (Asteraceae)	Lukumentok	Whole plant	Aesthetic value
25.	*Cremanthodium reniforme* (Asteraceae)	Lukumentok	Whole plant	Aesthetic value
26.	*Dactylorhiza hatagirea* (Orchidaceae)	Heart Orchid, Salampanja, Ambolakpa	Whole plant	Attractive flowers of aesthetic value
27.	*Delphinium brunonianum* (Ranunculaceae)	Himalayan Larkspur, Lunde-Kaon	Flower spike	Attractive flowers of aesthetic value
28.	*Delphinium cashmerianum* (Ranunculaceae)	Himalayan Larkspur, Lunde-Kaon	Flower spike	Attractive flowers of aesthetic value

Contd..

Table 7. Contd...

29.	*Delphinium vestitum* (Ranunculaceae)	Himalayan Larkspur, Lunde-Kaon	Flower spike	Attractive flowers of aesthetic value
30.	*Dianthus crinitus* (Caryophyllaceae)	Liamtnang	Flowers	Ornamental
31.	*Dianthus incaca* (Caryophyllaceae)	Mulanku	Flowers	Ornamental
32.	*Dracocephalum heterophyllum* (Lamiaceae)	Zinkzer / Ziptsi	Whole plant	As hanging basket
33.	*Echinops cornigerus* (Asteraceae)	Eckzema	Flower-heads	Having attractive flowers
34.	*Echinops niveus* (Asteraceae)	Eckzema	Flower-heads	Having attractive flowers
35.	*Elaeagnus angustifolium* (Elaeagnaceae)	Oleaster, Sershing	Whole plant	An ornamental small tree having fragrant flowers
36.	*Epilobium angustifolium* (Onagraceae)	Willow herb	Whole plant	As a hedge
37.	*Epilobium latifolium* (Onagraceae)	Willow herb	Whole plant	As a hedge
38.	*Epilobium roseum* (Onagraceae)	Willow herb	Whole plant	As a hedge
39.	*Euphorbia stracheyi* (Euphorbiaceae)	Tharno	Whole plant	Ornamental
40.	*Gagea dschungarica* (Amaryllidaceae)	Yellow star of Bethlehem	Flowers	Ornamental
41.	*Gagea lutea* (Amaryllidaceae)	Yellow star of Bethlehem	Flowers	Ornamental
42.	*Geranium pratense* (Geraniaceae)	Meadow cranesbil, Gadur	Flowers	Ornamental
43.	*Glechoma nivalis* (Lamiaceae)	Gaumej	Whole plant	Ornamental

Contd..

Table 7. Contd...

44.	*Impatiens amplexicaulis* (Balsaminaceae)	—	Whole plant	Attractive flowers of aesthetic value
45.	*Impatiens glandulifera* (Balsaminaceae)	—	Whole plant	Attractive flowers of aesthetic value
46.	*Inula obtusifolia* (Asteraceae)	—	Whole plant	Attractive flowers of aesthetic value
47.	*Iris kumaonensis* (Iridaceae)	Tesmamentok	Whole plant	Attractive flowers of aesthetic value, suitable for border
48.	*Iris lactea* (Iridaceae)	Tesmamentok	Whole plant	Attractive flowers of aesthetic value, suitable for border
49.	*Juniperus communis* (Juniperaceae)	Juniper, Sukpa	Whole plant	An ornamental shrub of aesthetic value,
50.	*Juniperus macropoda* (Juniperaceae)	Juniper, Sukpa	Whole plant	An ornamental shrub of aesthetic value,
51.	*Juniperus recurva* (Juniperaceae)	Juniper, Sukpa	Whole plant	An ornamental shrub of aesthetic value,
52.	*Kochia indica* (Chenopodiaceae)	—	Whole plant	Attractive ornamental herb
53.	*Leontopodium leontopodium* (Asteraceae)	Himalayn Edelweiss, Palu	Whole plant	Attractive alpine ornamental herb
54.	*Leontopodium naum* (Asteraceae)	Himalayn Edelweiss, Palu	Whole plant	Attractive alpine ornamental herb
55.	*Lomatogonium carinthiacum* (Gentianaceae)	Tikta	Whole plant	Alpine ornamental herb
56.	*Malaxis muscifera* (Orchidaceae)	—	Whole plant	Attractive alpine ornamental herb
57.	*Malva verticillata* (Malvaceae)	Chinese mallow, Sukchel	Flowers	Ornamental
58.	*Meconopsis aculeata* (Papaveraceae)	Blue poppy, Achat sarmon	Flower spike	Attractive flowers of aesthetic value

Contd..

Table 7. Contd...

59.	*Meconopsis bikramii* (Papaveraceae)	Blue poppy, Achat sarmon	Flower spike	Attractive flowers of aesthetic value
60.	*Oxygrphis polyepetala* (Ranunculaceae)	—	Whole plant	Ornamental
61.	*Papaver nudicaule* (Papaveraceae)	Orange poppy	Whole plant	An ornamental herb
62.	*Pedicularis pycnantha* (Scrophulariaceae)	—	Whole plant	An ornamental herb
63.	*Pedicularis pyramidata* (Scrophulariaceae)	Lugrumarpo	Whole plant	An ornamental herb
64.	*Physalis alkekengi* (Solanaceae)	Winter cherry, Shoklo	Fruits	Attractive fruits of aesthetic value used by locals
65.	*Polygonum polycnemoides* (Polygonaceae)	Lamtso	Whole plant	An ornamental herb
66.	*Primula denticulata* (Primulaceae)	Sulumentok	Whole plant	An ornamental herb
67.	*Primula elliptica* (Primulaceae)	Sulumentok	Whole plant	An ornamental herb
68.	*Primula involucrata* (Primulaceae)	Sulumentok	Whole plant	An ornamental herb
69.	*Primula macrophylla* (Primulaceae)	Sulumentok	Whole plant	Flowers of ornamental value
70.	*Primula reptans* (Primulaceae)	Sulumentok	Whole plant	Flowers of ornamental value
71.	*Primula rosea* (Primulaceae)	Prime rose, Sulumentok	Whole plant	Flowers of ornamental value
72.	*Rhodiola imbricata* (Crassulaceae)	Rose root, Shrolo	Whole plant	Flowers of ornamental value, suitable for hang ing basket
73.	*Rhodiola tibetica* (Crassulaceae)	Rose root, Shrolo	Whole plant	Flowers of ornamental value, suitable for hang ing basket

Contd..

Table 7. Contd...

74.	*Rosa brunonii* (Rosaceae)	Wild rose, Siah	Flowers	Flowers of aesthetic value
75.	*Rosa ecae* (Rosaceae)	Wild rose, Siah	Flowers	Flowers of aesthetic value
76.	*Rosa eglanteria* (Rosaceae)	Wild rose, Siah	Flowers	Flowers of aesthetic value
77.	*Rosa macrophylla* (Rosaceae)	Wild rose, Siah	Flowers	Flowers of aesthetic value
78.	*Rosa webbiana* (Rosaceae)	Wild rose, Siah	Flowers	Flowers of aesthetic value
79.	*Rosularia alpestris* (Crassulaceae)	—	Flowers	Alpine ornamental herb
80.	*Salix babylonica* (Salicaceae)	Weeping willow	Whole plant	Tree of aesthetic value
81.	*Saussurea bracteata* (Asteraceae)	Pseudo Brahma kamal, Span-rtso-bo	Whole plant	Pseudo Brahma Kamal of aesthetic value
82.	*Saussurea gossypiphora* (Asteraceae)	—	Whole plant	An ornamental herb, just look like a snowball
83.	*Saussurea obvallata* (Asteraceae)	Brahma kamal, Span-rtso-bo	Whole plant	Brahma Kamal of aesthetic value
84.	*Sedum ewersii* (Crassulaceae)	Stone crop, Gomin	Whole plant	An ornamental herb
85.	*Silene edgeworthii* (Caryophyllaceae)	Globe flower	Whole plant	An ornamental herb
86.	*Silene moorcroftiana* (Caryophyllaceae)	—	Whole plant	An ornamental herb
87.	*Syringa emodi* (Oleaceae)	Lilac	Whole plant	Tree of aesthetic value
88.	*Thermopsis inflata* (Fabaceae)	Olmango	Whole plant	An ornamental herb

Contd..

Table 7. Contd...

89.	*Thylacospermum caespitosum* (Caryophyllaceae)	Tagarcan	Whole plant	Attractive in fresh as well as in dry form
90.	*Trollius acaulis* (Ranunculaceae)	Globe flower	Whole plant	Having ornamental value
91.	*Tulipa stellata* (Liliaceae)	Wild tulip, Kapichang	Whole plant	Having ornamental value

Table 8 : List of Magico-religious plants of Ladakh and Lahaul Spiti

S. No	BOTANICAL NAME WITH FAMILY	COMMON NAME	REMARKS
1.	*Aconitum violaceum* (Ranunculaceae)	Aconite, Patees, Bona-nagpo	Keep away bad evils
2.	*Arnebia euchroma* (Boraginaceae)	Rattanjot, Demok	Considered religious
3.	*Berberis pachyacantha* (Berberidaceae)	Berberis	Religious ceremonies
4.	Carex melanantha (Poaceae)	Grass	As a good luck
5.	*Cedrus deodara* (Pinaceae)	Deodar	Religious ceremonies
6.	*Delphinium brunonianum* (Ranunculaceae)	Himalayan Larkspur, Lunde-kaon	Religious ceremonies
7.	*Delphinium cashmerianum* (Ranunculaceae)	Himalayan Larkspur, Lunde-kaon	Religious ceremonies
8.	*Delphinium vestitatum* (Ranunculaceae)	Himalayan Larkspur, Lunde-kaon	Religious ceremonies
9.	*Elaeagnus angustifolium* (Elaeagnaceae)	Oleaster, Sersing	Religious ornament used in monastry

Contd..

Table 8. Contd...

10.	*Epilobium angustifolium* (Onagraceae)	Roseway, Willow herb, Wanpo	Religious places
11.	*Epilobium latifolium* (Onagraceae)	Roseway, Willow herb, Wanpo	Religious places
12.	*Epilobium roseum* (Onagraceae)	Roseway, Willow herb, Wanpo	Religious ceremonies
13.	*Juniperus communis* (Juniperaceae)	Juniper, Sukpa	Dhoop, essence
14.	*Juniperus macropoda* (Juniperaceae)	Juniper, Sukpa	Dhoop, essence
15.	*Juniperus recurva* (Juniperaceae)	Juniper, Sukpa	Dhoop, essence
16.	*Peganum harmala* (Zygophyllaceae)	Indian hermal, Sepan	Bad-eye effect.
17.	*Physalis alkekengi* (Solanaceae)	Chinese lanten, Shoklo	Relegious ornament used by locals
18.	*Physochlaina praealta* (Solanaceae)	Langthang	Halluginosic properties
19.	*Primula macrophylla* (Primulaceae)	Prime rose, Sulumentok	Religious ceremonies
20.	*Primula rosea* (Primulaceae)	Prime rose, Sulumentok	Religious ceremonies
21.	*Rosa brunonii* (Rosaceae)	Wild rose	Religious ceremonies
22.	*Rosa ecae* (Rosaceae)	Wild rose, Siah serpo	Religious ceremonies
23.	*Rosa eglanteria* (Rosaceae)	Wild rose	Religious ceremonies
24.	*Rosa macrophylla* (Rosaceae)	Wild rose	Religious ceremonies

Contd..

Table 8. Contd...

25.	*Rosa webbiana* (Rosaceae)	Wild rose, Siah marpo	Religious ceremonies
26.	*Verbascum thapsus* (Scrophulariaceae)	Flennel Mullein, dandashal	Against bad evils
27.	*Waldheimia glabra* (Asteraceae)	Palu	Religious ceremonies as essence
28.	*Waldheimia stoliczkai* (Asteraceae)	Palu	Religious ceremonies as essence
29.	*Waldheimia tomentosa* (Asteraceae)	Palu	Religious ceremonies as essence

Table 9 : List of fuel plants of Ladakh and Lahaul Spiti

S. No	BOTANICAL NAME WITH FAMILY	COMMON NAME	REMARKS
1.	*Acantholimon lycopodioides* (Plumbaginaceae)	Longe	A very important fuel plant for the winters
2.	*Androsace aizoon* (Primulaceae)	Sga-tig	Used for fire producing
3.	*Androsace villosa* (Primulaceae)	Fuzung	Used for fire producing
4.	*Anthemis cotula* (Asteraceae)	—	Used as fuel
5.	*Arenaria festucoides* (Caryophyllaceae)	Pianzo	Used as fuel
6.	*Arenaria kashmirica* (Caryophyllaceae)	Khilimentok	Used as fuel
7.	*Artemisia brevifolia* (Asteraceae)	Worm seed/ 'Burtse	Widely used as fuel
8.	*Artemisia dracunculus* (Asteraceae)	Tarragon, Sanche	Used for fuel
9.	*Artemisia gmelinii* (Asteraceae)	Khampa-Shirdi	Used as firewood
10.	*Artemisia nilagirica* (Asteraceae)	Khampa-Shirdi	Used as firewood
11.	*Artemisia parviflora* (Asteraceae)	Khamang	Used as fuel
12.	*Artemisia persica* (Asteraceae)	Burnak	Used as fuel
13.	*Artemisia salsoloides* (Asteraceae)	Khamchu	Used as fuel
14.	*Artemisia sieversiana* (Asteraceae)	Khamchu	Used for fire producing

Contd..

Table 9. Contd...

15.	*Artemisia tournefortiana* (Asteraceae)	Khamchu	Used as fuel
16.	*Astragalus cicerifolius* (Fabaceae)	Yakpo	Used as fuel
17.	*Astragalus frigidus* (Fabaceae)	Skarmalo	Used as fuel
18.	*Atriplex nitens* (Chenopodiaceae)	Lichiang	Used as fuel
19.	*Atriplex rosea* (Chenopodiaceae)	Rukpa	Used as fuel
20.	*Berberis ulicina* (Berberidaceae)	Khizer	Used as fuel
21.	*Betula utilis* (Betulaceae)	Bhojpatra	Fuelwood
22.	*Caragana brevifolia* (Fabaceae)	Dama	Used as firewood
23.	*Caragana versicolor* (Fabaceae)	Dama or Kama	Widely used as firewood
24.	*Cedrus deodara* (Pinaceae)	Deodar	Used as fuelwood
25.	*Cirsium arvense* (Asteraceae)	Creeping thistle	Used as fire
26.	*Colutea multiflora* (Fabaceae)	Bladder sena	Used firewood
27.	*Cotoneaster falconeri* (Rosaceae)	Kumchung	Widely used as firewood
28.	*Cotyledon thyrsiflora* (Crassulaceae)	—	Used as fuel
29.	*Crataegus songarica* (Rosaceae)	Hawthorn	Used as firewood

Contd..

Table 9. Contd...

30.	*Cynanthus lobatus* (Campanulaceae)	—	Fuel plant
31.	*Cynoglossum petiolatum* (Boraginaceae)	Ematso	Fuel plant
32.	*Cynoglossum zeylanicum* (Boraginaceae)	Hounds' tongue	Fuel plant
33.	*Dianthus angulatus* (Caryophyllaceae)	—	Fuel plant
34.	*Dipsacus mitis* (Dipsaceae)	Sichimo	Fuel plant
35.	*Dioscorea deltoidea* (Discoriaceae)	Sangli-mangli	Fuel plant
36.	*Ephedra gerardiana* (Ephedraceae)	Chhapat	Fuel plant
37.	*Erigeron multicaulis* (Asteraceae)	—	Fuel plant
38.	*Eriocycla thomsonii* (Apiaceae)	Temo	Fuel plant
39.	*Eritrichium fruticulosum* (Boraginaceae)	—	Fuel plant
40.	*Eritrichium nanum* (Boraginaceae)	—	Fuel plant
41.	*Erysimum repandum* (Brassicaceae)	Chharpa	Fuel plant
42.	*Galium tricorne* (Rubiaceae)	—	Fuel plant
43.	*Halogeton glomeratus* (Chenopodiaceae)	Phyang	Fuel plant
44.	*Haloxylon thomsonii* (Chenopodiaceae)	—	Fuel plant

Contd..

Table 9. Contd...

45.	*Hippophae rhamnoides* ssp. *turkestanica* (Elaeagnaceae)	Seabuckthorn, Tsermang	Fuel wood
46.	*Hippophae salicifolia* (Elaeagnaceae)	Seabuckthorn, Tsermang	Fuel wood
47.	*Hippophae tibetana* (Elaeagnaceae)	Seabuckthorn, Tsermang	Fuel plant
48.	*Indigofera heterantha* (Fabaceae)	Kieche	Fuel plant
49.	*Juniperus communis* (Juniperaceae)	Juniper, Sukpa	Fuel wood
50.	*Juniperus macropoda* (Juniperaceae)	Juniper, Sukpa	Fuel wood
51.	*Leontopodium leontopodium* (Asteraceae)	Palu	Fire producing plant
52.	*Lonicera asperifolia* (Caprifoliaceae)	Honeysuckle	Fuel plant
53.	*Lonicera glauca* (Caprifoliaceae)	Lamora	Fuel plant
54.	*Lonicera microphylla* (Caprifoliaceae)	Honeysuckle	Fire wood
55.	*Lonicera rupicola* (Caprifoliaceae)	Lamora	Fire wood
56.	*Lonicera spinosa* (Caprifoliaceae)	Lamora	Fire wood
57.	*Myricaria germanica* (Tamaricaceae)	Umbu	Fire wood
58.	*Myricaria squamosa* (Tamaricaceae)	Umbu	Fire wood
59.	*Oxytropis mollis* (Fabaceae)	Dhambokoya	Fire plant

Contd..

Table 9. Contd...

60.	*Picea smithiana* (Pinaceae)	Himalayan Spruce, Tosh	Fire wood
61.	*Pinus wallichiana* (Pinaceae)	Chir	Fire wood
62.	*Populus caspica* (Salicaceae)	Poplar, Changpa	Fire wood
63.	*Populus ciliata* (Salicaceae)	Himalayan Poplar	Fire wood
64.	*Populus euphratica* (Salicaceae)	Indian Poplar, Changp	Fire wood
65.	*Potentilla bifurca* (Rosaceae)	—	Fire wood
66.	*Potentilla cuneata* (Rosaceae)	Chantsu	Fire wood
67.	*Ribes alpestre* (Rosaceae)	Asian gooseberry, Zasot	Fire plant
68.	*Ribes himalense* (Rosaceae)	Asian gooseberry, Zasot	Fire wood
69.	*Ribes orientale* (Rosaceae)	Asian gooseberry, Zasot	Fire wood
70.	*Rosa ecae* (Rosaceae)	Wild rose, Siah	Fire wood
71.	*Rosa webbiana* (Rosaceae)	Wild rose, Siah	Fire wood
72.	*Salix alba* (Salicaceae)	Willow, Mulchang	Fire wood
73.	*Salix angustifolia* (Salicaceae)	Willow, Mulchang	Fire wood
74.	*Salix caesia* (Salicaceae)	Willow, Mulchang	Fire wood

Contd..

Table 9. Contd...

75.	*Salsola collina* (Chenopodiaceae)	—	Fuel plant
76.	*Sibbaldia cuneata* (Rosaceae)	—	Fuel plant
77.	*Sisymbrium loeselii* (Brassicaceae)	—	Fuel plant
78.	*Sophora alopecuroides* (Fabaceae)	Singtik	Fuel wood
79.	*Sophora moorcroftiana* (Fabaceae)	Singtik	Fuel wood
80.	*Tamarix gallica* (Tamaricaceae)	Tamarisk, Agrelta	Fuel wood
81.	*Thylacospermus caespitosum* (Caryophyllaceae)	Tagaracan	Fuel plant
82.	*Ulmus wallichiana* (Ulmaceae)	Elm, Umbuk	Fuel wood

Table 11 : List of potential forest plants of Ladakh and Lahaul Spiti

S. No	BOTANICAL NAME WITH FAMILY	COMMON NAME	REMARKS
1.	*Betula utilis* (Betulaceae)	Bhojpatra	A very useful forest tree
2.	*Caragana versicolor* (Fabaceae)	Dama / Kama	A high alpine scrub
3.	*Cedrus deodara* (Pinaceae)	Deodar / Deyor	Source of timber, etc.
4.	*Colutea multiflora* (Fabaceae)	Bladder Sena	A potential forest shrub
5.	*Cotoneaster falconeri* (Rosaceae)	Kumchung	-do-
6.	*Crataegus songarica* (Rosaceae)	Hawthorn	Source of timber, etc.
7.	*Elaeagnus angustifolium* (Elaeagnaceae)	Oleaster, Shersing	An ornamental tree
8.	*Fraxinus xanthoxyloides* (Oleaceae)	Ash	An ornamental tree
9.	*Hippophae rhamnoides* ssp. *turkestanica* (Elaeagnaceae)	Seabuckthorn, Tsermang	A potential forest shrub
10.	*Hippophae salicifolia* (Elaeagnaceae)	-do-	-do-
11.	*Hippophae tibetana* (Elaeagnaceae)	-do-	A high alpine forest scrub
12.	*Indigofera heterantha* (Fabaceae)	Kieche	Alpine forest shrub
13.	*Juglans regia* (Juglandaceae)	Walnut, sTarga	A horticulture forest tree
14.	*Juniperus communis* (Juniperaceae)	Juniper, Shukpa	An ornamental shrub of economic importance

Contd..

Table 11. Contd...

15.	*Juniperus macropoda* (Juniperaceae)	Juniper, Shukpa	An ornamental shrub of economic importance
16.	*Juniperus recurva* (Juniperaceae)	Juniper, Shukpa	An ornamental shrub of economic importance
17.	*Morus alba* (Moraceae)	Mulberry, Sahtoot	A horticulture forest tree of high value
18.	*Myricaria elegans* (Tamaricaceae)	Umbo	A common shrub along river belts
19.	*Myricaria germanica* (Tamaricaceae)	Umbo	A common shrum along river belts
20.	*Myricaria squamosa* (Tamaricaceae)	Umbo	A common shrum along river belts
21.	*Picea smithiana* (Pinaceae)	Himalayan Spruce, Tosh	The important & common forest tree for cold desert region
22.	*Pinus wallichiana* (Pinaceae)	Blue Pine, Chir	A valuable timber tree
23.	*Populus caspica* (Salicaceae)	White Poplar, Yarpa	A medium sized tree of economic value
24.	*Populus ciliata* (Salicaceae)	Himalayan Poplar, Yarpa	-do-
25.	*Populus deltoides* (Salicaceae)	Caroline Poplar, Yarpa	-do-
26.	*Populus euphratica* (Salicaceae)	Indian Poplar, Yulat	-do-
27.	*Populus nigra* (Salicaceae)	Black Popular, Italian Poplar, Yulat	Commonly planted forest tree of great economic value
28.	*Populus sylvestris* (Salicaceae)	Poplar, Yulat	A valuable forest tree
29.	*Rhododendron anthopogon* (Ericaceae)	Rhododendron, Balu	A potential forest shrub of economic value

Contd..

Table 11. Contd...

30.	*Rhododendron campanulatum* (Ericaceae)	Rhododendron	A high alpine scrub
31.	*Ribes alpestre* (Saxifragaceae)	Asian gooseberry, Zasot	A potential forest shrub
32.	*Ribes orientale* (Saxifragaceae)	Asian gooseberry, Zasot	-do-
33.	*Robinia pseudacacia* (Fabaceae)	Pseudo Acacia, Kikar	A medium forest tree, suitable for cold desert areas
34.	*Rosa webbiana* (Rosaceae)	Wild Rose, Siah	A ornamental shrub
35.	*Salix alba* (Salicaceae)	White willow, Mulchang	A common forest tree of high altitude areas.
36.	*Salix angustifolia* (Salicaceae)	Willow, Mulchang	-do-
37.	*Salix babylonica* (Salicaceae)	Weeping willow	An ornamental forest tree.
38.	*Salix caesia* (Salicaceae)	Willow	A small shrub
39.	*Salix daphnoides* (Salicaceae)	Violet Willow	A medium sized tree.
40.	*Salix denticulata* (Salicaceae)	Willow, Tookchang	-do-
41.	*Salix excelsa* (Salicaceae)	Crack willow, Red wood willow, Mulchang	A very useful forest tree.
42.	*Salix hastata* (Salicaceae)	Willow, Selchang	A small forest shrub
43.	*Salix sclerophylla* (Salicaceae)	Willow, Solchang	-do-
44.	*Salix tetrasperma* (Salicaceae)	Willow, Solchang	-do-

Contd..

Table 11. Contd...

45.	*Sophora moorcroftiana* (Fabaceae)	Singtik	A potential leguminous shrub
46.	*Syringa emodi* (Oleaceae)	Lilac	A potential ornamental tree
47.	*Tamarix gallica* (Tamaricaceae)	Tamarisk, Agrelta	A very useful shrub along river belts
48.	*Ulmus wallichiana* (Ulmaceae)	Elm, Umbuk	A very useful forest tree

List of Colour Plates

Fig. 414. *Dactylorhiza hatagirea* (vegetative propagation)

Fig. 415. *Dactylorhiza* sp (cultivation)

Fig. 416. *Dracocephalum* sp (nursery)

Fig. 417. *Dracocephalum heterophyllum* (in full bloom)

Fig. 418. *Ephedra gerardiana* (nursery)

Fig. 419. *Ephedra gerardiana cultivation*

Fig. 420. *Hippophae rhamnoides* (vegetative propagation)

Fig. 421. *Hippophae tibetana* (nursery)

Fig. 422. *Inula racemosa* (nursery)

Fig. 423. *Inula racemosa* (cultivation)

Fig. 424. *Origanum* sp (nursery)

Fig. 425. *Origanum vulgare* (in cultivation)

Fig. 426. *Physochlaina praealta* (nursery)

Fig. 427. *Podophyllum* sp (nursery)

Fig. 428. *Podophyllum hexandrum* (ready for harvest)

Fig. 429. *Rheum webbianum* (vegetative propagation)

Fig. 430. *Rheum webbianum* in cultivation

Fig. 431. *Rhodiola imbricata* (Cultivation)

Fig. 432. *Rubia cordifolia* (in fruiting stage)

Fig. 433. *Salvia sclarea* (in full bloom)

Fig. 434. *Saussurea* sp (Nursery)

Fig. 435. *Saussurea lappa* (In fruiting stage)

Fig. 436. *Urtica hyperborea* (ready for harvest)

Fig. 437. *Valeriana jatamansi* (cultivation)

Fig. 438. *Verbascum thapsus* (nursery)

❑❑❑

Fig. 1. The Maitreya Budha

Fig. 2. Trans Himalayan cold desert

Fig. 3. Nun Kun peak 7,135 m

Fig. 4. Khardungla (5,520 m) the world's highest moterable pass

Fig. 5. Confluence of Indus and Zanskar river

Fig. 6. Onset of summer (agriculture)

Fig. 7. Pangongtso lake (4,218 m)

Fig. 8. Kee monastry (the most sacred place)

Fig. 9. Panamik hot spring (Nubra)

Fig. 10. A drogpa feature

Fig. 11. A traditional kitchen

Fig. 12. Temperate vegetation

Fig. 13. Alpine vegetation

Fig. 14. Oasitic vegetation

Fig. 15. High alpine desert vegetation

Fig. 16. Cotton ball type (*Saussurea* spp.)

Fig. 17. A solid mat alpine plant (*Thylacospermum* spp)

Fig. 18. A spiny cushion (*Acantholimon* spp.)

Fig. 19. An alpine spiny bush (*Caragana* spp.)

Fig. 20. A soft cushion plant (*Rhodiola* spp.)

Fig. 21. Alpine plants thriving subzero temperature

Fig. 22. A traditional library

Fig. 23. A traditional ploughing practice

Fig. 24. A traditional winnowing practice

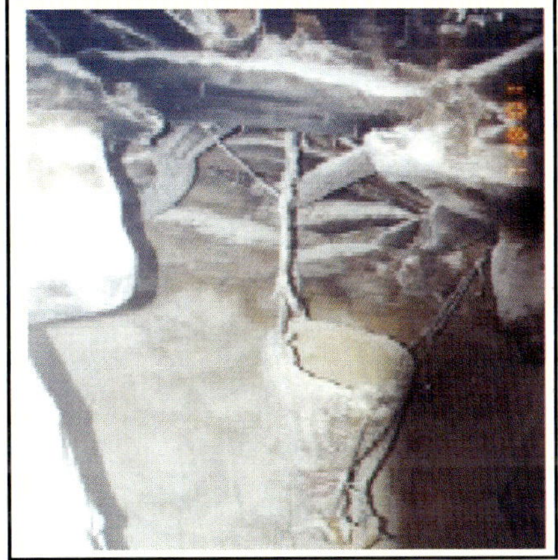

Fig. 25. A traditional floor mill

Fig. 26. Frying barley for making sattu

Fig. 27. A traditional storage of medicinal plants

Fig. 28. Wild edible dry leaves

Fig. 29. Preparing chhang (local beer)

Fig. 30. A traditional collection of milk

Fig. 31. Collection of wild fodder plants

Fig. 32. Weaving wool

Fig. 33. Collection of fuel plant

Fig 34. Exploring trans Himalaya

Fig 35. Indus valley (a panaromic view)

Fig 36. Indus river (a lifeline)

Fig 37. Nubra valley (a cold desert sand dune)

Fig 38. Changthang valley

Fig 39. Tsomoriri lake (Changthang)

Fig 40. Black neck crane

Fig 41. Suru valley (a panaromic view)

Fig 42. Zanskar valley

Fig 43. Spiti valley (a panaromic view)

Fig 44. Kunzam Pass (Spiti, 4,500 m)

Fig 45. Lahaul valley (a panaromic view)

Fig 46. Chandra Bhaga river (a lifeline)

Fig 47. Amchi (Herbalist)

Fig 48. Amchi Preparing herbal medicine

Fig 49. Traditional drying of herbs

Fig 50. *Aconitum heterophyllum*

Fig 51. *Aconitum heterophyllum* (dried roots)

Fig 52. *Aconitum violaceum*

Fig 53. *Anaphalis busua*

Fig 54. *Anaphalis triplinervis* var *monocephala*

Fig 55. *Anaphalis triplinervis* var *intermedia*

Fig 56. *Anemone rivularis*

Fig 57. *Anemone rupicola*

Fig 58. *Arctium lappa*

Fig 59. *Areneria bryophylla*

Fig 60. *Arenaria grifithii*

Fig 61. *Arnebia euchroma*

Fig 62. *Arnebia guttata*

Fig 63. *Astragalus rhizanthus*

Fig 64. *Berberis kunawarensis*

Fig 65. *Berberis pachyantha*

Fig 66. *Berberis ulicina*

Fig 67. *Berberis zabellina*

Fig 68. *Bergenia ligulata*

Fig 69. *Bergenia stracheyi*

Fig 70. *Cassiope fastigata*

Fig 71. *Cerastium cerateoides*

Fig 72. *Circium wallichii*

Fig 73. *Codonopsis clematidea*

Fig 74. *Codonopsis ovata*

Fig. 75. *Colchicum luteum*

Fig 76. *Colchicum luteum* (fresh bulbs)

Fig 77. *Corydalis crasissima*

Fig 78. *Corydalis flabellata*

Fig 79. *Corydalis govaniana*

Fig 80. *Corydalis meifolia*

Fig 81. *Corydalis thyrsiflora*

Fig 82. *Cousinia falconeri*

Fig 83. *Cousinia thomsonii*

Fig 84. *Cremanthodium ellisii*

Fig 85. *Crepis flexuosa*

Fig 86. *Cuscuta europaea*

Fig 87. *Cuscuta reflexa*

Fig 88. *Dactylorhiza hatagirea*

Fig 89. *Dactylorhiza hatagirea* (dried roots)

Fig 90. *Delphinium brunonianum*

Fig 91. *Delphinium cashmerianum*

Fig 92. *Dianthus anatolicus*

Fig 93. *Ephedra gerardiana*

Fig 94. *Ephedra intermedia*

Fig 95. *Epipactis helleborine*

Fig 96. *Euphorbia stracheyi*

Fig 97. *Euphrasia officinalis*

Fig 98. *Galium aparine*

Fig 99. *Galium pauciflorum*

Fig 100. *Gaultheria trichophylla*

Fig 101. *Gentiana algida*

Fig 102. *Gentianella moorcroftiana*

Fig 103. *Gentianella paludosa*

Fig 104. *Geranium collinum*

Fig 105. *Geranium pretense*

Fig 106. *Gypsophylla ceratoides*

Fig 107. *Heracleum pinnatum*

Fig 108. *Herminium monorchis*

Fig 109. *Hippophae rhamnoides* ssp *turkestanica* (yellow)

Fig 110. *Hippophae rhamnoides* ssp *turkestanica* (red)

Fig 111. Ladakh Berry

Fig 112. DRDO HERBAL TEA

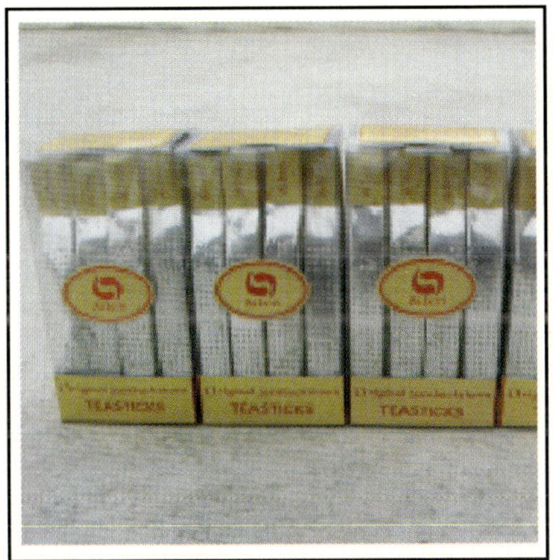

Fig 113. DRDO herbal tea (Si-berry-sticks)

Fig 114. *Hippophae salicifolia*

Fig 115. *Hippophae tibetana*

Fig 116. *Hyocyamus niger*

Fig 117. *Hyoscyamus pusillus*

Fig 118. *Hypecoum leptocarpum*

Fig 119. *Hypericum perforatum*

Fig 120. *Inula obtusifolia*

Fig 121. *Inula racemosa*

Fig 122. *Inula racemosa* (dried roots)

Fig 123. *Inula rhizocephala* var *rhizocephaloides*

Fig 124. *Jurinea ceratocarpa*

Fig 125. *Lagotis cashmeriana*

Fig 126. *Lancea tibetica*

Fig 127. *Lloydia serotina*

Fig 128. *Malaxis muscifera*

Fig 129. *Minuartia kashmirica*

Fig 130. *Morina longiflora*

Fig 131. *Onosma hispidum*

Fig 132. *Orobanchae hansii*

Fig 133. *Pedicularis cheilanthifolia*

Fig 134. *Pedicularis longiflora var tubiformis*

Fig 135. *Pedicularis mollis*

Fig 136. *Phlomis bracteosa*

Fig 137. *Physochlaina praealta*

Fig 138. *Picrorhiza kurrooa*

Fig 139. *Plantago depressa*

Fig 140. *Podophyllum hexandrum*

Fig 141. *Podophyllum hexandrum* (dried rhizome)

Fig 142. *Polygonatum multiflorium*

Fig 143. *Polygonatum verticillatum*

Fig 144. *Potentilla anserina*

Fig 145. *Potentilla atrosanguinea*

Fig 146. *Potentilla fruticosa*

Fig 147. *Ranunuculus laetus*

Fig 148. *Ranunculus tricuspis*

Fig 149. *Rheum moorcroftianum*

Fig 150. *Rheum speciforme*

Fig 151. *Rheum webbianum* (fresh roots)

Fig 152. *Rhodiola heterodonta*

Fig 153. *Rhodiola imbricata*

Fig 154. *Rubia cordifolia*

Fig 155. *Saussura bracteata*

Fig 156. *Saussurea glacialis*

Fig 157. *Saussurea gnaphaloides*

Fig 158. *Saussurea jacea*

Fig 159. *Saussurea lappa*

Fig 160. *Saussurea obvallata*

Fig 161. *Saxifraga jaquemontii*

Fig 162. *Saxifraga pulvinaria*

Fig 163. *Saxifraga stenophylla*

Fig 164. *Scrophularia koelzii*

Fig 165. *Sineceo chrysanthemoides*

Fig 166. *Solidago virg-aurea*

Fig 167. *Statice macrorrhabdos*

Fig 168. *Swertia petiolata*

Fig 169. *Thallictrum foetidum*

Fig 170. *Thermopsis barbata*

Fig 171. *Thlaspi arvense*

Fig 172. *Tribulus terrestris*

Fig 173. *Verbascum thapsus*

Fig 174. *Vincetoxicum hirundinaria*

Fig 175. *Viola kunawarensis*

Fig 176. *Youngia tenuifolia*

Fig. 177. Aromatic oil

Fig. 178. *Achillea millefolium*

Fig. 179. *Artemisia annua*

Fig. 180. *Artemisia brevifolia*

Fig. 181. *Artemisia dracunculus*

Fig. 182. *Artemisia gmelinii*

Fig. 183. *Artemisia sieversiana*

Fig. 184. *Artemisia tournefortiana*

Fig. 185. *Biebersteinia odora*

Fig. 186. *Chrysanthemum pyrethroides*

Fig. 187. *Dracocephalum heterophyllum*

Fig. 188. *Elsholtzia densa*

Fig. 189. *Elsholtzia eriostachya*

Fig. 190. *Ferula jaeschkeana*

Fig. 191. *Humulus lupulus*

Fig. 192. *Hyssopus officinalis*

Fig. 193. *Juniperus indica*

Fig. 194. *Mentha longifolia*

Fig. 195. *Nepeta coerulescens*

Fig. 196. *Nepeta floccosa*

Fig. 197. *Nepeta glutinosa*

Fig. 198. *Nepeta longibracteata*

Fig. 199. *Nepeta podostachys*

Fig. 200. *Origanum vulgare*

Fig. 201. *Oxytropis macrophylla*

Fig. 202. *Perovskia abrotenoides*

Fig. 203. *Salvia sclarea*

Fig. 204. *Tanacetum dollicophyllum*

Fig. 205. *Tanacetum gracile*

Fig. 206. *Tanacetum tibeticum*

Fig. 207. *Thymus serpyllum*

Fig. 208. *Valeriana jatamansi*

Fig. 209. *Waldheimia glabra*

Fig. 210. *Waldheimia stoliczkai*

Fig. 211. *Waldhemia tomentosa*

Fig. 212. *Allium carolinianum*

Fig. 213. *Allium humile*

Fig. 214. *Allium prezewalskianum*

Fig. 215. *Amaranthus spinosus*

Fig. 216. *Barbaria vulgaris*

Fig. 217. *Bidens pilosa*

Fig. 218. *Bunium persicum*

Fig. 219. *Bunium persicum* (fresh corms)

Fig. 220. *Capparis spinosa*

Fig. 221. *Carum carvi* var *gracile*

Fig. 222. *Chaerophyllum villosum*

Fig. 223. *Chenopodium album*

Fig. 224. *Chenopodium botrys*

Fig. 225. *Chenopodium foliosum*

Fig. 226. *Christolea crassifolia*

Fig. 227. *Draba altaica*

Fig. 228. *Eremurus himalaicus*

Fig. 229. *Erigeron multiradiatus*

Fig. 230. *Fagopyrum esculentum*

Fig. 231. *Fragaria vesca*

Fig. 232. *Gentiana tianschanica*

Fig. 233. *Lactuca dolichophylla*

Fig. 234. *Lactuca lassertiana*

Fig. 235. *Lepidium latifolium*

Fig. 236. *Malva verticillata*

Fig. 237. *Morus alba*

Fig. 238. *Oxyria digyna*

Fig. 239. *Pedicularis bicornuta*

Fig. 240. *Polygonum affine*

Fig. 241. *Prunus armeniaca*

Fig. 242. *Psychrogeton andryaloides*

Fig. 243. *Rheum webbianum*

Fig. 244. *Rheum webbianum* (juicy leaf stalks)

Fig. 245. *Rumex nepalensis*

Fig. 246. *Rumex patientia*

Fig. 247. *Sedum ewersii*

Fig. 248. *Silene setisperma*

Fig. 249. *Sonchus oleraceus*

Fig. 250. *Sonchus oleraceus* (Fresh leaves)

Fig. 251. *Taraxacum officinale*

Fig. 252. *Taraxacum sikkimenis*

Fig. 253. *Urtica dioca*

Fig. 254. *Urtica hyperborea*

Fig. 255. Sheep & goats on alpine pastures

Fig. 256. Herd of Yak

Fig. 257. Double hump camel of Nubra valley

Fig. 258. Zanskar ponies

Fig. 259. High altitude alpine pastures

Fig. 260. Harvesting of wild fodder plants

Fig. 261. Loose storage of fodder plants

Fig. 262. Traditional storage of fodder plants

Fig. 263. *Aconogonum alpinum*

Fig. 264. *Aconogonum tortusum*

Fig. 265. *Agropyron repens*

Fig. 266. *Astragalus munorii*

Fig. 267. *Astragalus subliformis*

Fig. 268. *Astragalus webbianus*

Fig. 269. *Astragalus zanskarensis*

Fig. 270. *Chesnea cuneata*

Fig. 271. *Cicer microphyllum*

Fig. 272. *Clematis orientalis*

Fig. 273. *Clematis tibetana*

Fig. 274. *Convolvulus arvensis*

Fig. 275. *Galinsoga parviflora*

Fig. 276. *Hippophae* sp. grazed by animals

Fig. 277. *Iris kumaonensis*

Fig. 278. *Iris lactea*

Fig. 279. *Lathyrus humillis*

Fig. 280. *Lindelofia anchusoides*

Fig. 281. *Lindelofia stylosa*

Fig. 282. *Lotus corniculatus*

Fig. 283. *Medicago falcata*

Fig. 284. *Medicago media*

Fig. 285. *Medicago sativa*

Fig. 286. *Millilotus officinalis*

Fig. 287. *Oxytropis cachemiriana*

Fig. 288. *Oxytropis lapponica*

Fig. 289. *Oxytropis tatarica*

Fig. 290. *Phragmitis australis*

Fig. 291. *Pleurospermum candollii*

Fig. 292. *Pleurospermum stellatum*

Fig. 293. *Potentilla microphylla*

Fig. 294. *Potentilla sericea*

Fig. 295. *Prangos pabularia*

Fig. 296. *Pseudomertensia achioides*

Fig. 297. *Rosa* sp grazed by animals

Fig. 298. *Stachys tibetica*

Fig. 299. *Trifolium pretense*

Fig. 300. *Trifolium repens*

Fig. 301. Collection of wild flowers as decoration

Fig. 302. *Aquilegia fragrans*

Fig. 303. *Aquilegia pubiflora*

Fig. 304. *Aster diplostephioides*

Fig. 305. *Aster flaccidus*

Fig. 30116. *Aster tibeticus*

Fig. 307. *Braya alpina*

Fig. 308. *Caltha palustris*

Fig. 309. *Campanula latifolia*

Fig. 310. *Choriospora sabulosa*

Fig. 311. *Cremanthodium reniforme*

Fig. 312. *Echinops cornigerus*

Fig. 313. *Epilobium angustifolium*

Fig. 314. *Epilobium latifolium*

Fig. 315. *Epilobium roseum*

Fig. 316. *Glechoma nivalis*

Fig. 317. *Impatiens amplexicaulis*

Fig. 318. *Leontopodium leontopodium*

Fig. 319. *Leontopodium nanum*

Fig. 320. *Meconopsis aculeata*

Fig. 321. *Meconopsis bikramii*

Fig. 322. *Papaver nudicaule*

Fig. 323. *Primula denticulata*

Fig. 324. *Primula elliptica*

Fig. 325. *Primula involucrata*

Fig. 326. *Primula macrophylla*

Fig. 327. *Primula reptans*

Fig. 328. *Rosa brunonii*

Fig. 329. *Rosa ecae*

Fig. 330. *Rosa webbiana*

Fig. 331. *Rosularia alpestris*

Fig. 332. *Saussurea gossipifora*

Fig. 333. *Silene edgeworthii*

Fig. 334. *Thermopsis inflata*

Fig. 335. *Tulipa stellata*

Fig. 336. *Arnebia* sp (dried roots)

Fig. 337. *Betula* sp. (bark as writting materials)

Fig. 338. *Juniperus* sp. (as incense)

Fig. 339. *Peganum harmala*

Fig. 340. *Physalis alkekengi*

Fig. 341. Dried flowers of *Physalis* sp as a traditional culture

Fig. 342. Collection of wild fuel plants

Fig. 343. Storage of wild fuel plants

Fig. 344. *Acantholimon lycopodioides*

Fig. 345. *Anthemis cotula*

Fig. 346. *Arenaria festucoides*

Fig. 347. *Caragana brevifolia*

Fig. 348. *Caragana versicolor*

Fig. 349. *Circium arvense*

Fig. 350. *Cotoneaster falconeri*

Fig. 351. *Erigeron multicaulis*

Fig. 352. *Eritrichium fruticulosum*

Fig. 353. *Eritrichium nanum*

Fig. 354. *Halogeton glomeratus*

Fig. 355. *Juniperus* sp (as fuel)

Fig. 356. *Lonicera asperifolia*

Fig. 357. *Lonicera spinosa*

Fig. 358. *Oxytropis mollis*

Fig. 359. *Potentilla cuneata*

Fig. 360. *Ribes alpestre*

Fig. 361. *Ribes himalense*

Fig. 362. *Ribes orientale*

Fig. 363. *Sibbaldia cuneata*

Fig. 364. *Sophora moorcroftiana*

Fig. 365. *Thylacospermum caespitosum*

Fig. 366. *Betula utilis*

Fig. 367. *Cedrus deodara*

Fig. 368. *Colutea multiflora*

Fig. 369. *Elaeagnus angustifolium*

Fig. 370. *Hippophae rhamnoides* ssp. *turkestanica*

Fig. 371. *Hippophae tibetana*

Fig. 372. *Juglans regia*

Fig. 373. *Juniperus communis*

Fig. 374. *Juniperus macropoda*

Fig. 375. *Juniperus macropoda* (branches)

Fig. 376. *Juniperus recurva*

Fig. 377. *Morus alba*

Fig. 378. *Myricaria elegans*

Fig. 379. *Myricaria germaniaca*

Fig. 380. *Myricaria squamosa*

Fig. 381. *Picea smithiana*

Fig. 382. *Pinus wallichiana*

Fig. 383. *Populus euphratica*

Fig. 384. *Populus nigra*

Fig. 385. *Rhododendron campanulatum*

Fig. 386. *Robinia pseudacacia*

Fig. 387. *Salix babylonica*

Fig. 388. *Salix fragilis*

Fig. 389. *Syringa emodi*

Fig. 390. *Tamarix gallica*

Fig. 391. *Ulmus wallichiana*

Fig. 392. In situ conservation

Fig. 393. Natural regeneration (*Meconopsis* sp)

Fig. 394. Filed gene bank (in winter)

Fig. 395. Field gene bank (in summer)

Fig. 396. Germplasm maintenance

Fig. 397. Seed germination under control

Fig. 398. Germplasm evaluation

Fig. 399. Micropropagation of endangered plants

Fig. 400. Training to amchis

Fig. 401. Training to NGO's

Fig. 402. Training to farmers

Fig. 403. Troops training

Fig. 404. Alpine Herbal garden (3,454 m.)

Fig. 405. *Achillea millefolium* (nursery)

Fig. 406. *Achillea millefolium* (cultivation)

Fig. 407. *Aconitum heterophyllum* (Vegetative propagation)

Fig. 408. *Aconitum heterophyllum* (nursery)

Fig. 409. *Artemisia annua* (ready for harvest)

Fig. 410. *Artemisia dracunculus* (ready for harvest)

Fig. 411. *Bidens pilosa* (cultivation & harvesting)

Fig. 412. *Carum carvi* var. *gracile* (nursery)

Fig. 413. *Cicer microphyllum* (cultivation)

Fig. 414. *Dactylorhiza hatagirea* (vegetative propagation)

Fig. 415. *Dactylorhiza* sp (cultivation)

Fig. 416. *Dracocephalum* sp (nursery)

Fig. 417. *Dracocephalum heterophyllum* (in full bloom)

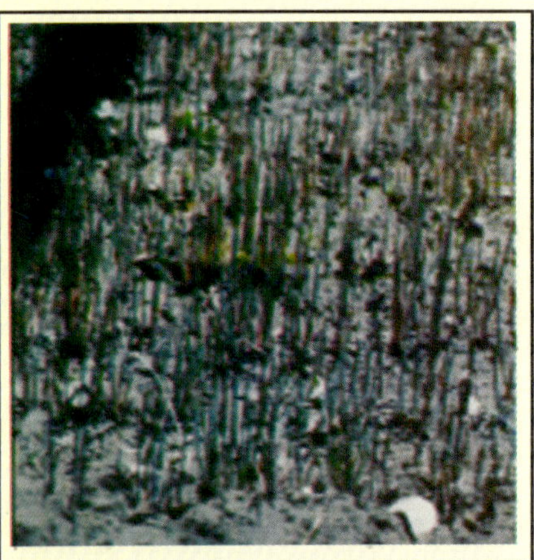

Fig. 418. *Ephedra gerardiana* (nursery)

Fig. 419. *Ephedra gerardiana cultivation*

Fig. 420. *Hippophae rhamnoides* (vegetative propagation)

Fig. 421. *Hippophae tibetana* (nursery)

Fig. 422. *Inula racemosa* (nursery)

Fig. 423. *Inula racemosa* (cultivation)

Fig. 424. *Origanum* sp (nursery)

Fig. 425. *Origanum vulgare* (in cultivation)

Fig. 426. *Physochlaina praealta* (nursery)

Fig. 427. *Podophyllum* sp (nursery)

Fig. 428. *Podophyllum hexandrum* (ready for harvest)

Fig. 429. *Rheum webbianum* (vegetative propagation)

Fig. 430. *Rheum webbianum* in cultivation

Fig. 431. *Rhodiola imbricata* (Cultivation)

Fig. 432. *Rubia cordifolia* (in fruiting stage)

Fig. 433. *Salvia sclarea* (in full bloom)

Fig. 434. *Saussurea* sp (Nursery)

Fig. 435. *Saussurea lappa* (In fruiting stage)

Fig. 436. *Urtica hyperborea* (ready for harvest)

Fig. 437. *Valeriana jatamansi* (cultivation)

Fig. 438. *Verbascum thapsus* (nursery)

Bibliography

Agarwal, V.S. (1986). Economic Plants of India. Bishen Singh Mahendra Pal Singh, Dehra Dun & Kailash Prakashan, Calcutta.

Agharkar, S.P. (1953). Medicinal Plants – Pt I of Bombay, Gazetter of Bombay State (Revised Series), Volume : Genral – A, The Govt Central Press, Bombay.

Ahlgren, G.H. (1956). Forage Crops. McGraw-Hill Book Co., Inc., New York. 2nd Edn. 1959.

Ahluwalia, H. P.S. (1980). Hermit Kingdom Ladakh. Vikas Pub., New Delhi.

Ahmad, S. & R.R. Stewart. (1958 & 1959). Grasses of West Pakistan. *The Biological Society of Lahore*, Monograph No. 3, Pt I Panniocoideae, 1958; Pt II Pooideae, 1959.

Ali, S.I. (1959). Revision of the genus *Oxytropis* from W. Pakistan and N.W. Himalayas. *Phyton* (Austria), **8** : 49-61.

Ali, S.I. (1961). Revision of the genus *Astragalus* Linn. from W. Pakistan and N.W. Himalayas. *Biologia*, **7** : 7-92.

Ali, S.I. (1965). Contribution to the genus *Astragalus* Linn. from W. Pakistan and N.W. Himalayas. *Bot. Notiser*, **118**(1) : 187-196.

Anonymous, (1948-76). The Wealth of India- Raw Materials., I-IX. C.S.I.R., New Delhi.

Anonymous, (1984). All India Coordinated Research Project on Ethnobiology (Annual Report). Department of Environment, New Delhi.

Anonymous, (1986). The Useful Plants of India. Publication & Information Directorate, C.S.I.R., New Delhi.

Anonymous, (1990). Ethnobiology in India : A status Report. Ministry of Environment & Forests, Govt. of India, New Delhi.

Anonymous, (1993). Ethnobotany in India. Sci. Pub., Jodhpur (Raj.).

Arora, R.K. & Anjula Pandey. (1996). Wild Edible Plants of India Diversity Conservation and Use. I.C.A.R., N.B.P.G.R., Pusa, New Delhi.

Arora, R.K., M.L. Maheshwari, K.P.S. Chandel, & R. Gupta. (1980). Mano (*Inula racemosa*) Little Known Aromatic Plants of Lahaul Valley, India. *Econ. Bot.* **34**(2) : 175-180.

Aswal, B.S. & B.N. Mehrotra. (1979). New Records of Plants from Himanchal Pradesh, India. *ibid.*, **2**(4) : 322.

Aswal, B.S. & B.N. Mehrotra. (1980). Contribution to the Flora of Lahaul Valley (North West Himalayas) –II. Some New Plant records. *J. Econ. Taxon. Bot.*, **1** : 115-117.

Aswal, B.S. & B.N. Mehrotra. (1981). Contribution to the Flora of Lahaul Valley (North West Himalayas) –III. A Note on the Nomenclature of some Plants. *Ibid*, **2** : 236.

Aswal, B.S. & B.N. Mehrotra. (1983). Contribution to the Flora of Lahaul Valley (North West Himalayas) –IV. Some New Plant records. *Indian J. For.*, **6**(4) : 314-318.

Aswal, B.S. & B.N. Mehrotra. (1983). New Records of Plants from Himanchal Pradesh, India –III. *J. Econ. Taxon. Bot.*, **4**(3) : 991-994.

Aswal, B.S. & B.N. Mehrotra. (1985). Contribution to the Flora of Lahaul Valley (North-West Himalayas) –V. Phytogeographical Aspects. *J. Econ. Taxon. Bot.*, **7**(2) : 299-307.

Aswal, B.S. & B.N. Mehrotra. (1987). Ethnobotanical Studies on the Flora of Lahaul Valley (North-West Himalayas). In : *Recent Advances in Plant Sciences*. (Eds. M. R. Sharma and B. K. Gupta). Pp: 116-130.

Atal, C.K. & Kapur B.M. (Eds. 1982). Cultivation & Utilization of Medicinal Plants. Regional Research Lab. Jammu- Tawi. India.

Atal, C.K. & Kapur B.M. (Eds. 1982). Cultivation & Utilization of Aromatic Plants, Regional Research Lab. Jammu- Tawi. India.

Atkinson, E.T. (1882). The Himalayan Gazetteer. , 1-3, Cosmo Publications, New Delhi (Reprint 1973).

B.P. (1980). British Pharmacopoeia. (Her Majesty's Stationary Office, University Press, Cambridge), 1973; 2nd, 1980; Addendum, 1983.

B.P.C. (1949). The British Pharmaceutical Codex The Pharmaceutical Press, London.

B.P.C. British Pharmaceutical Codex (1963 & 1968). The Pharmaceutical Press London.

Badhwar, R.L & R.R. Fernandez. (1964). Edible Wild Plants of the Himalayas". Printed at the Survey of India Offices 103 (P.Z.O.) Printing group, Dehradun. Distributed by Manager of Publications, Delhi.

Ballabh, B. (2002). Ethnobotany of Boto Tribe of Ladakh Himalaya. Ph. D. Thesis, Kumaun University, Nainital.

Bhandari, M.M. (1978). Flora of Indian Desert. Sci. Pub., Jodhpur (Raj.).

Bhattacharyya, A. (1991). Ethnobotanical observation in the Ladakh Region of Northern J&K State, India, *Econ. Bot.*, **45** (3) : 305-308.

Bhattacharyya, P.K. and S.K. Mukerjee. (1978). Indian Cuscutaceae. *Indian J. For.*, **1**(2) : 156-162.

Binu, S., T. S. Nayar & P. Pushpangadan, (1992). An Outline of Ethnobotanical Research in India. *J. Econ. Taxon. Bot.*, Addl. Ser., **10** : 405-428.

Blatter, E. (1927-29. Beautiful Flowers of Kashmir, 2 Vols. John Balc, Danielsson, London.

Bobrov, E. G. (1967). A review of the genus *Myricaria* Desv. and its History. *Bot. J. USSR*, **52** : 924-936.

Caius, J.F. (1986). The Medicinal and Poisonous Plants of India. Sci. Pub., Jodhpur.

Champion, H.G. & S.K. Seth. (1968). A Revised Survey of the Forest Type of India. Govt. of India, Publ. New Delhi.

Chaurasia, O.P. & B. Singh, (1996-2001). Cold Desert Flora , **(I-V)**. Field Research Laboratory, Leh - Ladakh. (J&K).

Chaurasia, O.P., B. Singh & S.K. Sareen, (1999). Ethno-medico-botaniical Survey of Nubra Valley (Ladakh). *J. Econ. Taxon. Bot.*, **23** (1) : 167-172.

Chaurasia, O.P., B. Singh, D.V. Singh & B.B. Bhatt, (1999). Rare, Endangered & Threatened Medicinal Flora of Ladakh Himalaya & Their Conservation Strategies. In : South East Asian Seminar on Herbs and Herbal Medicines 16-19 Jun. pp : 132.

Chaurasia, O.P., B.B. Bhatt & D.V. Singh. (2000). Ethnobotany of Ladakh Himalaya. In : Ladakh : Past & Present. (Ed. P.S. Jina). Gyan Publishing House, New Delhi.

Chaurasia, O.P., Brahma Singh & Basant Ballabh. (2001). Potential Aromatic Flora of Himalayan Cold Desert : Ladakh & Lahul - Spiti. *J. Econ. Taxon. Bot.*, **25** (1) : 91-97.

Chaurasia, O.P., Brahma Singh & S.K. Sareen. (1999). Ethno-Medicinal Plants of Arctic Desert - Ladakh Used in Veterinary Practices. *J. Econ. Taxon. Bot.*, **23**(1) : 161-166.

Chaurasia, O.P., Narendra Singh & Brahma Singh. (2000). Potential wild Leafy Vegetables from Ladakh. *Indian Horticulture*. pp 4-5.

Chaurasia, O.P. & Padma Gurmeet, (2004). A checklist on medicinal & aromatic plants of trans Himalayan cold desert (Ladakh & Lahaul-Spiti). Field Research Laboratory, Leh Ladakh (J&K).

Chohan, A.S. (1983). Historical Study of Society and Culture in Dardistan and Ladakh. New Delhi.

Chopra, R.N., I.C. Chopra & B.S. Verma. (1969). Supplement to Glossory of Indian Medicinal Plants. I.C.A.R. Publications & Information Directorate, New Delhi.

Chopra, R.N., R.L. Badhwar & S. Ghos. (1949). Poisonous Plants of India. I.C.A.R., Calcutta, 2nd Edn. Revised and Enlarged, 2nd Vols, 1965.

Chopra, R.N., R.L. Badhwar & S. Ghosh. (1984). Poisonous Plants of India. Academic Publishers, Jaipur.

Chopra, R.N., S.L. Nayar & I.C. Chopra. (1956). Glossory of Indian Medicinal Plants. I.C.A.R., New Delhi.

Choudhury, B. (1967). Vegetables : India - The Land and People. National Book Trust, India, New Delhi.

Chowdhery, P.K. and B.M. Wadhwa. (1984). Flora of Himanchal Pradesh Analysis, **(1-3)**, Howrah.

Clarke, C. B. (1876). Compositae Indicae. London.

Claus, E.P. (1961). Pharmacognosy. Henry Kimpton, London. 4th Edn. 98.

Collett, H. (1902). Flora Simlensis, Thacker, Spink, Calcutta and Shimla.

Cowen, D.V. (1950). Flowering trees and Shrubs in India. Thacker & Co. Ltd. Bombay.

Cunningham, Alexander: (1977). Ladakh (Physiacal, Statistical and Historical with notices of the Surrounding Countries), Sagar Pub., New Delhi, Reprint.

Dar, G.H., J. Vir, P. Kachroo & G.M. Buth (1983). Ethnobotany of Kashmir-I, Sind Valley, *J. Econ. Taxon. Bot.* **5** (3) : 668-675.

Das, Pflanzenreich : (1900). Regni Vegetabilis Conspectus. A. Engler. H. R. Engelmann. J. Craner, Weinhein, Haft.

Dastur, J.F. (1951). Useful Plants of India and Pakistan. D.B. Taraporevala Sons & Co. Ltd. Bombay.

Deva, Som & H.B. Naithani. (1986). The *Orchid* Flora of North West Himalaya, New Delhi.

Dhar, Uppendra & Kachroo, P. (1983). Alpine Flora of Kashmir. Sci. Pub., Jodhpur, India.

Dhar, Uppendra & Kachroo, P. (1989). Alpine Flora of Kashmir – Phytogeographic Assessment. In : "The Vegetational Wealth of the Himalayas". (Ed. G.S. Paliwal). Puja Pub., Delhi., Puja Pub., Puja Publ, Delhi. pp : 327-332.

Dorjai, Gawai (1995). Thungs-Pai-Te Maid-Shail-Gi-Melong, Me-Regs Pai Tun Khang Lhasa Tibet.

Duhoon, S. S. (1996). Seabuckthorn, a less-known wonder plant of ethnobotanical importance in cold deserts of India. *J. Econ. Taxon. Bot.* (Additional Series). 12 : 43-45.

Duthie, J.F. (1893). Report on a Botanical Tour in Kashmir, 1892. Rec. *Bot. Surv. Ind.,* **1** : 1-18.

Duthie, J.F. (1903-1929). Flora of the Upper Gangatic Plain and of the Adjacent Siwalik and Sub-Himalayan Tracts. Government Press, Calcutta, 3 Vols. Reprinted Edn. *Bot. Sur. of Ind.,* Calcutta, 2 Vols, 1960.

Duthie, J.F. (1906). The *Orchids* of North West Himalayas. Calcutta.

Dutta, U. C. (1989). The Materia Medica of the Hindus, Mittal Publ. New Delhi.

Dymock, W., C. J.H. Warden & D. Hooper. (1893). Pharmacographia Indica. Trubner & Co., London. **3 :** 149.

Dymock, W., C.J.H. Warden & D. Hooper. (1976). Pharmacographia Indica : A History of principal Drugs of Vegetable Origin, met with British India. Tribebner & Co. Ltd, London, 3 Vols. 1889-91; Index and Appendix, 1993; reprint Bishen Singh Mahendra Pal Singh & Periodical experts, New Delhi, **3** : 1976.

Encyclopaedia Americana. (1962). American Croporation, New Yark, 28 Vols.

Encyclopaedia Britinnaca. (1967). Encyclopaedia Britinnaca.Inc. Chicago, 24 Vols.

Engler, H.G.A. (1900-1953). Das Pflanzenreich Regni Vegetabilis Conspectus. Hefte. Berlin, 1-107.

Francke, A.H. (1905 & 1980). 'Ladakh : The Mysterious Land', Ess. Pub. Co. New Delhi, pp 20-21.

Francke, A.H. (1977). A History of Ladakh. With critical introduction and annotation by S S Gergan and F M Hassnain, Sterling pub., Pvt. Ltd. New Delhi.

Francke, A.H. (1977). A History of Western Tibet. (Rep. Edn.) New Delhi.

Gadgil, M & Devasia, P. (1995). Intellectual Property Rights and biological resources : specifying geographical origins and prior knowledge of users, Curr. Sci. 69 : 637-639.

Gardner, C.A. & H.W. Bennetts. (1956). The Toxic Plants of Western Australia. West Australian News papers Ltd. Perth.

Ghosh, R. (1957). R. Ghosh's Pharmacology, Materia Medica and Therapeutics, including applied Pharmacology. (Ed. S.K. Biswas). Hilton & Co., Calcutta.

Goel, Anil K. & B.S. Aswal. (1987). *Juncus rohtangensis* Goel *et.* Aswal (Juncaceae) - A New Species from Himanchal Pradesh, India. *Indian J. For.*, **10**(4): 262-263.

Gohil, P.N. & M.A. Quardi (1992). Ethnobotany of Kargil- Medicinal Plants used by Balti, Dard, Bokpa Races, *J. Econ. Taxon. Bot.* Addl. Ser. **10** : 301-306.

Good, R. (1929). The Taxonomy and Geography of the Sino-Himalayan Genus *Cremanthodium* Benth. *J. Linn. Soc. Bot.*, **48** : 259-316.

Goyal, H.R., P. Kishore, R.K. Issar, S.T. Phuntsog & G. Namgyal. (Compiled & Eds. 1996-1998). Research on Amchis System of Medicine - A Research Profile. Amchi Research Unit Leh-Ladakh. Central Council for Research in Ayurveda & Siddha; (Ministry of Health and Family Welfare) Govt. of India, New Delhi.

Grierson, A.J.C. (1964). A Revision of the Asters of Himalayan area. *Notes Royal Bot. Gard., Edinb.*, **26**(1) : 66-163.

Gupta, O.P., T.N. Srivastava, S.S. Gupta & O.P. Badola. (1980). An Ethnobotanical & Phytochemical Screening of High Altitude Plants of Ladakh Part- I, *Bull. Medico-Ethnobotanical Research*, **1** (3) : 301-317.

Gupta, O.P., T.N. Srivastava, S.S. Gupta & O.P. Badola. (1981). An Ethnobotanical & Phytochemical Screening of High Altitude Plants of Ladakh Part- II, *Bull. Medico-Ethnobotanical Research*, **2** (1) : 67-68.

Gupta, R.K. (1967). Medicinal Plants of Western Himalaya, *J. Agrl. Trop. Bot. Applied*, **9** : 1-55.

Gupta, R.K. (1989). Mediterranean Influence in the Flora of Western Himalayas. In : "The Vegetational Wealth of the Himalayas". (Ed. G.S. Paliwal). Puja Pub., Delhi pp 175-193.

Gurmet Padma, O.P. Chaurasia, Brahma Singh & D.V. Singh (1998). Some traditional medicinal plants of Khardungla and Changla (Ladakh) Used in Amchi medicine. *B.M.E.B.R.*, XIX, **3-4** : 93-101.

Hara, H. (1963). Spring Flora of Sikkim Himalaya, Hoikusha, Japan.

Hara, H. (1966). Flora of Eastern Himalaya. The University of Tokyo Press, Japan.

Hara, H. (1968). Photo-Album of Plants of Eastern Himalaya, Inoue Book Co., Tokyo.

Hara, H. (1975). New of Noteworthy Flowering Plants from Eastern Himalaya (18). *Ibid*, **51**(5) : 129-135.

Hara, H. and L.H.J. Williams. (1979). An Enumeration of the Flowering Plants of Nepal, **2** : London.

Hara, H., Chater, C.A. & L.H.J. Williams. (1982). An Enumeration of the Flowering Plants of Nepal, **3** : London.

Harshberger, J.W. (1895). Some New Ideas. Philadelphia Evening Telegraph.

Homoeopathic Pharmacopoeia of India. (1971). Govt of India, Min of Health, New Delhi.

Hooker, J.D. (1872-1897). A Sketch to the Flora of British India, Oxford.

Hooker, J.D. (1872-1897). The Flora of British India Vols. 1-7 Reeves & Co. London (Secretary of States for India, London).

Hutchinson, J. (1955). British Wild Flowers, Pegnguin Books Ltd. Middlesex, 2 ,

I.P. (1966). Pharmacopoeia of India (The Indian Pharmacopoeia). Manager of Publications, Delhi.

Ilyina, M.M. (1974). Plant Raw Materials of U.S.S.R. *Academy of Sciences of U.S.S.R.* Moscow, **2** : Natural Plants.

Irshad A., A. Navchoo & G.M. Buth. (1989). Medicinal System of Ladakh, India. *J. Ethnopharmacology*, **26** : 137- 146.

Irshad A., A. Navchoo & G.M. Buth. (1990). Ethnobotany of Ladakh, India. *Econ. Bot.*, **44**(3) : 318-321.

Jafari, S.M.H. (1955-1956). Some Cruciferae of W. Pakistan, Afganistan and N.W. Himalayas. *Note Royal Bot. Gdn., Edinb.*, **22**(1-2) : 1-199.

Jain, S.K. & R.R. Rao, (1977). A Hand Book of Field and Herbarium Methods. Today and Tomorrow's Prints & Pub. New Delhi.

Jain, S.K. & R.R. Rao, (1997). A Hand Book of Field and Herbarium Methods. Today and Tomorrow's Prints & Pub. New Delhi.

Jain, S.K. & R.R. Rao. (1983). Ethnobotany in India - An overview, *Bot. Surv. Ind.* Hawarah.

Jain, S.K. (1963). Medicinal Plants Lore of the tribal of Bastar, *Econ. Bot.*, **19** : 236-250.

Jain, S.K. (1963). Plants used in medicine by Madhya Pradesh. *Bull. Regional Research Lab.*, Jammu, **1** : 126-128.

Jain, S.K. (1964). The Role of a Botanist in folklore Research. *Folk-lore,* **5** : 145-150.

Jain, S.K. (1967). Ethnobotany : Its scope in study. *Indian Mus. Bull.*, **2** : 39-43.

Jain, S.K. (1989). Methods of Approaches in Ethnobotany. *Society of Ethnobotanists*, Lucknow.

Jain, S.K. (1991). Dictionary of Indian Folk- Medicine and Ethnobotany. Deep Publications, New Delhi.

Jain, S.K. (1994). Medicinal Plants. National Book Trust. New Delhi.

Jain, S.K. (Ed. 1981). Glimpses of Indian Ethnobotany. Oxford and IBH Publishing Co. New Delhi.

Jain, S.K. (Ed. 1987). A Manual of Ethnobotany. (Proceeding of the Training Course and Workshop on Ethnobotany held at Lucknow, 10-15 March, 1986). Sci. Pub. Jodhpur.

Jain, S.K. (Ed. 1996). Ethnobotany in Human Welfare (Man Plant Relationships & Conservation Diversity). Deep Publication, New Delhi.

Jain, S.K., Minnis, Paul E. & N.C. Shah. (1986). A World Dictionary of Ethnobotanists. *Society of Ethnobotanists*, Lucknow.

Jain, S.K. & Srivastava, S. (2001). Indian ethnobotanical literature in last two decades - a graphic review and further directions, Ethnobotany. 13 : 1-8.

Jina P.S. (1983-84). Boudh Bikshus of Ladakh. In: Proceed of All India Conference on Buddhist & Jaina Monasticism, Deptt. Of Buddhist Studies, Delhi University, Delhi.

Jina, P.S. (1995). Famous Western Explorers to Ladakh. Indus Pub. Co. New Delhi.

Jina, P.S. (1996). High Pasturelands of Ladakh Himalaya. Indus Pub. Co. New Delhi.

Jina, P.S. (1996). Ladakh the Land & the people. Indus Pub. Co. New Delhi.

Joshi, Pramila. (1993). Ethnobotany of the Tribal Communities of the Kumaun Himalaya. Ph. D. Thesis, Kumaun University, Nainital. (Unpublished).

Kachroo, P., B.L. Sapru & U. Dhar, (1977). Flora of Ladakh. Bishen Singh Mahendra Pal Singh 23-A Connaught Place, Dehra Dun.

Kant, Shashi & Pushp Gupta. (1998). Algal Flora of Ladakh. *J. Econ. Taxon. Bot.* Addl. ser, 15. Sci. Pub. Jodhpur.

Kapoor, L.D., B.K. Abrol & I.C. Chopra, (1957). Medicinal Plants of High Altitude. *Bull. Nat. Inst. Sci. Ind.* **9** : 223-229.

Kapur, T.B. (1987). AVSM Brig: Ladakh the Wonder Land. Mittal Publishers New Delhi.

Kaul, M.K. (1997). Medicinal Plants of Kashmir & Ladakh (Temperate & Cold Himalaya). Indus Publishing Company, New Delhi.

Kaul, M.K., A.K. Bhatia & C.K. Atal. (1985). Ethnobotanical studies in Notrthwest & Trans-Himalaya. Contribution to the wild food plants of Ladakh. *J. Econ Taxon. Bot.*, **6** : 523-527.

Kaul, M.K., P.K. Sharma & V. Singh. (1998). Crude Drugs of Zanskar (Ladakh) : Conservation Assessment and Management Plan Workshop - Kullu (H.P.)16-18 April.

Kaul, Shridhar & HN. Kaul. (1995). Ladakh Through The Ages - Towards a New Identity, Indus Pub. Co., New Delhi.

Khan, Reayat. (1989). Utilization & Genetic Improvement of the Wild & Cultivated Plants of the Himalaya. In : "The Vegetational Wealth of the Himalayas". (Ed. G.S. Paliwail). Puja Pub. Delhi. Pp. 521-526.

Khare, C.P. (2004). Encyclopedia of Indian Medicinal plants. Springler. Germany.

Koelz, Walter N. (1979) Notes on the Ethnobotany of the Lahaul a Province of the Punjab. Quart. *J. Crude Drug Res.*, **17**(1) : 1-56.

Kritikar, K.R. and B.D. Basu. (1935). & An ICS retd. Eds. Revised by E. Blatter, J. F. Caius and K.S. Mhaskar. Indian Medicinal Plants. Lalit Mohan Basu, Allahabad, 2nd Edn, 4 Vols, (with 1-1033 Plates). Rep. 1995-1999.

Kumar, A.M. & A.R. Naqshi, (1990). Ethnobotany of Jammu- II Banihal. *J. Econ. Taxon. Bot.*, **14**(1) : 67-74.

Lambert, Cowley. (1877). Trip of Kashmir and Ladakh, Henry & King, London.

Maheshwari, J.K. (1987). Interdisciplinary Approaches in Ethnobotany. In : A Manual of Ethnobotany. (Ed. S.K. Jain). Sci. Pub. Jodhpur. Pp. 23-32.

Maheshwari, J.K. (1989). Some Thoughts on the Endangered Flora of the Himalayas. In : "The Vegetational Wealth of the Himalayas", (Ed. G.S. Paliwal). Puja Pub., Delhi, Pp. 256-268.

Maheshwari, P. & Umrao Singh. (1965). Dictionary of Eonomic Plants in India. I.C.A.R., New Delhi.

Mani, M.S. (1978). Ecology and Phytogeography of High Altitude Plants of Northwest Himalaya. Oxford & IBH Pub. Co. New Delhi, Mumbai, Calcutta.

Mani, M.S. (1990). Fundamentals of high altitude Biology. Oxford and IBH, Publishing Company Pvt. Ltd.

Martin, J. Gary, (1995). Ethnobotany "A Method Manual". Chapman & Hall. London.

Mashelkar, R.A. (2001). Intellectual Property Rights and the Thirld world. Curr. Sci. 81 : 955-965.

Menon, A.K. (1960). Indian Essential Oils : *A Review. C.S.I.R.* New Delhi.

Moorcroft, W. & G. Trebeck. (1841). Travels in Himalayan Provinces of Hindustan. 1-2. London.

Murti, K.S. (1996). Biodiversity of Indian Cold Desert Ecosystem Status and Conservation, A Book by Bot. Sur. of Ind., Dehradun, India (Unpublished).

Nadkarni, K.M. (1954). Indian Materia Medica. Popular Book Depot, Bombay & Dhootapapeshwar Prakashan Ltd. Panvel, 2 Vols, 3rd Edn, Revised and Enlarged by A.K. Nadkarni.

Nair, N.C. 1964. Some Plant Records for Lahaul and Spiti. *ibid.* **6** : 151-153.

Nakao, S. Living Himalayan Flowers. (1964). The Mainichi Newspapers, Tokyo.

Namgial, Gurmet, Phuntsog, ST. (1990). Amchi Pharmaco-Theraputic. Amchi Research Centre Leh-Ladakh, CCRAS. New Delhi. 110066.

Nasir, E. & S.I. Ali. (1972-1976). Flora of West Pakistan. Eds. An Annotated Catalogue of the Vascular Plants of West Pakistan and Kashmir. R.R. Stewart. Fakhri Printing Press, Karachi.

Nasir, E. and S.I. Ali. (1971-1982). Flora of West Pakistan. Eds. Nos. **8** : (1971) Karachi.

Navchoo I.A. & G.M. Buth, (1992). Ethnobotany of Ladakh, J&K State. *J. Econ Taxon. Bot.*, Addl. Ser., **10** : 251-258.

Navchoo, I.A. & G.M. Buth. (1990). Ethnobotany of Ladakh, India, Bevarages, Narcotics, Food. *Econ. Bot.*, **44** (3) : 318-321.

Nayar, M.P. & A.R.K. Sastry, (1987). Red Data Book of Indian Plants., **I.** Bot. Sur. Ind.

Om Prakash Vidyarthi. (1997). Wild & Cultivated plants of Jammu, Kashmir & Ladakh. Directorate of Social Forestry Project Jammu & Kashmir Govt.

Pande, P.C. & S S Samant. Plant diversity of the Himalaya. 2001. Gyan Publishers, Nainital.

Pande, P.C., D.S. Pokhariya & J.C. Bhatt. (1999). Ethnobotany of Kumaun Himalaya. Sci. Pub. Jodhpur (India).

Pengelly A. (2004). The constituents of medicinal plants. CABI Publishing. Cambridge.

Polunin Oleg. & Adam Stainton. (1989). Flowers of the Himalaya. Oxford University Press. New Delhi.

Prajapati, N.D. & etal. (2004). A handbook of Medicinal plants. Agrobios. Jodhpur.

Pushpangadan, P. (1990). Health status of tribals of India. Key note address in National seminar on health of tribal people. Indian Society for health administration, 21-26 pp. 57-61.

Raghunathan, K. (1987). Medico-Ethno-Botanical Surveys and their role in Research in Ayurveda in Manual of Ethnobotany, Sci. Pub. Jodhpur, (Raj.).

Raghunathan, K. (Ed. 1976). Preliminary Techno Economical Survey of Natural Resources and Herbal Wealth of Ladakh. CCRIMH, New Delhi.

Rao, R.R. & R.K. Hajara. (1987). Methods of Research in Ethnobotany : In : A Manual of Ethnobotany. (Ed. S.K. Jain). Sci. Pub. Jodhpur. Pp : 33-41.

Rau, M.A. (1960). On a collection of Plants from Lahaul, Bull. *Bot. Surv. Ind.*, **2** : 45-56.

Rau, M.A. (1974). Vegetation and Phytogeography of the Himalaya In: Ecology and Biogeography in India. (Ed. M.S. Mani). The Hague, 247-280.

Rau, M.A. (1975). High Altitudes Flowering Plants of Western Himalaya. Bot. Sur. Ind. Hawrah.

Rau, M.A. (1981). Western Himalayan Flora, In: "The Himalaya"-Aspects of Change. (Ed. J.S. Lall). Oxford & Delhi, 50-63.

Raven, P.H. (1962). The Genus *Epilobium* in the Himalayan Region. Bull. Brit. Mus. *Nat. Hist. Bot.*, **2** : 325-382.

Rizvi, J. (1983). Ladakh, Crossroads of High Asia. New Delhi.

Sahni, K.C. (1989). Himalayan Flora and Physiography - A Study in Contrasts. In: The Vegetational Wealth of the Himalayas. (Ed. G.S. Paliwal), Puja Pub. Delhi. Pp : 194-200.

Samant, S.S., U. Dhar & R.S. Rawal. (2001). Diversity and Distribution of Wild Edible Plants of the Indian Himalaya. In : Plant Diversity of the Himalaya. (Eds. P.C. Pande and S.S. Samant). Pp. 421-482.

Santapau, H. & A.N. Henry. (1973). A Dictionary of Flowering Plants of India. Publications & Information Directorate (CSIR), New Delhi.

Santapau, H. (1966). Common Trees : India - The Land and People. National Book Trust, India, New Delhi.

Schultes, Richard Evans. (1996). The Plant Kingdom – A Thesaurus of Biodynamic Constituents. *Ethnobotany*, **8** : 2-13.

Schults, R.E. (1960). Tapping Our Heritage of Ethnobotanical Folklore. *Econ. Bot.*, 257-262.

Schults, R.E. (1962). The Role of Ethnobotanists in the Search for New Medicinal Plants. *Lloydia*, **25** : 257-266.

Shah, N.C. (1987). Ethnobotany in the Mountainous region of Kumaon Himalaya. Thesis submitted to the Kumaon University, Nainital for the degree of Doctor of Philosphy in Botany. 1-255.

Sharma, D.D. (2000). Brog-Yul : The Land and the People. In : Ladakh : Past & Present. (Ed. Prem Singh Jina), Gyan Publihing House, New Delhi.

Sharma, P.K. (1991). Herbal Remedies for Treating Rheumatic Pains in Jammu & Kashmir, *Indian J. For.*, **14** (3) : 206-210.

Sharma, P.S. & V. Singh. (1989). Ethnobotanical Studies in Northwest & Trans-Himalaya V. Ethno-veterinary Medicinal Plants Used in Jammu & Kashmir State, India. *J. Ethnopharmacology*, **27** (1-2) : 63-70.

Singh B. & O.P. Chaurasia, (1998). Medicinal Plants of Cold Desert of India. In Golden Jubilee National Sym. On species, Med., Aromatic Plants Biodiversity, Conservation and Utilization, Calicut. 10-12 Aug 98. Abstract p.15.

Singh B. & O.P. Chaurasia. (2000). Medicinal Flora of Indian Cold Desert. *Acta Hort.*, ISHS, Pub. Belgium. **523**(13) : 65-72.

Singh B., Chaurasia, O.P. & D.V. Singh. (1997). Plants Biotechnology and Herbal Flora of Ladakh. In : Proceedings of National Sym. on Herbal Medicines : Application of Biotech. and Futuristic Approach. 27-29 Mar. 97.

Singh B., O.P. Chaurasia & Basant Ballabh. (2001). Edible wild plants of Trans-Himalayan Cold Desert. In : Plant Diversity of the Himalaya. (Eds. P.C. Pande and S.S. Samant). Ynodaya Prakashan, Nainital. Pp. 483-512.

Singh, B. & O.P. Chaurasia. (1995). Comparative Morpho-physiological Studies in Wild and Cultivated Gram (Cicer spp.) at Ladakh. *Adv. Plant. Sci.*, **8**(2): 308-311.

Singh, B., O. P. Chaurasia & K. L. Jadav. (1995). An Ethnobotanical Study of Indus. Valley (Ladakh). *J. Econ. Taxon. Bot.*, Addl Ser., **12** : 92-101.

Singh, B., O.P. Chaurasia & K.L. Jadav. (1996). Botanical Curiosities in the Alpine Flora of Khardungla. . *J. Econ. Taxon. Bot.*, **20**(1) : 135-142.

Singh, B., O.P. Chaurasia, & DV. Singh. (2000). Medicinal and aromatic plants resources of cold arid zones and their utilization. Dynamics & Cold Arid agriculture. Kalyani Pub. New. Delhi, 147-165.

Singh, H.B. & R.K. Arora. (1978). Wild Edible Plants of India. I.C.A.R., New Delhi.

Singh, Harbhajan & T.A. Thomas. (1978). Grain Amaranths, Buckwheat and Chenopods. I.C.A.R., New Delhi.

Singh, Unrao, A.M. Wadhwani & B.M. Johri (1965). Dictionary of Economic Plants of India.

Indian Council of Agricultural Research, New Delhi.

Singh, V. (1997). Lesser Known Wild Edible Plants of Zanskar, (Ladakh) Himalaya, *J. Econ. Taxon. Bot.*, **21** (1) : 175-178.

Singh, Y.K. (1982). A Note on *Rhodiola imbricata* Edgew. (Crassulaceae). *ibid.*, **3**(3) : 303-306.

Srivasatava, T.N. & O.P. Gupta. (1982). Medicinal Plants Used by Amchis in Ladakh. In : Cultivation & Utilization of Medicinal & Aromatic Plants. (Eds. Atal, C.K. & B.M. Kapur)., pp. 103-106.

Srivasatava, T.N., O.P. Gupta, M.R. Uniyal & N.C. Shah. (1976). Preliminary Technoeconomic Survey of Natural Resources and Herbal Wealth of Ladakh. (Ed. K. Raghunathan). CCRIMH Publication, 21 (1st Edn).

Srivasatava, T.N., O.P. Gupta, M.R. Uniyal & N.C. Shah. (1996). Preliminary Technoeconomic Survey of Natural Resources and Herbal Wealth of Ladakh. (Ed. K. Raghunathan). C.C.R.I.M.H., Pub., 21 (2nd Edn).

Srivastava, T.N.. D.P. Badola & O.P. Gupta. (1981). Medicinal herbs Used by Amchis of Ladakh. *Bull. Ethno-Medico Bo.t Res.*, **2** (2) : 145-166.

Stewart, J.L. (1869). Punjab Plants, comprising Botanical and Vernacular Names and Uses of most of the Trees, Shrubs and Herbs of Economical Value, Growing within the Province. Govt. Press, Lahore.

Stewart, R.R. (1972). Flora of West Pakistan. An Annonated Catalogue of the Vascular Plants of West Pakistan and Kashmir. Fakhri Press, Karachi.

Stewart, R.R. (1972-1976). Flora of West Pakistan. Edited by E. Nasir & S. I. Ali. An Annotated Catalogue of the Vescular Plants of West Pakistan and Kashmir. Fåkhri Printing Press, Karachi.

Stewart, Ralph Randles. (1916-17). The Flora of Ladakh & Western Tibet. *Bull Bot. Tor. Club.*, **43** : 571-590.

Thomas, S. C. Li. (2000). Medicinal plants (Culture, Utilization and Phytopharmacology). Technomic Pub Co. USA.

Thomson, T. (1852). Western Himalaya and Tibet, a narrative of a Journey through the mountain of Northern India, 1847-48.

Uniyal. M.P. & R.K. Issar. (1988). Utility of HItherto Unknown Herbal Drugs Traditionally Used in Ladakh & Possible Alternative Medicine, *Bull. Medico- Ethnobot. Res.*, **9** (3-4) : 96-105.

Upholf, J.C. (1968). The Dictionary of Economic Plants. Verlag von J. Cramer, Lehre, 2nd Edn.

Usher, G. (1984). A Dictionary of Plants Used by Man. CBS Publishers & Distributors, New Delhi.

Vidyarthi, O.P. (1997). Wild & Cultivated Plants of Jammu, Kashmir & Ladakh. Pub. by Dte. of Social Forestry Project J&K Govt.

Virjee, J., G.H. Dar, P. Kachroo & G.M. Bhat. (1984). Taxo-Ethnobotanical Studies of the Rural Areas in Distt. Rajouri Jammu, *J. Econ. Taxon. Bot.*, **5** (4) : 831-834.

Viswanathan, M.V. & N.R. Mankad. (1984). Medicinal Plants of Ladakh. (J&K). *J. Econ. Taxon.*

Bot., **5** (2) : 401-407.

Viswanathan, M.V. (1999). Edible and Medicinal Plants of Ladakh (J&K). *J. Econ. Taxon. Bot.*, **23**(1) : 151-154.

Wallis, T.E. (1955). Text book of Pharmacognosy. J. & A. Churchill Ltd. London, 3rd; 4th 1960.

Ward, F.K. (1936). A Sketch of Vegetation and Geography of Tibet. *Proc. Linn. Soc. London.*, **148**(3) : 133-160.

Watt, G. (1808). The Commercial Products of India: An Abridgment of "The Dictionary of the Economic Products of India". G. Watt, John Murray, London.

Watt, G. (1881). Notes on the Vegetation of Chamba State and British Lahaul. *J. Linn. Soc. Bot.*, **18** : 368-382.

Watt, G. (1889-1893). D.E.P. : A Dictionary of the Economic Products of India. Government Press, Calcutta, 6 Vols. Index

Watt, G. (1989-96). A Dictionary of the Economic Products of India, (**1-6**), Govt. Press Calcutta, India.

❑❑❑

Index

BOTANICAL NAMES AND FAMILIES

D

E

F

G

H

I

J

K

L

M

S

T

U

V

W

X

Y

Z

❑❑❑

Index

OTHER NAMES

I

J

K

L

M

N

O

P

R

S

T

U

V

W

Y

Z

❑❑❑